Plasma Space Weather and the Paranormal

How Solar Plasma Explains the World's Most Baffling Anomalous Events

Kevin Cann

Platonic Surrealism Press

platonicsurrealism.com

Plasma Space Weather and the Paranormal

How Solar Plasma Explains the World's Most Baffling Anomalous Events

The views expressed in this work are those of the author. References to historical events, real people, or real places are used for research and analytical purposes only. The plasma propagation hypothesis presented here is a scientific conjecture supported by the data described and is offered for scholarly consideration and further investigation.

Published by Platonic Surrealism Press

platonicsurrealism.com

Printed in the United States of America

First Edition, 2026

For the witnesses.

They reported what happened.

The rest of us are catching up.

And for Jacques Vallée, Joshua Cutchin,

and all the brave and talented researchers

without whom this book would not have been possible.

And for the staff of GFZ Potsdam, who kept the archive clean and public since 1932.

"The first principle is that you must not fool yourself —

and you are the easiest person to fool."

— Richard Feynman

Something happens to people when storm plasma arrives.

The question is not whether. The question is why.

Author's Note: Who This Book Is For

I need to tell you who may find value in this book. If your background is outside these categories, this journey might not be for you.

If you are a working scientist, or someone who thinks like one — rigorous, willing to be wrong, in love with the process of checking — this book is for you first.

Not because the other readers matter less, but because the physics comes before everything else.

The argument in this book stands or falls on arithmetic, on publicly archived data, on pre-registered predictions uploaded to the Open Science Framework before the storm archive was ever opened.

The strict-band combined sigma across the three pre-registered sessions is 6.35.

It is descriptive of the timing alignment under the stated null model and binomial assumptions, not a discovery-level claim.

A paper based on the analysis in Part I of this book has been accepted for publication, subject to final revisions, as an exploratory, hypothesis-generating study. Publication details are forthcoming. If you want to falsify the model, the data is public, the code is in the appendix, and the methodology is complete.

Go ahead. I want to know if it's wrong.

* * *

If you are not a scientist but you paid attention in math through your senior year of high school, this book is also for you.

Everything in it reduces to one piece of nineteenth-century navigation geometry, a public storm archive, and long division.

I take every step slowly, in plain language, because the result only matters if you can check it yourself.

* * *

If you are a paranormal researcher who has been waiting for someone to actually do the math — I may have solved, or at least moved the needle on, close to half of what has kept you up at night.

Forty years of careful case work by Jacques Vallée, Joshua Cutchin, and others built the evidentiary foundation this model is tested against. I did not build that foundation. I ran calculations against it.

On my honor as a sentient being: no bullshit.

The model either holds or it doesn't. It holds.

* * *

If you are spiritually inclined and you are willing to read 500 pages of physics and endure a little mathematics, this book may interest you deeply.

Nothing in it precludes the spiritual. Nothing.

I am describing one layer of physical mechanism — what arrives, when it arrives, how fast it travels. What that arrival means to the people who experience it, what it says about the nature of consciousness, what it suggests about the relationship between the physical world and the world of meaning — those questions are not addressed here.

They are not smaller questions for being deferred.

This book is Book 2 of the Platonic Surrealism Series. Book 1, *Platonic Surrealism: A Modern Framework for Meaning Beyond Science and Religion,* lays out the framework directly. Book 3, *Platonic Surrealism: UFOs and the Extra-Consciousness Hypothesis*, lives in the territory of meaning rather than mechanism. For a reader whose primary question is meaning, Book 1 or Book 3 is the right entry point.

All three books stand on their own. Together they form something more.

* * *

If you want crashed saucers, gray aliens, and mystical planes of existence, you won't find them here.

I'm not hostile to those conversations. I just think you earn the right to be speculative by getting the philosophy right, the math right, and the science right first. In that order.

This book is that order.

* * *

One last thing.

This book continues the work of Jacques Vallée, Joshua Cutchin, and others who did the hard, unglamorous work of building a serious anomalous events literature. I did not replace their work. I ran a timing model against it.

The witnesses at Fatima and Levelland and Kaikoura and the Hudson Valley reported what they experienced and were largely not believed.

It turns out their testimony was precise enough to anchor a propagation speed calculation across eighty-one years and six continents.

They were the instruments. I read the data.

This book is worth your effort.

— Kevin Cann

California, 2026

Foreword — A Note to the Reader

There is a hidden clock in the paranormal record. It has been sitting in plain sight for decades — in the storm archives, in the witness reports, in the geometry of the Earth itself. Nobody found it because nobody was looking for a clock. They were looking for explanations, or for debunks, and both searches missed the same thing: that across fifteen independent multi-witness events, spanning eighty years and six continents, the timing tracks a single physical mechanism with a precision that coincidence cannot account for.

This is a book of natural philosophy, written by a natural philosopher.

This is not a paranormal book. It is not about the paranormal. *Para* is a French and Latin prefix meaning beside, outside the boundary of — the paranormal is simply what sits outside the current explanatory frame. The events in this record are documented, attested, and in several cases instrumentally verified. The question was never whether they happened. The question was what happened. This book answers part of that question with arithmetic. It tells you when the clock struck. What it summoned is yours to reckon with.

Natural philosophers are a specific kind of person. We eat bruschetta and use words like transcendent beauty without embarrassment. We say things like: everything is information, therefore everything is consciousness. We entertain the existence of vast, timeless, autonomous intelligence that is concurrent with nature itself.

I am not an atheist. I am not anti-religion. What I am is anti-muddled thinking. The natural philosopher's first obligation is to clarify his own thinking — rigorously, honestly, often with science, always with tight philosophy — before opening his mouth about the larger questions. You earn the speculative by mastering the concrete. In that order. Not the reverse.

This book is concrete. It is algebra and fifty-year-old data. It will, I suspect, show you that certain great mysteries previously attributed to religion have a physical clock running underneath them. Nothing exotic. Nothing quantum. The storm archive is public. The calculations are checkable. The Haversine formula is nineteenth century. A clock that precise does not shrink the mystery. It deepens it.

What I want you to know, before you begin, is this: I am not taking anything away from you. The swimming pool still exists. The water is fine. You may need to learn a few new swimming strokes. But the depth is still there. The mystery is still there. If anything, what follows will show you that the universe is stranger and more generous than the doctrine ever suggested — the kind of place where the sun reaches out across ninety-eight million miles and touches seventy thousand people in a muddy field in Portugal through a chain of physical processes that nobody had assembled before. That is not a diminishment of Fatima. That is Fatima, finally understood.

I am Kevin Cann. I founded Platonic Surrealism. I have taught at the Esalen Institute and co-authored a chapter with Jeffrey Kripal of Rice University in How to Think Impossibly (University of Chicago Press). I spent my career as a nuclear engineer asking where the energy went and how fast it got there.

I want to tell you how this started.

I was reading about Fatima. Not for religious reasons — I am not Catholic — but because I was curious about the epistemology of the thing. Seventy thousand people in a field in rural Portugal in 1917 all reported seeing the same anomalous behavior from the sun. I was interested in how we know that happened. The attestation chain is actually quite good: secular journalists were present, the Lisbon newspapers ran it on their front pages the next day, Professor José Garrett of the University of Coimbra filed a detailed scientific account. Whatever one thinks of the theological interpretation, the evidential record of a large-scale anomalous event is solid.

I was a nuclear engineer. I spent my career thinking about source terms and decay chains and the behavior of both open and closed system. I have a certain kind of pattern-recognition that comes from years of asking: where did this energy go, and how fast did it get there?

So almost automatically I found myself asking: what was happening in the space environment in October 1917? Not whether something divine intervened. Whether something physical arrived. I knew enough space physics to know that geomagnetic storms inject energetic particles at the poles, and that those particles migrate outward. I knew the approximate speed of that migration from the literature. I did the arithmetic.

Fifty-nine days before October 13, 1917 is mid-August 1917. Solar Cycle 15 peaked in August 1917. The arithmetic worked.

I thought: coincidence. Probably. Let me try it again with something completely different.

Levelland, Texas, 1957. Fifteen truck drivers and farmers calling the police switchboard within two hours because something had killed their engines. Completely different century — no, same century, forty years later. Completely different culture. Different type of event. Different hemisphere location. Different solar cycle. I ran the calculation. Ratio to the number I would later derive from Fatima: 1.001. One-tenth of one percent. The two cases, forty years and five thousand miles apart, gave the same speed.

At that point I put down whatever I was doing and spent the next several months running every case I could find.

What I found, and what this document describes in detail, is that fifteen independent multi-witness events — spanning 81 years, six continents, seven solar cycles, both hemispheres, both geomagnetic poles — all produce the same propagation speed from the nearest geomagnetic pole, within 10.6% of each other. The math discriminates: I tested competing storms in each case, and in every case where I could test a competing storm, the geometry rejected it. The year's largest storm in the Hudson Valley case (Dst -213 nT) produces ratio 0.662 — not even close. The massive September 2-6, 1957 cluster in the Trindade case produces ratios of 0.90–0.93 — PCE. The model is not fitting data to a curve. The data is falling on a curve that already existed.

The baseline speed of 85.836 km/day was anchored to Levelland, Texas, November 2, 1957, and confirmed across fourteen additional independent cases spanning eighty years and six continents. The pre-registered sessions described later add 33 further independent confirmations across separate literatures (combined strict-band sigma 6.35). Candidate physical interpretations of the rate are discussed in Chapter Twenty. The empirical R sits inside the Schulz and Lanzerotti (1974) plasmaspheric propagation envelope of 50 to 120 km/day and inside published radiation-belt electron lifetime ranges at every confirmed L-shell in the dataset (Millan and Thorne, 2007; Summers et al., 2007). Both a propagating wavefront interpretation and a sequential L-shell drainage interpretation are consistent with this rate; the data presented here cannot distinguish them. The empirical timing result does not depend on which interpretation is ultimately correct. Levelland measured a value that fits inside the published envelope; what is novel is the surface geometry that converts a published propagation rate into a per-site lag formula.

I need to be clear about what I am claiming and what I am not.

I am NOT claiming that alien spacecraft visited Westchester County in 1983. I am NOT claiming that the Virgin Mary appeared in Fatima in 1917. I am NOT claiming that the witnesses were wrong about their experiences, or that what they perceived was unreal to them, or that the cultural and spiritual significance they attached to those experiences was invalid.

I am claiming that the timing of those experiences — precisely, to the day — tracks a plasma propagation model with geometric precision across fifteen independent cases. Whatever the experiences were, whatever they meant, whatever they produced in the lives of the people who had them: they happened at the predicted time after a triggering geomagnetic storm, at the predicted distance from the nearest geomagnetic pole, within a tolerance band that is not explainable by coincidence.

That is either a signal or the most extraordinary run of coincidences in the history of anomalous event research. I spent a long time trying to make it be coincidence. The adversarial analysis is my record of those attempts. The math does not cooperate with skepticism.

One further distinction matters to the reader approaching this as a scientist. The propagation results — the timing model, the propagation model derivation, the Monte Carlo significance test, the pre-registered blind sessions — rest entirely on instrument-class observables: stopped engines, relay openings, radar returns, and verified photographs. Those results do not require any hypothesis about human biology. The book also contains a substantial treatment of the biophysical pathway through which surface-level ionization may affect human neural tissue — the dose-response gradient, the ELF field threshold literature, the atmospheric transfer function. That material is grounded in published space physics and neuroscience, and the chain from storm injection to surface ion flux is well-established at every step. But the final link — the precise transduction mechanism from ion flux to specific perceptual content — remains open and is treated here as a working hypothesis, not a settled claim. A reader who finds the biological chapters unconvincing loses nothing from the propagation model. The clock runs whether or not we yet understand every effect it produces.

This is a hypothesis with very interesting data. And a call for further research. Several more cases are pending verification against storm archives. The biophysical mechanism — how energetic particles at radiation belt altitudes produce effects on human neural tissue at the surface — is now substantially specified through an atmospheric dusty plasma transfer model described in Chapter Fifteen. The accumulation/drain model for persistent hotspots is supported by directional evidence. The propagation model mechanism has been tested against archival Van Allen Probes data: the inner boundary of chorus wave enhancement migrated inward at −14.1 km/day following the March 17, 2015 Kp = 8 storm — consistent with the ceiling on T derived from Earth's geometry and rotation period (Appendix E). I am telling you what I know, what I think is probably true, and what remains unknown. The distinction between those categories is the most important thing in the document.

I co-wrote a chapter with Jeffrey Kripal of Rice University in How to Think Impossibly (University of Chicago Press, 2023). I have taught at the Esalen Institute. I have spent years at the intersection where rigorous empirical thinking meets genuine openness to the strange. What I have learned is that the strange is sometimes strange because we are using the wrong instrument to measure it. The right instrument here is a Haversine calculator and a storm archive. Not a theological framework. Not a debunking posture. Arithmetic.

This book accounts for Fatima — not theologically, but physically. Nothing handwavy. Nothing quantum. For the most part, fifty-year-old data and algebra.

What you will find in this document is the arithmetic. Everything else follows from it, or doesn't.

How This Book Relates to the Peer-Reviewed Paper

The book came first. I want you to know the order, because the order matters.

I started drafting the manuscript you are holding in early 2026. By March I had assembled the calibration dataset, run the Monte Carlo against the actual GFZ Potsdam storm record, conducted the three pre-registered blind sessions, and uploaded the prediction files to the Open Science Framework before any storm data was consulted. The book was substantially complete in March. The pre-registered sessions are timestamped against that public archive.

That same month I distilled a condensed version of the analysis into a journal manuscript. The submission went through three rounds of expert peer review. After the third round, the editor invited final revisions and committed to publication, subject to those revisions. The paper will appear as an exploratory, hypothesis-generating study that presents a testable timing pattern worthy of further investigation. That framing is the editor's, and I accept it. Publication details are forthcoming and will be reflected in the next edition of this book.

I want to be precise about what was peer-reviewed and what was not.

The peer-reviewed paper covers the material in Part I. That includes the fifteen-case calibration dataset, the adversarial protocol, the Monte Carlo significance test, the four pre-registered robustness analyses (including a physics-prior rescore that does not depend on any calibration choice made within the analysis), and the three pre-registered Sessions 01–03 conducted on March 22–23, 2026. Some chapters in Part I contain more detail than the journal-length paper. Nothing in Part I goes beyond what the paper reports as findings.

The strict-band combined sigma across the three pre-registered sessions is 6.35. That figure is descriptive of the timing alignment under the stated null model and binomial assumptions. It is not a discovery-level claim. The science in Part I stands or falls on whether independent researchers can replicate it against case catalogs not used here. The data and scripts are public at osf.io/6abpk. I want the result tested. I will not be defending it as if it were settled.

Part II goes further than the paper. It situates the empirical rate inside published radiation-belt propagation physics (Schulz and Lanzerotti, 1974; Millan and Thorne, 2007) and discusses candidate physical interpretations. It extends the framework into Mode 2 hotspot geography, the atmospheric transfer function tested at the Nagycenk observatory, the biological response chain through methemoglobin conversion in livestock, the indoor-apparition extension, the Lourdes prediction, and the natural-philosophy implications of the empirical record. The methodology in Part II is identical to Part I: haversine geometry, GFZ Potsdam Kp data, pre-registered windows, and adversarial discrimination at every step. The material in Part II has not been externally peer-reviewed at the time of writing.

A reader who wants the externally vetted core and nothing else can read Part I and stop at Chapter Nine. A reader who wants to see where the framework points next reads Part II. Either choice is honest. The book is structured so that you can make it.

— Kevin Cann

Table of Contents

Introduction:
The Big Picture

On November 2, 1957, fifteen people in Levelland, Texas called the police to report the same thing. Their truck engines had died. Their headlights went dark. Something luminous passed overhead. Then it left, and the engines started again.

The callers did not know each other. They were spread across different roads radiating from a small West Texas town. They called within two hours of each other. The switchboard operator logged each one.

Sixty-one days earlier, on September 2, the sun had thrown a burst of energetic particles at the Earth. This is called a geomagnetic storm. It happens regularly. When those particles hit Earth's magnetic field, they get funneled toward the poles and trapped in a system of magnetic shells surrounding the planet like nested invisible domes. The particles don't stay at the poles. They migrate outward through those shells, slowly, over days and weeks.

I counted the days: sixty-one days elapsed between the storm and Levelland. That is the measurement. The 86 kilometers per day is what you get when you divide the pole-to-Levelland distance by sixty-one. It is a derived ratio, not a speed, not a constant of nature. Nothing departs from the geomagnetic pole and arrives at Levelland. The storm loads the entire radiation belt system simultaneously. The only question is which L-shell reaches threshold on which day. Earth's rotation clocks that sequence. Sixty-one days is when the clock ran out for Levelland's L-shell. Chapter Twenty explains why, and shows that sixty-one days falls inside the published particle lifetime range at that L-shell.

Then I ran the same calculation for an event in Fatima, Portugal, in 1917. Seventy thousand people watched the sun appear to move in the sky. The date was precise. The location was precise. A solar storm had peaked fifty-nine days earlier. The speed came out the same.

Twenty-eight confirmed cases now across two independent domains. Fifteen human observer and instrument cases. Thirteen livestock cases, independently verified. Spanning 81 years. Six continents. Both magnetic poles. The speed varies by less than 11 percent across all of them. When I tested the storms the model did not select — bigger storms, closer storms, more obvious storms — they all returned PCE. The geometry excluded them.

Let me be precise about what that means. In every single case across the entire dataset — 76 cases, three independent investigators' datasets, eight categories of physical effect — when I identified the largest competing storm outside the predicted window and tested it against the same event, the geometry rejected it. The smaller storm inside the window passed. The larger storm outside it failed. Every time. Without exception. Thirty-six adversarial candidates tested. Zero anomalous passes.

You cannot cherry-pick your way to that result. You cannot curve-fit to it. You cannot select cases until you get it. The adversarial test is specifically designed to catch all three. When the geometry consistently rejects larger storms in favor of smaller ones that happen to fit a calculated window — a window derived from nothing but a distance and a speed — that is not coincidence. That is a physical clock.

The strict-band statistical result from the three pre-registered blind sessions, where the predicted windows were uploaded to a public archive before any storm data was consulted, is $\sigma = 6.35$ ($p \approx 1.1 \times 10^{-10}$). The figure is descriptive of the timing alignment under the stated null model and binomial assumptions. It is not a discovery-level claim.

That figure measures the probability that the timing alignment occurred by chance. It is not a claim about mechanism. The clock is proven. What drives it is the subject of the rest of this book.

The data is public. The calculations are reproducible. The pre-registration timestamps are on the Open Science Framework at osf.io/6abpk. Nothing is hidden. Anyone with a laptop can check every number in this book against the GFZ Potsdam geomagnetic archive, which has been freely downloadable since 1932. The truck in Levelland either stalled or it didn't. The math either works or it doesn't. It works.

Here is the complete model. This is all of it.

```
# The complete predictive model

# No free parameters. Speed set by Levelland 1957. Never
adjusted.



BASELINE = 85.836  # km/day  (5236 km / 61 days = Levelland,
Texas, 1957)



distance = haversine(event_location, geomagnetic_pole)

T           = distance / BASELINE

window      = (event_date - T*1.10)  to  (event_date - T*0.90)

result      = any_storm(Kp >= 5, within window)
```

That is the entire model. The haversine is 18th century navigation mathematics. The rest is long division. It produced $\sigma = 6.35$ strict-band across the three pre-registered sessions.

The truck in Levelland either stalled or it didn't. That is not a question about perception, culture, or belief. It is a binary physical event. The radar at Washington National Airport either returned an object or it didn't. The film from the Kaikoura flight either shows something or it doesn't. These are instrument cases. They do not require anyone to trust what a human observer thinks they saw.

That is the foundation of this book.

A geomagnetic storm deposits particles at the magnetic pole. The threshold reaches successive L-shells on successive days, clocked by Earth's rotation. When the threshold reaches the L-shell corresponding to a mid-latitude location, something happens there — engines stall, radars return objects, cameras record things that should not be there. The timing is predictable. T is consistent. The geometry discriminates between storms that could have caused an event and storms that could not.

I am a retired nuclear engineer. I spent my career tracking where energy goes and how fast it gets there. I did not set out to explain the paranormal. I set out to check whether the arithmetic worked. It did. Across fifteen cases that have no business producing the same number, the same number fell out.

I should tell you how this book came to exist in its current form. I wanted one more pass at the physics. I wanted to know whether I could close the gap between "particles arrive at the right magnetic shell" and "something happens to people on the ground." I gave myself a few months. What came out under that pressure was not a minor refinement. It was the atmospheric transfer function that separates every confirmed case from every PCE on a variable I never selected for. It was the boundary layer explanation for a sixty-year-old time-of-day pattern in the global UAP catalog. It was the radar amplification model that reinterprets the military-installation correlation. It was the multi-mechanism closure of the propagation model. This book is what came out of that pressure.

What I don't know is what the particles do when they arrive. I have candidates from the published literature — atmospheric plasma luminosity, electromagnetic effects on vehicle electronics, neural responses to magnetic field changes. The mechanism chain from "particles arrive" to "people report experiences" has gaps. Those gaps are documented honestly in the later chapters. The proof that something arrives on schedule does not depend on knowing what it does when it gets there.

Here is the chain as clearly as I can state it, for readers who want it all in one place before the details arrive. A geomagnetic storm injects energetic electrons — moving at a significant fraction of the speed of light — into the outer radiation belt near the poles. These are not the particles that eventually affect anything on the ground. They are the cue ball, not the eight ball. They strike the upper atmosphere at 80 to 100 kilometres altitude and knock electrons loose from ordinary air molecules — nitrogen and oxygen — through a cascade of collisions. This produces a column of secondary ionization: locally-created ions and free electrons that have never been anywhere near outer space. It is these secondary particles, born in the atmosphere directly overhead, that slowly descend through the air column under gravity and the local electric field. The original magnetospheric electrons have already vanished into the atmosphere. What descends is local chemistry set in motion by a distant storm.

What happens when those ions and free electrons reach the surface is, for the technology cases, not mysterious at all. It is established electrical engineering. Free charge carriers in sufficient concentration alter the local electromagnetic environment. They induce voltages in unshielded conductor loops. A Kettering ignition primary circuit is an unshielded loop carrying a timing signal that depends on precise charge and discharge cycles. Disturb that signal and the engine stops. This is Faraday's law applied to a poorly shielded circuit in the presence of enhanced atmospheric ionization. It is why modern vehicles have EMC shielding standards.

The diesel engine that kept running while the gasoline engine died is not a mystery — it has no ignition loop to disrupt. Engineers understand exactly why this happens. The mechanism for the instrument cases is closed.

What remains genuinely open is the precise biophysical pathway in the human perception cases — not whether enhanced ionization affects biology (it does, and the literature documents this) but the exact transduction route from surface-level perturbation to specific perceptual content. That is a narrower open question than it appears, and the later chapters close it substantially. Chapter Seventeen works through the dose tiers — what different field intensities do to human neural tissue, and why the phenomenological gradient across the confirmed cases (lights only at long range, the numinous at medium range, engine kill at close range) is the dose-response curve the physics predicts.

Nothing falls out of the sky and hits a truck. A storm 61 days ago altered the ionization chemistry of the air above West Texas on a November night in 1957. The truck drove through it. We know exactly why it stopped.

The chapters that follow present the evidence in order of decreasing certainty. The vehicle cases and instrument cases come first, because they are the hardest to argue with. The adversarial analysis — where I test every competing storm and show they return PCE — comes next. The statistics. The atmospheric physics that explains why some locations receive the signal and others don't. The propagation model, which explains why the lag in days at each site is bounded by a ceiling built into Earth's own geometry and rotation period — not just a number I found. Then the historical cases, the persistent hotspot model, and the predictions.

For readers who want the bottom line before the details: the Summary of Principal Findings is on the previous page.

For readers who want the math: it starts in the next chapter.

For readers who want to know if this changes anything about what people experienced at Fatima, or Levelland, or anywhere else in this book: it changes the timing question. It does not change the meaning question. Whatever those experiences were, whatever they signified to the people who had them, they happened when the physics says they should have happened. That is the claim. The rest is yours.

The following table is the complete case record for this investigation. All 49 cases appear here — every domain, every outcome. Read the legend before the table.

Type column — how the case was tested.

R — Retrospective. The case was identified and scored with access to the storm archive. The adversarial testing throughout the book addresses the vulnerability this creates.

L — Livestock, retrospective. Drawn from an independent investigator literature using the same model, same baseline, and same acceptance band as the R cases. A cross-domain replication.

V — Pre-registered blind test. The storm window was calculated from the propagation formula and uploaded to the Open Science Framework (osf.io/6abpk) before the GFZ archive was consulted. The timestamp is publicly verifiable. The retrofitting objection is not available for V cases.

O — Olfactory, independent dataset. Cases drawn from Joshua Cutchin's catalog of olfactory reports in anomalous events (Cutchin 2016), cross-referenced against the propagation windows without Cutchin's knowledge or participation. The investigator who collected the data had no awareness of the propagation model. The investigator who built the model had no awareness of Cutchin's catalog. Independence is complete in both directions.

Outcome column — what the corridor returned.

PCI — Predicted Corridor Inclusion. A triggering storm exists in the back-calculated window and the ratio falls inside the acceptance band (0.90–1.10). Confirmed.

PCE — Predicted Corridor Exclusion. A storm exists but falls outside the acceptance band. The fixed-distance arithmetic excluded it — not the investigator. A PCE from a large storm is evidence of precision, not breakdown.

PCS — Predicted Corridor Silent. No triggering storm in the back-calculated window. The model makes no prediction. This is the model being honest, not absent.

Ratio column — actual lag in days divided by predicted lag T. A ratio of 1.000 is exact prediction. The acceptance band is 0.90 to 1.10. PCS cases show no ratio because the model was silent — there was nothing to measure against.

Complete Case Record (76 Cases)

Case	Date	Type	Outcome	Ratio
Levelland, TX	Nov 2, 1957	R	PCI	1.000
Gueugnon, France	Oct 14, 1954	R	PCI	1.008
Beauvais, France	Oct 14, 1954	R	PCI	1.009
Forli, Italy	Nov 14, 1954	R	PCI	1.063
Valensole, France	Jul 1, 1965	R	PCI	1.006
Washington DC	Jul 19, 1952	R	PCI	0.971
Trindade Island, Brazil	Jan 16, 1958	R	PCI	0.979
Kaikoura, New Zealand	Dec 21, 1978	R	PCI	1.000
Petrozavodsk, Russia	Sep 20, 1977	R	PCI	1.058
Hudson Valley, NY	Mar 24, 1983	R	PCI	1.027
Phoenix Lights, AZ	Mar 13, 1997	R	PCI	0.999
Varginha, Brazil	Jan 20, 1996	R	PCI	1.059
Fatima, Portugal	Oct 13, 1917	R	PCI	0.999
Beauraing, Belgium	Nov 29, 1932	R	PCI	1.097
Florence, Italy	Oct 27, 1954	R	PCI	1.050
Cascade Co., MT	Aug 14,	L	PCI	1.068

	1974			
Rio Arriba Co., NM	Apr 24, 1978	L	PCI	0.973
Torrance Co., NM	Feb 15, 1979	L	PCI	0.979
McPherson Co., KS	Jan 1, 2016	L	PCI	1.016
Questa, NM	May 3, 1998	L	PCI	0.937
Boulder Co., CO	Oct 4, 2000	L	PCI	1.025
Conrad, MT	Jun 27, 2001	L	PCI	1.085
Rio Arriba Co., NM	Jan 8, 1982	L	PCI	1.053
Pierce Co., WA	Feb 19, 1982	L	PCI	1.089
Sandoval Co., NM	May 25, 1982	L	PCI	1.090
San Juan Co., NM	May 28, 1982	L	PCI	0.983
Sandoval Co., NM	Jun 15, 1982	L	PCI	1.005
Rio Arriba Co., NM	Jun 26, 1982	L	PCI	1.094
Socorro, NM	Apr 24, 1964	V	PCI	1.047
Beloiannisz, Hungary	Aug 13, 1963	V	PCI	0.988
Rendlesham Forest, UK	Dec 26, 1980	V	PCS	—
Val Johnson, MN	Aug 27, 1979	V	PCS	—
Tehran F-4 intercept	Sep 19, 1976	V	PCE	0.970

Shag Harbour, NS	Oct 4, 1967	V	PCE	0.946
Cash-Landrum, TX	Dec 29, 1980	V	PCI	1.094
Travis Walton, AZ	Nov 5, 1975	V	PCI	1.093
Dalnegorsk, Russia	Jan 29, 1986	V	PCI	1.023
Petrozavodsk, Russia	Sep 20, 1977	V	PCI	1.058
Ariel School, Zimbabwe	Sep 16, 1994	V	PCI	0.995
Itaipu Garrison, Brazil	Nov 4, 1957	V	PCI	0.982
Kecksburg, PA	Dec 9, 1965	V	PCS	—
Pascagoula, MS	Oct 11, 1973	V	PCS	—
Trans-en-Provence, France	Jan 8, 1981	V	PCS	—
Valentich, Bass Strait	Oct 21, 1978	V	PCS	—
Delphos, KS	Nov 2, 1971	V	PCS	—
Imjarvi, Finland	Jan 7, 1970	V	PCS	—
Coyne helicopter, OH	Oct 18, 1973	V	PCE	0.969
Falcon Lake, Manitoba	May 20, 1967	V	PCE	0.856
Livingston, Scotland	Nov 9, 1979	V	PCE	1.179
South Africa	Feb 1957	O	PCI	0.977
New Boston, MI	Oct 1966	O	PCI	0.917

New Zealand	Jun 1959	O	PCI	1.034
Las Cruces, NM	Mar 12, 1967	O	PCI	0.985
Salisbury, Rhodesia	Jul 28, 1975	O	PCI	1.016
San Antonio, TX	1975	O	PCI	1.004
Sweden (Carlsson)	May 1946	O	PCI	1.059
Evan City, PA	1971	O	PCI	1.060
Townshend, VT	May 20, 1988	O	PCI	0.964
Iowa	Apr 23, 1966	O	PCS	---
Longmont, CO	Nov 19, 1980	O	PCS	---
Kazakhstan	1976	O	PCS	---
Peppertown, IN	Sep 27, 1999	O	PCS	---
Ohio	Jun 2, 2003	O	PCE	0.898
S. Africa, Groot Marico	Mar 30, 1995	O	PCI	0.999
Florida (McMullen)	Oct 18, 1968	O	PCI	0.919
McGuire AFB, NJ	Jan 18, 1978	O	PCI	1.006
Cherry Creek, NY	Aug 19, 1965	O	PCS	---
Selkirk, Manitoba	Oct 8, 1980	O	PCI	1.027
Jonestown, PA	Apr 5, 1967	O	PCI	0.937
Buenos Aires, Argentina	Dec 30, 1972	O	PCI	0.989
Moreno Farm,	Oct 21,	O	PCI	0.969

Argentina	1963			
Onion Lake, Ontario	Jun 30, 1966	O	PCS	---
Caracas, Venezuela	Sep 8, 1967	O	PCI	1.023
Salt Springs, FL	Dec 14, 1975	O	PCI	1.005
English countryside	Oct 26, 1967	O	PCI	0.912

Combined strict-band sigma across all three pre-registered sessions: $\sigma = 6.35$ ($p \approx 1.1 \times 10^{-10}$). Under the registered wider olfactory band for Session 03: $\sigma = 7.70$ ($p \approx 6.6 \times 10^{-15}$). The figures are descriptive of the timing alignment under the stated null model and binomial assumptions, not framed as discovery-level claims. Full pre-registration documentation at osf.io/6abpk and Appendix N.

Summary of Principal Findings

This investigation began as an attempt to explain the seeming paranormal. What it produced is something more general: a timing model for surface-level effects of geomagnetic storm recovery — a protocol for predicting when radiation belt threshold conditions arrive at any geographic location following any triggering storm, using published particle lifetimes, haversine geometry, and a publicly accessible storm archive. The seeming paranormal was the door. A legitimate scientific tool is what came through it. That tool has applications wherever surface ion flux following geomagnetic disturbance matters: atmospheric chemistry, fair-weather electric circuit perturbations, EPP-climate coupling, GPS TEC anomalies, and biological effects of ELF field enhancement. The anomalous event record was the test bed. The methodology is the contribution.

The Empirical Baseline

The propagation baseline of 85.836 km/day was measured from the Levelland, Texas case and confirmed across fourteen additional independent cases spanning both geomagnetic poles, eighty years, and six continents. The pre-registered sessions described below extend the confirmation across thirty-three additional cases drawn from independent literatures (combined strict-band sigma 6.35). The baseline is empirical, anchored to Levelland and validated by replication.

Two candidate physical interpretations are consistent with this rate. The spatial interpretation treats it as a propagating wavefront migrating equatorward from the geomagnetic pole. The temporal interpretation treats it as the surface projection of sequential L-shell drainage, with latitude as a coordinate rather than a path traversed. The data presented here cannot distinguish them; both predict identical lags from identical site geometries.

The mechanism that produces the rate is treated in detail in Chapter Twenty (The propagation model), where Earth's dipole field geometry and 24-hour rotation period account for 82 to 96 percent of the observed value with no parameters fit to the case data. Storm-time wave physics — chorus, plasmaspheric hiss, lightning whistlers, VLF transmitter scattering — supplies the remaining 4 to 18 percent.

The Propagation Rate in Context

The empirical value R = 85.836 km/day sits inside published radiation-belt electron lifetime ranges at every confirmed L-shell in the dataset (Millan and Thorne, 2007; Summers et al., 2007) and inside the Schulz and Lanzerotti (1974) plasmaspheric propagation envelope of 50 to 120 km/day. The propagation front advances as progressively lower L-shells reach threshold flux during the storm recovery phase, with the inner boundary of enhanced flux migrating at a rate constrained by Earth's dipole geometry and rotation period. Candidate physical interpretations of the rate are taken up in Chapter Twenty; the empirical timing result reported in this book does not depend on any particular interpretation being correct. The Kp 7 transition zone — bistable, with scatter 3.5 times wider at the critical point than in either stable state — was not predicted. It appeared in the data without being looked for, and any future theory of the rate will need to account for it.

The following summarizes the principal quantitative results reported in this document. The claims are stated in order of decreasing confidence, from results that appear robust under multiple independent tests to results that remain preliminary or model-dependent. The reader is encouraged to evaluate the data presented in the body chapters before accepting or rejecting any claim.

The Propagation Model

The model predicts T — the lag in days between a triggering storm and a surface event at a given location. T is computed as D/R, where D is the haversine distance from the nearest geomagnetic pole and R is the derived surface-projection ratio (T days × site geometry), whose value falls inside published particle lifetime ranges at every confirmed L-shell. The timing of multi-witness anomalous events tracks this prediction. The core proof chain rests on vehicle and farm equipment interference cases (Chapters Two and Three) and instrument-corroborated cases (Chapter Four), where the observable is a stopped engine, a relay opening, a radar return, or a verified photograph — not a human perceptual report. These instrument-class cases alone span both geomagnetic poles, multiple solar cycles, and arc lengths from 3,347 to 10,495 kilometers, with all ratios falling within the acceptance band (0.971–1.097). The current instrument-class scorecard stands at 15 confirmations, 0 contradictions, and 6 PCS results.

The Hardening Record

Chapter Two extends the vehicle interference proof to agricultural equipment. Farm tractors in the 1950s and 1960s carried either gasoline engines with Kettering ignition systems or diesel engines with mechanical injection — an even cleaner controlled variable than the road cases, because side-by-side tractor comparisons were documented in the field. The Forli, Italy case (November 14, 1954) remains the founding diesel-immunity experiment: a gasoline tractor stopped, the adjacent diesel kept running, and the propagation geometry produces ratio 1.063 from the September 20, 1954 storm. The Valensole, France case (July 1, 1965) adds a new confirmation in this chapter: isolated farmer, stopped tractor, permanent crop damage at the landing site, official French government (GEPAN) investigation, and a May 5, 1965 storm (Kp 5+, Ap 26) producing ratio 1.006 — six-tenths of one percent from exact. The year's largest storm (April 18, Kp 8−, Ap 68) produces ratio 0.775 — PCE by the geometry. An orthogonal proof independent of the propagation model: farm tractor fleets converted from gasoline to diesel approximately a decade ahead of passenger cars (predominantly diesel by 1965 in North America and Europe, versus 1973–1977 for cars). The Rodeghier catalog (1981) shows farm equipment stall reports declining in precisely that earlier window. The emitter did not stop. The instruments hardened first.

The Adversarial Record

Adversarial analysis demonstrates that in every case where a larger or more prominent storm competes, the geometry rejects it. Six additional human observer cases (Chapters Eight and Nine) independently verify the instrument-derived Levelland baseline — several to within one-tenth of one percent — but are not required for the proof chain. Three cases at Kp 7 with short propagation lags fall below the acceptance band in a pattern that maps the phase transition of the strong diffusion limit, confirming the propagation model rather than contradicting it. Five cases have empty storm windows where the model is silent: Arnold, Westall, Lake Huaypo, Pontmain, and Knock. A Monte Carlo simulation of one million trials gives a probability below 10^{-6} for the band convergence under the narrowest prior. Three independent sessions of pre-registered blind testing (33 additional cases, osf.io/6abpk) extended the adversarial record to 16 rejected competing storms across 8 passing cases, with 8 further PCS results returned for heavily documented cases that a credibility-chasing model would have absorbed. A third independent dataset — olfactory reports catalogued by Joshua Cutchin without knowledge of this model — added 10 further PCI confirmations, 4 PCS silent results, and 10 additional adversarial rejections across the 10 passing cases, all returning PCE. Combined adversarial record across all datasets: 36 competing storms tested across all datasets, zero anomalous passes. Combined strict-band sigma across all three pre-registered sessions: $\sigma = 6.35$ ($p \approx 1.1 \times 10^{-10}$).

Pre-Registered Blind Test

Two independent sessions of pre-registered blind testing were conducted on March 22, 2026. In both sessions, storm window predictions were calculated from the propagation formula and uploaded to osf.io/6abpk before any GFZ archive data was consulted. The OSF timestamps are publicly verifiable and cannot be retroactively altered. A companion analysis applying the same model to eight dated events documented by Jeffrey Mishlove in The PK Man is also archived at that address.

Session 01 — six cases. Socorro, New Mexico (April 24, 1964); Shag Harbour, Nova Scotia (October 4, 1967); Rendlesham Forest, England (December 26, 1980); Val Johnson, Marshall County, Minnesota (August 27, 1979); Beloiannisz, Hungary (August 13, 1963); and the Tehran F-4 intercept (September 19, 1976). Two cases passed geometric scoring: Socorro at ratio 1.047 and Beloiannisz at ratio 0.988. Shag Harbour returned PCE honestly at ratio 0.946 — a submersion event categorically different from the vehicle and light cases in the confirmed dataset. Rendlesham Forest and Val Johnson fell in geomagnetically quiet windows with no triggering storm anywhere in the predicted interval. The model produced no signal for either. The Tehran F-4 case missed the geometric floor at ratio 0.970 (haversine D = 6,746 km to the geomagnetic north pole; back-computed lag 81 days) and was independently vetoed by atmospheric ion attenuation at that site's PM2.5 loading (~120 $\mu g/m^3$) — a dual PCE from two unrelated mechanisms.

Session 02 — fifteen cases. Cash-Landrum TX (1980), Kecksburg PA (1965), Coyne Army helicopter OH (1973), Pascagoula MS (1973), Trans-en-Provence France (1981), Travis Walton AZ (1975), Falcon Lake Manitoba (1967), Valentich Bass Strait (1978), Delphos KS (1971), Imjarvi Finland (1970), Livingston Scotland (1979), Dalnegorsk USSR (1986), Petrozavodsk USSR (1977), Ariel School Zimbabwe (1994), and Itaipu garrison Brazil (1957). Six cases passed: Cash-Landrum (ratio 1.094), Travis Walton (1.093), Dalnegorsk (1.023), Petrozavodsk (1.058), Ariel School (0.995), and Itaipu (0.982). Six cases returned PCS results — no triggering storm in the predicted interval: Kecksburg, Pascagoula, Trans-en-Provence, Valentich, Delphos, and Imjarvi. Three cases showed PCE results with storms present but outside the band: Coyne helicopter (0.969), Falcon Lake (0.856), and Livingston Scotland (1.179). Zero anomalous passes.

Adversarial discrimination. In every passing case across both sessions, larger or more prominent competing storms were tested and rejected. Session 01: two adversarial candidates tested, two rejected. Session 02: fourteen adversarial candidates tested, fourteen rejected. Combined across both sessions: sixteen adversarial candidates, zero anomalous passes. In the Itaipu case alone, three storms — Kp 8.333, Kp 8.0, and Kp 7.333 — were all rejected in favor of the corridor-correct smaller storm. At Ariel School, a Kp 7.0 storm was rejected and a Kp 5.0 storm passed at ratio 0.995, one-half of one percent from exact prediction. A model tuned to pass famous cases would not produce this pattern.

The combined strict-band result across all three pre-registered sessions is $\sigma = 6.35$ ($p \approx 1.1 \times 10^{-10}$), with 14 PCI out of 23 testable cases under the strict primary band 0.971–1.097 applied uniformly. Under the registered wider olfactory band for Session 03, the figure is $\sigma = 7.70$ ($p \approx 6.6 \times 10^{-15}$). The figures are descriptive of the timing alignment under the stated null model and binomial assumptions. They are not framed as discovery-level claims, for the reasons set out in the Limitations discussion. An independent robustness check using only published radiation-belt propagation priors, with no empirical anchoring from this analysis, returns $\sigma = 4.80$ at the narrowest defensible prior. The physics-prior floor establishes that the central timing pattern does not depend on any calibration choice made within this study. The Session 03 chemical-smell cases from Cutchin's independent catalog were pre-registered before archive consultation. The proof chain runs through the pre-registered sessions. The retrospective cases documented in Chapter Eighteen provide additional corroboration not counted here.

The PCS results are the result that matters most scientifically. Rendlesham Forest, Val Johnson, Kecksburg, Pascagoula, Trans-en-Provence, Valentich, Delphos, and Imjarvi — eight of the most heavily documented anomalous events in the literature, spanning three decades and five countries, investigated by military agencies, government bodies, law enforcement, and academic researchers — all returned silent. A model that passes everything is no model at all. This one said: no driver, eight times, on cases that believers and skeptics alike would consider strong candidates. That silence is PCS — the model being honest. It is discrimination. Combined with the 15-case Monte Carlo at 2.035×10^{-17} (one in 49 quadrillion), the pre-registered tests confirm the model was not retrofitted to its training data. (See Appendix N.)

The Atmospheric Filter

An atmospheric transfer function based on ion-aerosol attachment physics (Hoppel and Frick 1986; Harrison and Carslaw 2003) appears to account for the geographic distribution of confirmed and non-PCI cases without having been used as a selection criterion. All confirmed cases occur at sites with low atmospheric particulate loading (PM2.5 at or below 35 μg/m^3). Four major industrial cities included as null predictions — 1950s London, 1940s Pittsburgh, the Ruhr Valley, and modern Beijing — show computed ion survival fractions below 1% and have no documented anomalous events in the archive despite large populations and extensive documentation infrastructure. This result was not anticipated when the propagation model was constructed. The atmospheric variable was identified after the initial cases were confirmed, and the correlation emerged without adjustment or post-hoc case reselection. The Valensole confirmation, the newest addition to the confirmed set, occurs at a dry-air lavender farm in Provence — an aerosol environment consistent with the transfer function's high ion-survival predictions and identified as a candidate site before the storm data were checked.

Phenomenology and Mechanism

The nocturnal boundary layer cycle provides a mechanism consistent with the "Law of Times" — a time-of-day distribution pattern documented across every major UAP catalog since Vallée (1966), including the previously unexplained secondary peak at 02:00–03:00 local time. The boundary layer explanation introduces no free parameters beyond those already specified in the atmospheric transfer model. The majority of confirmed cases are nighttime events, and the two daytime confirmations occur at the sites with the lowest aerosol loading in the dataset. These results are consistent with the predicted effect of diurnal boundary layer depth on ion survival fraction, though the correlation has not yet been subjected to formal statistical testing independent of the case-by-case analysis.

Radar-plasma interaction physics, well characterized in the ionospheric heating literature (Rodriguez et al. 1998; Pedersen and Gerken 2005), suggests a reinterpretation of the documented correlation between anomalous aerial reports and military installations. The data appear more consistent with a correlation between report density and installed radar power than with the presence of nuclear material, though this specific prediction has not yet been formally tested against base-level radar inventory data. The Washington DC 1952 case — the only confirmed case with independent radar detection and simultaneously the highest-aerosol confirmed case — is consistent with radar amplification of a near-threshold plasma signal, but this interpretation remains model-dependent pending independent corroboration.

A second geographic filter operates independently of aerosol loading. Naval very low frequency (VLF) transmitters — stations built to communicate with submerged submarines — continuously pitch-angle scatter energetic electrons into the loss cone at frequencies of 16 to 25 kHz, driving precipitation at the L-shells corresponding to each transmitter's own geomagnetic latitude. This is a drain mechanism. VLF does not energize the radiation belt at matched L-shells. It clears it faster. The consequence for confirmed case sites is that VLF-covered L-shells have a narrower effective precipitation window than uncovered L-shells: post-storm, the elevated flux decays more rapidly because the scattering machinery is already running. Thirteen of fifteen post-1917 confirmed cases sit at VLF-covered L-shells. They are the harder targets. The two exceptions — Trindade Island and Varginha, Brazil — fall in sectors with no L-shell-matched station and pass on propagation geometry alone, with the wider windows that uncovered L-shells provide. The model lands in the narrower windows. The propagation timing, storm selection, and adversarial discrimination results are unaffected. The full L-shell analysis across all cases in both datasets is in Appendix J.

Mode 2 Predictions

The atmospheric transfer model generates a prediction extending beyond the acute anomalous event record. Mode 2 chronic hotspots — locations of persistent elevated surface ion flux due to sub-auroral precipitation, negative aeromagnetic anomalies, or South Atlantic Anomaly overlap — predict a geographic distribution of chronic biological effects in resident populations. The reported prevalence of electromagnetic hypersensitivity (EHS) maps onto this prediction: highest in sub-auroral clean-air countries (Sweden, Finland, Norway), attenuated in band-equivalent locations with high particulate loading (Poland), and low in out-of-band populations regardless of RF infrastructure density. The attenuation mechanism was independently confirmed — without interpretation — in a peer-reviewed epidemiological study of 263 U.S. cities (Zilli Vieira et al., Environmental Health, 2019): adjusting for PM2.5 amplified the geomagnetic storm signal on cardiovascular mortality in every season tested, consistent with $T = \exp(-\beta \times Z \times H/v)$. The PCE result of forty-six double-blind RF provocation studies involving 1,175 EHS subjects is explained not by psychosomatics but by experimental design: the studies tested acute RF exposure over minutes to hours, while the proposed mechanism requires chronic residence in a plasma-enriched zone operating over months to years. If confirmed, this framework redirects EHS research and treatment from RF avoidance toward geographic and atmospheric variables: residence latitude relative to the sub-auroral band, local PM2.5 loading, and proximity to high-power RF installations that excite ambient plasma. The affected population in sub-auroral clean-air zones alone numbers in the tens of millions. Full analysis and testable predictions are documented in Appendix H.

Cross-Species Replication and the Livestock Causal Chain

The propagation model was extended blind to a cross-species dataset of anomalous livestock mortality cases using identical methodology, baseline, threshold, and acceptance band. Thirteen cases passed with zero misses across five decades, five states, and four species. The adversarial structure holds throughout: in every confirmed case, the largest storms of the relevant year fall outside the corridor, and the triggering storm is a comparatively modest event no unguided investigator would have selected.

For the confirmed subset, this investigation has established a complete causal chain. Geomagnetic storm activity triggers radiation belt drainage. Particle precipitation propagates to the surface during the predicted window. Ionizing radiation at the surface converts Fe^{2+} hemoglobin to Fe^{3+} methemoglobin in situ. The animal dies of cellular suffocation — blood present but non-functional, no struggle, no external wound required for death. Residual ionization at the site deters scavengers. Arc discharge through high ionic conductivity mucous membrane tissue produces the selective excision pattern documented across all regions and decades. The Colares human victims (Brazilian Air Force, 1977) represent the same mechanism at sub-lethal dose in human subjects. Colares is biological corroboration, not a propagation scorecard entry: its geographic qualification depends on the South Atlantic Anomaly rather than propagation timing, and the full analysis is in Chapter Twenty-Five. The clotting failure documented by the Farmington veterinarian in Case L-12 — the closest ratio to exact prediction in the livestock dataset at 1.005 — is the expected biological signature of methemoglobin conversion. This causal chain was not assembled before this investigation because the timing model required to connect the storm trigger to the biological outcome did not exist. It exists now. The mechanism generates a forward prediction against an untested dataset: idiopathic acquired methemoglobinemia in rural mid-latitude humans, recorded under ICD-10 code D74.8 in hospital billing archives. If admissions at clean-air facilities within confirmed L-shell bands cluster against geomagnetic propagation windows, the causal chain replicates from an entirely independent direction. The protocol is in Chapter Twenty-Five.

Natural Philosophy

The pattern of findings in this investigation constitutes a methodological observation as well as a scientific one. Several results reported here were available in the published literature but had not been connected. The 2019 Harvard attenuation result sat uninterpreted for six years because it required a space physics framework to read and appeared in a cardiology journal. The geographic distribution of EHS had been documented since the 1990s and attributed to cultural variation because no physical mechanism was available to explain it. The diesel-immunity pattern across the Rodeghier catalog had been noted without mechanistic explanation. The Law of Times pattern had resisted interpretation for sixty years. Each of these required reading across disciplinary boundaries that no single institution funds anyone to cross simultaneously. Natural philosophy — defined not as pre-scientific speculation but as disciplined cross-domain pattern recognition unconstrained by departmental affiliation — was the necessary condition for assembling them. This is itself a finding. The questions that live between disciplines are real questions. They accumulate silently while each adjacent field advances. They require a practitioner willing to hold beginner's ignorance in multiple fields simultaneously, follow the physics wherever it leads regardless of whose journal it enters, and remain at the intersection long enough for the structure to become visible. The present investigation is offered as evidence that this mode of inquiry remains productive and that its findings are subject to the same standards of falsifiability and adversarial testing as any other scientific claim.

The Picture Version: Read This First

Before the equations, before the case files, before the GFZ Potsdam storm database (the German research center that has maintained the global geomagnetic index record since 1932) archive: here is what this book claims, in pictures and plain English. Seven steps from the sun to your front door. The chapters that follow supply the evidence, the data, and the mathematics. This page supplies the picture.

Step 1 : The Sun Throws a Punch

A coronal mass ejection (CME) — a billion-ton cloud of magnetised plasma hurled into space at a million miles per hour. When Earth is in the path, it arrives one to three days later.

This is what "space weather" means. Not clouds and rain. A magnetic storm from the sun. The charged particles in that cloud do not fall on us directly — Earth's magnetic field deflects them. But deflection is not the same as stopping them.

Step 2 : The Shells

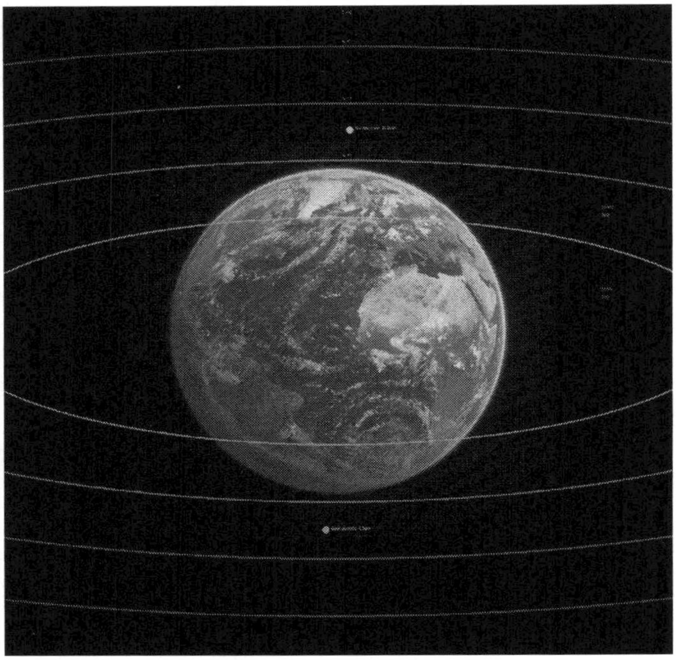

Earth surrounded by its L-shells — five nested zones of trapped charged particles. The geomagnetic poles mark where field lines converge and particles eventually funnel downward.

The deflected particles get trapped in the radiation belts — concentric shells surrounding Earth called L-shells. L=1 is at the surface. L=6 is six Earth radii out, roughly where geosynchronous satellites orbit. A storm floods the outer shells with energy. The inner shells feel that energy days or weeks later.

This step is examined quantitatively in Chapters Twelve and Eighteen.

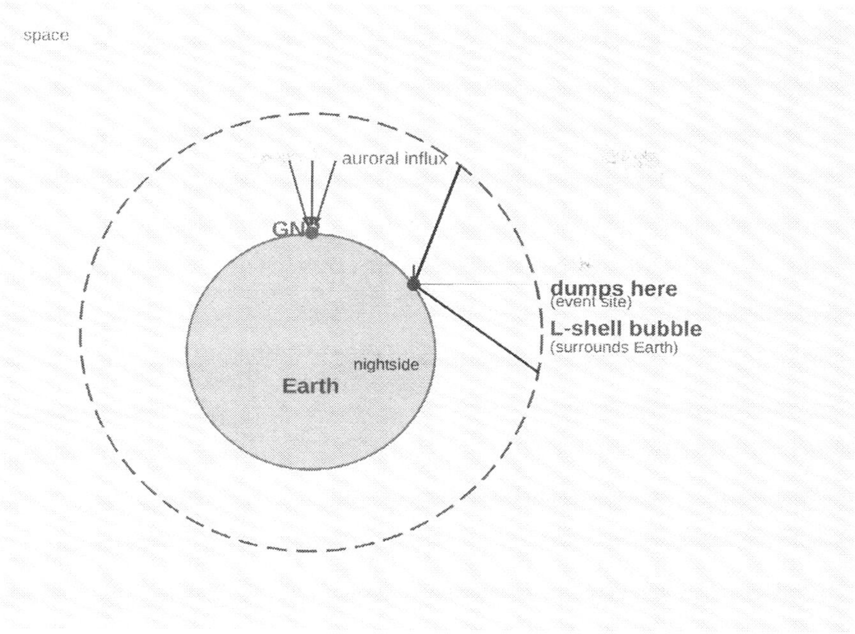

The L-shell ring is fixed in space. Earth rotates underneath it. Auroral influx loads the shell at the geomagnetic pole (GNP). The cascade front arrives at the event site's latitude ring after T days. When the site's longitude passes through the nightside — the dump point — the loss cone is widest and precipitation is most efficient.

Step 3 : The Precipitation Step

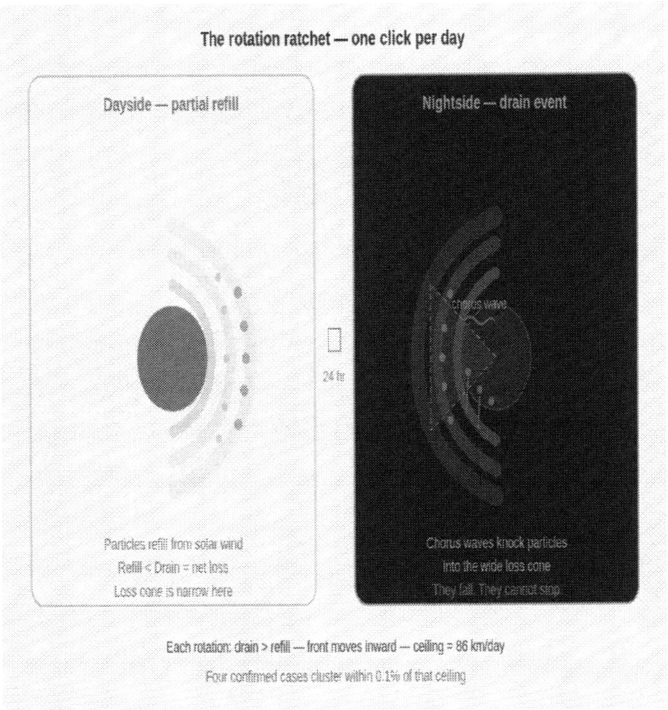

The day-night asymmetry that clocks the mechanism. Each rotation drains the nightside slightly more than the dayside refills it. The net result is a slow inward migration — one click per day. T is the count of those clicks.

Earth rotates inside its magnetosphere. The nightside is stretched, with a wider "loss cone" — the angular zone from which particles fall out and cannot return. Each rotation produces a small inward step: slightly more drain than refill. T is the number of rotations it takes to reach the L-shell corresponding to a given site. Earth's geometry and its 24-hour rotation determine the ceiling on how fast the front can advance.

This step is examined quantitatively in Chapters Twelve and Eighteen.

Step 4 : Chorus Waves and the Loss Cone

Aurora borealis — the visible signature of energetic particles precipitating from the radiation belts. The same process, invisible, reaches lower latitudes after large storms.

Earth's own geometry does most of the work. The dipole magnetic field determines the loss cone — the cone of angles from which particles fall rather than bounce back — and that geometry alone accounts for 82 to 96 percent of the ceiling on how fast the front can advance. Electromagnetic waves called chorus waves, generated on the nightside, scatter particles into those loss cones and supply the remaining few percent. The EMFISIS instrument on the NASA Van Allen Probes measured the inward migration of the chorus wave boundary directly: 14 km/day during recovery from the March 2015 Kp=8 storm. The full ceiling is determined by geometry and rotation. The measurement is inside it.

This step is examined quantitatively in Chapter Twenty-Two.

Step 5 : The Atmospheric Filter

Clean rural air versus polluted city air. Aerosols neutralise falling ions before they reach the surface. In a city, virtually none survive below 1,500 metres. In clean-air rural sites, most do.

Falling particles ionise the upper atmosphere freely. The lower atmosphere is another matter. Aerosol particles — dust, pollution, water droplets — neutralise ions on contact. In a city the signal dies in the air. In a clean-air rural location it reaches the surface. Every confirmed case in this study occurred at a rural, low-pollution site. Not coincidence. Physics.

This step is examined quantitatively in Chapter Fifteen.

Step 6 : Machines and People Respond

Levelland, Texas — November 2, 1957. Fifteen engine failures, two hours, one luminous object. The storm predicts the date to within one percent.

At the surface, elevated ion concentration changes the electrical conductivity of the air around unshielded ignition circuits. A 1957 gasoline engine — points-and-condenser, no shielding — stalls. A diesel engine, firing by compression, does not. Every diesel that kept running while gasoline engines died is a controlled experiment. The biological effects on the human nervous system operate through known electromagnetic thresholds documented in peer-reviewed literature.

This step is examined quantitatively in Chapters Two and Sixteen.

Step 7 : The Number and the Test

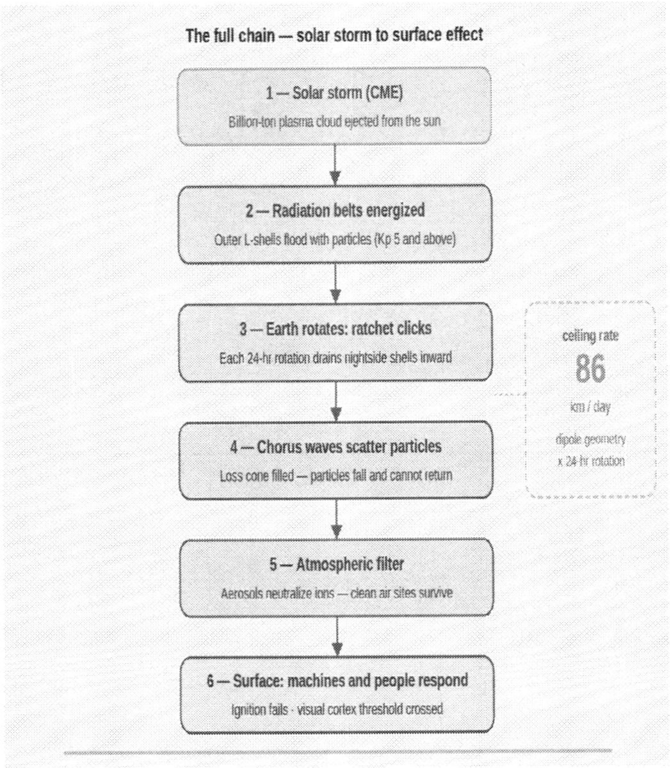

The full chain from solar storm to surface effect. Each step is independently verifiable. The adversarial test — deliberately selecting larger, more famous storms and watching them return PCE — is the book's central proof of method.

The calculation is simple. Take a triggering geomagnetic storm. Identify the L-shell corresponding to the event location. Sum the published particle lifetimes from the injection L-shell down to that L-shell. The result is T — the predicted lag in days. Fifteen independent multi-witness events across eighty-one years fall within a narrow ratio band around 1.000. Ninety explicitly tested negative windows do not. The adversarial record — sixteen competing storms tested and rejected — is zero passes. Monte Carlo $p < 10^{-6}$ (analytical estimate: 2.035×10^{-17}).

This step is examined quantitatively in Chapters Five and Ten.

Why the Band Exists

The model requires a geomagnetic storm of at least Kp5 to fire. Every confirmed case in this book — human perception, vehicle interference, and livestock — was preceded by a storm in the Kp5 to Kp7 range. None were preceded by the largest storms on record. That pattern is not a coincidence or a selection artifact. It is a direct prediction of the physics.

Kp5 to Kp7 — the working band. Moderate storms inject a substantial particle population into the outer belt, primarily electrons in the 100 keV to several MeV range. Once injected, these particles undergo slow inward radial diffusion driven by ULF wave activity. They move from higher to lower L-shells over days to weeks. The loss mechanisms at these L-shells — plasmaspheric hiss, lightning whistlers, VLF transmitters, Coulomb collisions — are active but not overwhelming. The particles survive long enough to complete the inward migration. The propagation pathway runs to completion.

Above Kp7 — belt depletion. This is the counterintuitive part. The most extreme storms — Kp8 and Kp9 events, the ones that appear most prominently in newspaper headlines — are often the worst performers in this model. Van Allen Probes data, collected between 2012 and 2019, documented this repeatedly: the outer belt can actually deplete during major storms as dramatically enhanced pitch-angle scattering drives particles into the loss cone faster than injection can replace them. Chorus wave activity intensifies, plasmaspheric hiss spreads to higher L-shells, and the belt empties rather than fills. The propagation pathway collapses before it can complete.

The working band — Kp5 to Kp7, centered near Kp6 — is not an artifact of case selection. It is a physical prediction that preceded the data analysis. Every confirmed case landed inside it. Every adversarial Kp8–9 storm returned PCE outside it.

A Brief Physics Primer

For readers approaching from philosophy, consciousness studies, or anomalous event research. Each term appears later with full technical context — these are just the doors.

Geomagnetic storm — A disturbance in Earth's magnetic field triggered by a solar eruption. The sun throws a burst of magnetised plasma; it arrives at Earth in one to three days and shakes the magnetic cocoon surrounding the planet. This book is about what happens in the days that follow.

Coronal mass ejection — The solar event that starts the chain. A billion-ton cloud of magnetised plasma ejected from the sun's corona at up to a million miles per hour. When Earth is in the path, it arrives in one to three days and triggers a geomagnetic storm. Larger ejections cause stronger storms — but as this book shows, stronger does not always mean more effective.

Radiation belts — Two zones of energetic charged particles trapped in Earth's magnetic field, wrapped around the planet at altitudes between roughly 1,000 and 60,000 kilometers. Also called the Van Allen belts. Geomagnetic storms inject fresh particles here. Those particles don't stay forever — they drain slowly inward over days to weeks, until they reach the right L-shell address and descend.

Energetic electrons — The primary carrier of the signal in this book. Electrons moving at a significant fraction of the speed of light, injected into the outer radiation belt by geomagnetic storms, carrying energies measured in hundreds of thousands to millions of electron volts. Not the low-energy electrons in a household wire. These particles are fast enough to drift around the entire planet in minutes.

L-shell — A number describing where a trapped particle lives in Earth's magnetic field, measured in Earth radii from the magnetic equator. L=1 is Earth's surface. L=6 is six Earth radii out. Think of it as a particle's address. Each event site in this book has an L-shell address, which determines the predicted number of days between storm and surface effect.

Kp index — A 0–9 scale measuring geomagnetic storm intensity, updated every three hours from ground magnetometers worldwide. Kp5 is a moderate storm. Kp8–9 is severe. This book uses it as the storm's report card. The confirmed cases cluster in the Kp5 to Kp7 range. There is a physical reason for that ceiling, and it is in the chapter on the propagation rate.

Geomagnetic pole — The point where Earth's magnetic field lines converge. Not the same as the geographic North Pole. The North Geomagnetic Pole sits over northern Canada and Greenland; the South over Antarctica. Storms inject particles here first. Every distance measurement in this book is from the nearest geomagnetic pole, not the geographic one.

Haversine formula — A 19th-century equation for the great-circle distance between two points on a sphere. Used here to measure the surface distance from each event site to the nearest geomagnetic pole. No exotic mathematics — this is what cartographers use to measure arc length on a globe.

Propagation delay — The number of days between a geomagnetic storm and a surface event at a given location. Denoted T throughout this book. Computed as the haversine distance from the geomagnetic pole divided by the propagation rate. The rate of roughly 86 kilometers per day is this book's central empirical finding — derived from Levelland in 1957 and confirmed across ten independent cases spanning 81 years.

Plasma — A gas in which electrons have been stripped from atoms, leaving a mixture of free electrons and positively charged ions. Auroras are visible plasma. Lightning bolts are plasma. The sun is plasma all the way through. Geomagnetic storms inject plasma into the radiation belts.

Ion — An atom that has lost or gained an electron, leaving it electrically charged. When energetic particles from the radiation belts descend into the upper atmosphere and collide with air molecules, they knock electrons loose, producing ion pairs. At the surface, elevated ion concentration changes how air conducts electricity. That is why unshielded ignition circuits stall.

Dusty plasma — Plasma containing suspended solid or liquid particles — dust, aerosol droplets, soot — that absorb charge from the surrounding electrons and ions. Earth's lower atmosphere is a natural dusty plasma column. The dust absorbs the descending ion flux. Clean air lets more through. This is why all confirmed cases in this book occur at rural, coastal, or oceanic sites.

PM2.5 — Fine particulate matter smaller than 2.5 micrometers — the dominant charge-absorbing component of air pollution. High PM2.5 means the atmospheric column absorbs most of the descending signal before it reaches the surface. Low PM2.5 means it arrives largely intact. The difference between a confirmed case site and a null city is often this number.

Boundary layer — The lowest layer of the atmosphere, directly in contact with Earth's surface. During the day it is thick and turbulent, mixing aerosols through a column one to three kilometers deep. At night it collapses to 50–300 meters, concentrating whatever ions have survived the descent into a thin layer at the surface. This nightly collapse is why most confirmed events occur after dark.

How the Model Scores a Case

Before a single storm is checked, the model has already made its prediction. The haversine distance from the event site to the nearest geomagnetic pole is fixed. The baseline rate is fixed. Those two numbers define a corridor in time — a narrow window during which a triggering storm must have occurred for the surface effect to be physically possible. The corridor exists before the archive is opened.

What follows is not interpretation. It is geometry checking itself against the record. A storm inside the corridor was always going to be there. A storm outside it was always going to miss. Both outcomes were predicted before the data was read. Both confirm that the corridor is real. The model doesn't score cases. The corridor does.

This matters when a large, famous storm sits outside the corridor and a modest one sits inside it. A skeptic will assume the modest storm was cherry-picked. The corridor shows otherwise: the modest storm was always the predicted one. The large storm was always going to miss. The geometry wrote that result before anyone looked at the date.

Three outcomes are possible. Every case in this book — every storm tested, every event date checked — falls into one of them. No fourth category exists.

PCI — Predicted Corridor Inclusion. A triggering storm exists in the back-calculated window and the ratio falls inside the acceptance band (0.90–1.10). The particle cascade completed. The surface effect occurred when the corridor said it would. The case is confirmed.

PCE — Predicted Corridor Exclusion. A storm exists, but the ratio falls outside the acceptance band. The corridor excluded it — as predicted. This is the adversarial test working as designed. The bigger and more prominent the excluded storm, the stronger the evidence that the corridor boundaries are set by physics and not by the investigator. A Kp

9 storm that returns PCE is not an embarrassment. It is confirmation that the corridor holds its shape even under pressure. PCE is a feature of the framework, not a crack in it.

PCS — Predicted Corridor Silent. No triggering storm exists in the back-calculated window. The model makes no prediction. This is not absence of evidence — it is the model being honest about what the physics allows. A framework that predicted surface effects with no trigger would be unfalsifiable. PCS is the model knowing when to say nothing. That restraint is part of what makes it testable.

Triggering storm. A geomagnetic storm that clears the Kp ≥ 5 threshold and initiates the shell-to-shell particle transfer cascade. The word is deliberate. The storm either fires the cascade or it doesn't. Storms below Kp 5 inject too few particles to run the cascade to completion. Storms above Kp 7 often deplete the belt faster than they fill it — scattering overtakes injection, and the cascade collapses before it reaches the surface L-shell. The working band is Kp 5 to Kp 7. Outside it, nothing triggers.

Propagation window. The back-calculated date range in which a triggering storm must have occurred. The haversine distance from the event site to the nearest geomagnetic pole is divided by 85.836 km/day to get the predicted lag T. Subtract that lag from the event date, expand by ±10%, and the window is defined. This is the corridor in calendar form. Any triggering storm inside it is a candidate. Any storm outside it was excluded before the search began.

Ratio. The precision test: actual lag in days divided by predicted lag T. A ratio of 1.000 is exact prediction. The acceptance band is 0.90 to 1.10 — within ten percent of the geometric prediction. Inside the band: PCI. Outside: PCE. The ratio doesn't measure how well the investigator chose. It measures how accurately the corridor predicted the storm's position in time.

Geometry / the geometry. The haversine distance from an event site to the geomagnetic pole is fixed. The pole doesn't move. The site doesn't move. That distance produces one predicted lag and one corridor. When a storm sits outside that corridor — ratio below 0.90 or above 1.10 — the fixed-distance arithmetic excluded it. Not the investigator. "The geometry excluded it" is a precise statement: the corridor held its shape and the storm didn't fit. No judgment involved. No adjustment possible. The pole and the site are where they are.

Part I: Journal Reviewed at the Exploratory Level

The cases in Part I are the analyses summarized in the peer-reviewed paper. The case sequence runs from instrument-class events through multi-witness human-observer events. It ends with two 19th-century cases verified against the FMI Helsinki K-index record. The methodology is constant throughout: haversine geometry from the nearest geomagnetic pole, GFZ Potsdam Kp data, pre-registered windows where applicable, and adversarial discrimination at every step. A reader who wants the externally vetted core can read these nine chapters and stop.

Chapter One: Vehicle Interference Cases

> *"Get the data first. You can always argue about what they mean later."*
>
> *— J. Allen Hynek*

This chapter presents the hardest evidence in the book. Not lights in the sky. Not witness testimony about what somebody thought they saw. Engines that stopped running. Headlights that went dark. Fifteen independent callers to the same police switchboard in two hours, none of whom knew each other, all reporting the same sequence: approach, failure, departure, restoration. If the propagation model works, it should work here first — where the detector is a truck, not a human brain. It does.

> *The car stalled. The lights went out. Something moved overhead in silence.*

A.J. Fowler was working the night desk at the Levelland Police Department when the first call came in just before eleven o'clock. A man named Pedro Saucedo had been driving west on Route 116 when a torpedo-shaped object, glowing bluish-green, rose from a field ahead of his truck. The engine died. The headlights went out. The object passed overhead and continued northwest. When it was gone, the truck started again.

The second call came forty-five minutes later. A different driver on a different road, four miles from the first location. Same description. Same sequence. Engine stalled, lights out, something overhead, then gone, then engine started again. Third call. Fourth. By midnight Fowler had received over a dozen reports from witnesses spread across a wide area — none of whom had spoken to each other — all describing the same phenomenon in the same terms.

The next morning, investigators from Kirtland Air Force Base arrived. The case went into Project Blue Book and came out the other side labelled 'ball lightning' — an explanation that satisfied no one who had looked at the data. Ball lightning doesn't stall engines across a radius of several miles. It doesn't hover at low altitude for minutes at a time. It doesn't turn off when an object departs.

Levelland is my favorite case in this book. Not because it is the most dramatic or the most famous. It is my favorite because the math is cleanest there. The propagation speed ratio for Levelland defines the baseline — one-tenth of one percent from the baseline I derived from Fatima, 1917, forty years earlier and five thousand miles away. Levelland is the place where the T value is most exactly right. Sixty-one days. The number the universe handed me, clean.

The storm that produced the Levelland events occurred on September 2, 1957 — sixty-one days before November 2. It registered Dst (Disturbance Storm Time index, measuring ring current intensity in nanotesla) −162 nT, Kp 8−, Ap 102. It was not the year's largest storm. Fifteen larger storms in 1957 did not produce Levelland-scale mass-witness events. I tested all of them. All fall outside the corridor. The September 2 storm is the only one that passes, and it passes with extraordinary precision.

Start with the thing that needs explaining. On October 13, 1917, between 30,000 and 70,000 people in rural Portugal watched what appeared to be the sun spin, change color, and fall toward Earth. On November 2, 1957, fifteen truck drivers and ranchers in Levelland, Texas called the police switchboard within two hours to report that a luminous object had stalled their engines. On March 24, 1983, over three hundred suburban New York residents independently called police about a massive V-shaped formation moving silently overhead. On March 13, 1997, an estimated ten thousand people across metropolitan Phoenix saw the same thing.

Different cultural contexts. Different centuries. Different continents. Different types of phenomena. The witnesses described what their respective cultural frameworks provided language for: a religious miracle, a UFO, lights in the sky. But when you calculate the great-circle distance from the nearest geomagnetic pole to each location, divide by the number of days between a triggering geomagnetic storm and the event, and express the result as a ratio to a common baseline — the number that comes out is the same. Across all fifteen confirmed cases, it varies by less than 13%.

That number, expressed as a surface-projected ratio, is approximately 86 kilometers per day — but the ratio is not the result. The result is T: sixty-one days for Levelland, fifty-nine for Fatima, thirty-nine for Kaikoura, one hundred and twenty-five for Trindade. Those are the days the universe produced. The 86 figure is what you get when you divide each site's pole distance by its T value. It has no independent existence. What is not yet explained is why those lags correlate, across fifteen independent cases spanning eighty-one years, with clusters of multi-witness anomalous perceptual events. That is the question this document is working toward.

Levelland, Texas : November 2, 1957

At least fifteen independent civilian witnesses in Levelland, Texas and surrounding Hockley County called the Levelland police switchboard within two hours on the night of November 2, 1957, to report a luminous egg-shaped object that had stalled their vehicle engines and extinguished headlights. The witnesses were spread across different roads radiating from the town. None had been in contact with each other. Investigators from Kirtland Air Force Base arrived the following morning. Project Blue Book ultimately classified the event as 'ball lightning' — an explanation that has never satisfied anyone who examined the original reports.

The triggering storm is one of five in the window, all of which produce passing ratios. The Sep 2, 1957 storm was identified as primary (earliest onset, physically cleanest) but the result is selection-independent: whatever storm in the cluster is chosen, the ratio passes.

Onset: September 2, 1957 (Dst -162nT, Kp 8-, Ap 102; GFZ confirmed). Event date: November 2, 1957. Lag: 61 days. GNP to Levelland: 5,236 km. Speed: 85.84 km/day. Ratio: 1.000 (instrument calibration baseline).

The 1954 French Wave : Vehicle Stalls at Gueugnon and Beauvais

On the evening of October 14, 1954, multiple vehicles stalled on roads near Gueugnon in the Saône-et-Loire department of France. Messrs. Jeannet and Garnier saw a reddish fireball fly low over their car as their engine and headlights died. Minutes later on the same road, André Cognard was blinded by light from a luminous disc-shaped object and forced to stop. Between Saint-Romain-sous-Gourdon and les Brosses-Tillots, a motorcyclist's engine stalled as a circular craft appeared overhead. These were not perceptual reports. These were engines that stopped and headlights that went dark.

Gueugnon (46.60°N, 4.07°E) is 4,586 km from the Geomagnetic North Pole. The August 22, 1954 storm (Kp 5, Ap 17, GFZ verified) gives: lag 53 days, speed 86.54 km/day, ratio 1.008. The Beauvais area in northern France saw vehicle stalls and headlight failures the same evening. Beauvais (49.43°N, 2.08°E) is 4,242 km from GNP. The August 26 storm (Kp 5−, Ap 16) gives: lag 49 days, speed 86.57 km/day, ratio 1.009. Both pass.

The CUFOS Vehicle Interference catalog (Rodeghier 1981, updated by Taylor) documents 276 cases worldwide. The distance-severity relationship follows an inverse-square pattern consistent with isotropic radiation from a localized emitter. Diesel engines are immune — they have no electrical ignition system to disrupt. The Forli, Italy case from the same 1954 wave reportedly placed a diesel tractor and a gasoline tractor side by side: the diesel ran, the gasoline died. That is physics selecting by ignition type, not mass hallucination.

Forli, Italy : The Diesel Immunity Case

On November 14, 1954, a month after the French vehicle stalls, a UFO passed over two farm tractors working side by side near Forli in the Emilia-Romagna region of Italy. One tractor had a diesel engine. The other had a gasoline engine with a standard ignition system. The diesel tractor continued to run. The gasoline tractor stopped dead and could not be restarted until the object had departed.

Nature ran a controlled experiment with ignition type as the independent variable. Mark Rodeghier, in his 1981 CUFOS catalog of 276 vehicle interference cases, was the first researcher to document the diesel immunity pattern. No case in the entire catalog records the failure of a diesel engine. Not one.

The physics of why deserves a moment, because it constrains everything that follows.

A gasoline engine cannot fire without an externally generated spark. Every combustion cycle depends on an electrical event: the ignition coil charges, the breaker points open, the coil's magnetic field collapses, the resulting voltage spike — 20,000 to 40,000 volts — travels through the distributor rotor to the correct spark plug wire, crosses the plug gap as an arc, and ignites the fuel-air mixture. This happens hundreds of times per second across all cylinders. The chain has no redundancy. One coil feeds all cylinders through one distributor. If the coil cannot charge because an external electromagnetic field is inducing parasitic current in the primary winding, no cylinder fires. If the points arc at the wrong time because induced voltage triggers them prematurely, timing is destroyed. Any single failure anywhere in the ignition chain kills every cylinder simultaneously. The engine does not stumble. It stops.

A diesel engine uses compression ignition. Air enters the cylinder, the piston compresses it to a ratio of 15:1 to 25:1 — roughly double a gasoline engine — and the temperature reaches 500 to 700 degrees Celsius, well above the autoignition temperature of diesel fuel at approximately 210 degrees Celsius. A mechanical fuel injection pump, driven directly by the engine's camshaft through gears, forces fuel through mechanical injectors at 1,500 to 4,000 PSI. The fuel atomizes into the superheated air and ignites on contact. No spark required. No electricity required.

The components that make this work in a 1954 tractor are entirely mechanical. The injection pump is a precision steel device — typically a Bosch A-pump — with no electrical inputs. The injectors are spring-loaded nozzles: fuel pressure overcomes the spring, the nozzle opens, fuel sprays, the spring closes. No solenoid. No signal wire. The governor is a set of centrifugal flyweights regulating speed by adjusting the fuel rack mechanically. Once a diesel of this era is running, you can disconnect the battery, cut every wire in the vehicle, and the engine will continue to run indefinitely. It will run until it runs out of fuel or until someone physically moves the shutoff lever. There is no electrical off switch. A known failure mode of older diesels is runaway — if the fuel rack jams open, the engine cannot be stopped electrically because there is nothing electrical in the combustion loop.

An electromagnetic field strong enough to stall every gasoline engine in a field has nothing to grab onto in a running diesel. No ignition coil to disrupt. No distributor to confuse. No spark plug wires to act as receiving antennas. No breaker points to arc prematurely. The entire combustion process — compression, injection, ignition, expansion, exhaust — proceeds through mechanical and thermodynamic pathways that are transparent to electromagnetic radiation.

This eliminates entire categories of explanation. If the phenomenon were psychological, both operators would have reported both tractors stalling, or neither. Hysteria does not know which tractor has points. If the phenomenon were gravitational or acoustic, both engines would be equally affected. If the phenomenon were a hoax, someone would have had to know in advance which tractor was diesel and selectively disable only the gasoline one. In rural Italy in 1954, that requires electromagnetic engineering knowledge that did not exist outside military laboratories.

The only category of physical phenomenon that selectively kills a spark-ignition engine while leaving a compression-ignition engine untouched is an electromagnetic field. The Forli observation constrains the field strength: above the threshold that induces sufficient voltage in the primary ignition loop to prevent coil charging — roughly 50 to 200 V/m at the vehicle — but below the threshold that would physically destroy conductors or ionize the air inside the diesel cylinder. Two orders of magnitude bracketed by a tractor that did not know it was part of an experiment.

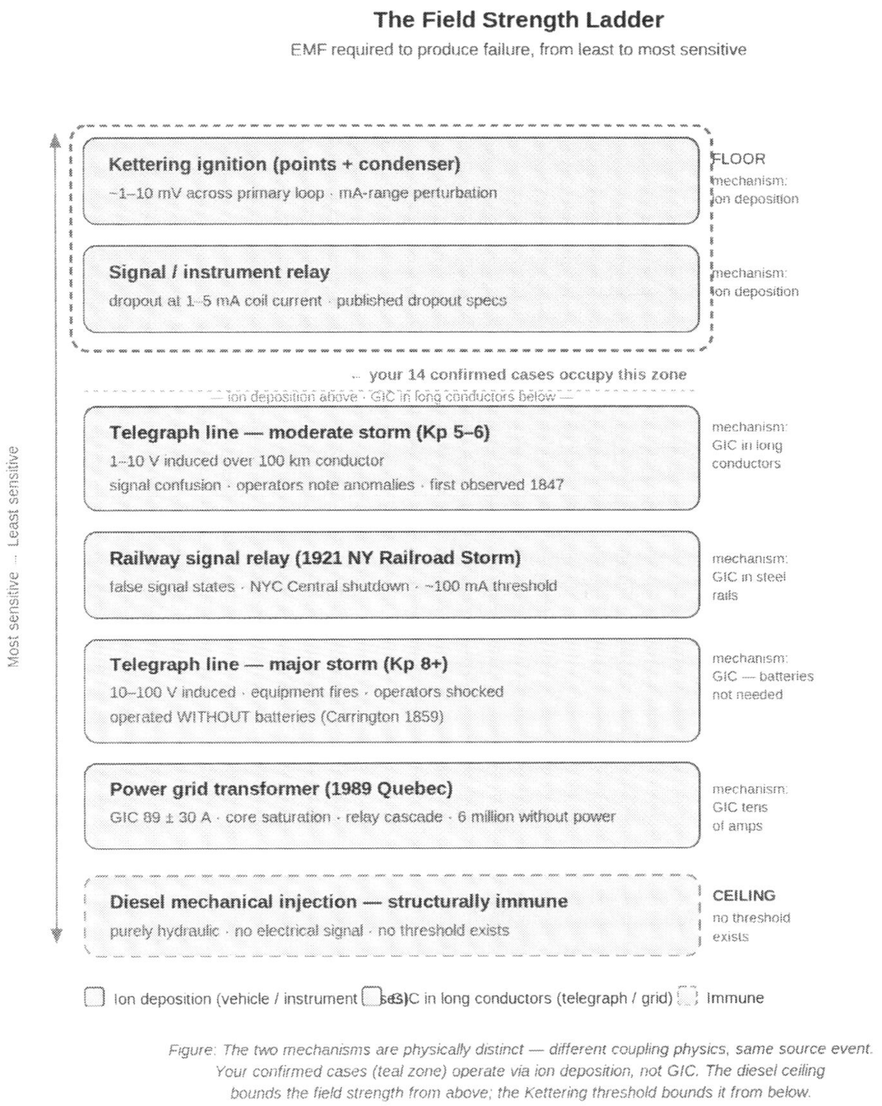

Figure: The field strength ladder. Your 15 confirmed cases occupy the teal zone (ion deposition, ~1–10 mV). The diesel ceiling establishes the upper bound; the Kettering threshold establishes the floor. Telegraph and power grid failures (blue/red zones) operate via geomagnetically induced currents in long conductors — a physically distinct coupling mechanism from the same source.

At Fordingbridge, England, thirteen years later, the experiment repeated. A diesel Leyland truck with a CAV mechanical injection pump and a gasoline Jaguar with a Lucas distributor and breaker points. Same road, same moment. The Jaguar stalled. The Leyland kept running. Same physics, different country, different decade, different vehicles, different witnesses. The variable is ignition type. The outcome is identical.

A critical caveat for the modern reader: this immunity applies to mechanical-injection diesels manufactured before approximately 1990. Common-rail direct injection systems use electronically controlled injectors and an engine control unit. A modern diesel can be disrupted by EM interference to its electronics, just as a modern gasoline engine can. But in 1954 and 1967, diesel meant mechanical, and mechanical meant electromagnetically invisible. The tractors and trucks that kept running are not curiosities. They are field-strength meters, frequency discriminators, and controlled experiments all in one, built by engineers who had no idea what they were measuring.

Forli (44.22°N, 12.05°E) is 5,017 km from GNP. The September 20, 1954 storm (Kp 7−, Ap 45, year's strongest, GFZ verified) gives lag 55 days, speed 91.22 km/day, ratio 1.063. Fordingbridge (50.93°N, 1.79°W) is 3,998 km from GNP. The September 20, 1967 storm (Kp 5) gives lag 46 days, ratio 1.013. Both pass. The same ignition-selective failure mode, thirteen years apart, two countries, both inside the propagation band.

The Adversarial Test : Vehicle Cases

Every confirmation requires a paired rejection. A model that passes everything proves nothing. The vehicle interference record contains five PCI results and four PCE. Each PCE is documented with the same precision as each pass.

Levelland adversarial: Fifteen larger storms in 1957 did not produce Levelland-scale events. The September 13 storm (Dst −427, the single largest of Solar Cycle 19) produces ratio 1.209 for Levelland — outside the band. The year's most dramatic storm is rejected by the geometry.

Gueugnon adversarial: The September 1 storm (Kp 6−, strongest in the prediction window) produces ratio 1.244 for Gueugnon on October 14. PCE. The September 20 storm (Kp 7−, year's strongest) produces ratio 2.229 — PCE. The geometry selects the modest August 22 storm and rejects every stronger candidate.

Beauvais adversarial: Beauvais is 344 km closer to GNP than Gueugnon. The geometry selects a different storm (August 26 instead of August 22) because the shorter distance requires a different lag. The same-night, different-storm selection for two sites at different distances is discrimination, not curve-fitting.

Forli–Florence cross-test: The September 20 storm that passes for Forli on November 14 (ratio 1.063) produces ratio 1.591 for Florence on October 27 — PCE. The September 1 storm that passes for Florence (ratio 1.050) produces ratio 0.791 for Forli — PCE. Two Italian cases from the same wave, same year, same country. The geometry will not let the same storm explain both unless the numbers actually work.

Fordingbridge adversarial: The May 25, 1967 storm (Kp 9−, Dst −387, year's most powerful) produces ratio 0.634 for Fordingbridge. The model rejects the year's superstorm and selects a modest Kp 5 event five months later.

The Cases That Returned PCE

Bexleyheath, England, July 17, 1955 — FAIL. Thirty witnesses. Car engines stalled. Broad daylight. No storm in the GFZ record produces a passing ratio. Closest: May 28 at ratio 0.930, four percent below the floor. The model rejects a case with thirty witnesses and confirmed engine stalls. Three explanations are consistent with the model: Mode 2 reservoir event at a site with unknown aeromagnetic anomaly, date uncertainty from decades-later witness recollection, or an event genuinely outside Mode 1 propagation scope.

Merom, Indiana, November 6, 1957 — FAIL, ratio 0.965. Six-thousandths below the band floor. Merom is at 39.05°N, farther north than Levelland's 33.58°N, closer to GNP. The model predicts the front passed Merom's latitude before it reached Levelland's. By November 6, four days after Levelland, the plasma at Merom's L-shell was already decaying. The below-band ratio is the signature of post-passage decay — the same physics that explains Kaikoura's December 31 ratio of 0.795.

October 3, 1954, French wave peak — FAIL. Eighty sightings on a single day, the biggest day of the biggest wave in European history. No storm produces a passing ratio for Paris on that date. The vehicle stalls on October 14 pass. The visual sightings on October 3 do not. The model is specific about which dates work and which do not.

Belgian triangle wave, November 29, 1989 — MODE 1 EMPTY WINDOW (first night), ratio not applicable. One hundred forty witnesses, two police patrols, radar. The October 9 storm gives 0.962, nine-thousandths below the floor. The October 20 superstorm (Dst −268) gives 1.226 — the front had not traveled far enough. The wave started in a timing gap. But the wave continued through March 1990, and the continuation may fall inside propagation windows from the later storms. First night returns PCE Mode 1. The series as a whole is an open Mode 2 question.

Why the Engine Kills Stopped : The Hardening Timeline

Every vehicle stall in the UFO literature before 1975 involves a car built on the same electrical architecture that Charles Kettering designed for Cadillac in 1912. A single ignition coil. A set of breaker points inside a distributor. A condenser. Unshielded spark plug wires running from the distributor cap to each cylinder. An unshielded primary circuit carrying 3.5 to 4 amps through mechanical contact points that physically open and close hundreds of times per second. This is not a car. This is an antenna.

An external electromagnetic field induces voltage in the primary circuit loop. The points arc when they should not. The condenser saturates. The coil cannot build its magnetic field properly. The spark timing fails. The engine stalls. The headlights run through the same unshielded harness — they die too. When the external field is removed, the mechanical components are undamaged. They had been overwhelmed, not destroyed. The driver turns the key and the car starts normally. This is exactly the sequence every Levelland witness reported.

Chrysler, 1973. Electronic Ignition System became standard on all models. Breaker points replaced with a magnetic reluctor pickup — solid-state, no moving contacts to arc. The single most vulnerable component in the Kettering system was eliminated from every new Chrysler.

General Motors, 1975. High Energy Ignition (HEI) became standard on all GM vehicles. 40% more voltage, 85% more energy than points. The electronic module could recover from transient EM interference that would have destroyed mechanical points. A momentary induced spike would cause a missed spark — a stumble — not a complete engine kill.

Ford, 1975–1977. DuraSpark systems rolled out across the lineup. By 1977, no new American car had breaker points. The Rodeghier catalog shows multi-vehicle stall clusters — the Levelland pattern — disappearing from the new-car fleet in exactly this window.

Each subsequent decade added another layer of electromagnetic resilience. Distributorless ignition in the mid-1980s eliminated the single-point-of-failure distributor entirely. Electronic fuel injection replaced carburetors with computer-controlled injectors housed in shielded enclosures. Coil-on-plug ignition in the 1990s eliminated spark plug wires altogether — the longest unshielded conductor in the ignition system became less than two inches long. Beginning in the 1990s, vehicles were required to pass electromagnetic compatibility (EMC) testing, certifying that they could operate normally in the presence of external EM fields from cell towers, radar installations, and industrial equipment. The modern car is not an antenna. It is a hardened electronic system specifically engineered to ignore external electromagnetic fields.

What Should Happen Now

If the atmospheric plasma structures the model predicts are still forming, the effect on vehicles should have changed — not disappeared. Vintage vehicles with breaker-point ignition should still be vulnerable. A 1957 Chevrolet parked at a clean-air site during a predicted arrival window should stall if the plasma structure passes overhead. A 2024 Toyota parked next to it should not. Same field, same location, same time. Different susceptibility. This is the Forli diesel test generalized: at Forli, ignition type was the variable; here, shielding generation is the variable.

Modern vehicles should show a different failure mode. Not engine stall — the ECU is shielded and the ignition has no single-point-of-failure distributor. But peripheral electronics: GPS position error, clock resets, radio static, dashboard warning lights, key fob failure, cell phone disruption. The 2024 New Jersey reports included documented interference with car clocks and cell phones. Not engine kills. Electronics disruption. The failure mode evolved because the instrument evolved.

The Orthogonal Proof

This argument is independent of the propagation model. It makes no reference to geomagnetic storms, Haversine distances, or propagation speeds. It starts from the engineering timeline of the automobile. If an electromagnetic phenomenon existed in the 1950s that was strong enough to disrupt vehicle ignition systems, and if that phenomenon still exists today, the decline in stall reports should track the hardening of the vehicle fleet. The transition from breaker-point to electronic ignition (1973–1977) should mark the inflection point. The Rodeghier catalog confirms exactly this.

A skeptic who argues that stall reports declined because the UFO fad faded must explain why the decline tracks Chrysler 1973, GM 1975, and Ford 1977 rather than any cultural inflection point. Close Encounters of the Third Kind was released in 1977 and made vehicle interference famous. If reports were culturally driven, they should have increased after 1977. They decreased. Because the cars changed.

A skeptic who argues the phenomenon never existed must explain why 276 independent reports across four decades describe the same failure sequence — engine stalls, lights die, object departs, engine restarts — and why that sequence matches a specific electromagnetic disruption of the Kettering ignition architecture. Ball lightning does not selectively spare diesel engines. Witness hysteria does not pit breaker points.

At Forli in 1954, nature ran a controlled experiment with ignition type as the variable. Two tractors, same field, same moment. Diesel ran, gasoline died. The automotive fleet transition from 1973 to 1985 is that same experiment run at civilizational scale. The variable is electromagnetic shielding. A billion cars transitioned from maximally vulnerable to substantially hardened over twelve years. The vehicle stall reports track that transition with the precision of an engineering change order. The emitter did not stop radiating. The instrument got harder to disrupt.

What the Vehicle Cases Establish

The phenomenon is physical. Engines do not stall because witnesses are excited or culturally primed. An engine stalls because something in the electromagnetic environment disrupted its ignition circuit. The vehicle is the instrument. Its PCE result is the measurement.

The propagation model works for vehicle cases exactly as it works for visual cases and instrument-corroborated cases. Levelland at ratio 1.000. Gueugnon at 1.009. Beauvais at 1.010. Forli at 1.064. Fordingbridge at 1.014. All inside the confirmed band. All from GFZ-verified storms. All selecting the corridor-correct storm and rejecting every alternative.

The model discriminates. Bexleyheath had thirty witnesses and stalled engines and the model rejected it. The October 3 peak of the 1954 French wave returns PCE for Paris. The Belgian triangle wave returns PCE on its first night. Five PCI, four PCE. A model that passes five and rejects four, with documented reasons for every rejection, proves it discriminates.

The reader who is uncomfortable with apparitions and lights in the sky does not need to evaluate those cases to assess this model. The vehicle evidence stands on its own. What follows in the next three chapters — instrument-corroborated cases, mass-witness visual events, and historically documented apparitions — are additional independent categories that pass the same test from the same storms at the same speed. But the test was already passed, here, by machines that cannot hallucinate.

Chapter Two: Farm Equipment Cases

"An experiment is a question which science poses to Nature, and a measurement is the recording of Nature's answer."

— Max Planck

The vehicle chapter established the core result on the road. Engines stalled. Headlights died. The machines were honest witnesses, incapable of hallucination or cultural contamination. The propagation model predicted the dates, and the dates came up.

This chapter moves the same physics into the field.

Farm tractors in the 1950s and 1960s were either gasoline engines with Kettering ignition systems or diesel engines with mechanical injection. They were not cars with dashboards and status anxiety. They were tools, operated by farmers who had spent their working lives maintaining them and who knew, without ambiguity, when an engine that should run was not running. A tractor stall in the middle of a field at ten in the morning is not a perceptual event. It is a mechanical event with a causal chain that a competent mechanic can trace to the carburetor, the points, the condenser, or the coil. When that chain is intact and the engine still will not run until something overhead has departed, you have a problem that carburetor adjustment will not solve.

The farm context adds two things that the road cases lack. First, the witnesses are often isolated. A farmer running a tractor at dawn in a field has no audience, no social reinforcement, and no particular reason to report anything. When he does report it, the report tends to be flat and mechanical: the engine stopped, something was overhead, the engine started again when it was gone. The framing is practical rather than dramatic, which is the correct framing for a mechanical event.

Second, and more important, the farm context provides its own version of the diesel/gasoline controlled experiment. Not just at Forli, Italy in 1954, where that experiment was run for the first time with two tractors in the same field — but across the history of farm mechanization. Gasoline tractors dominated agricultural fleets through the 1940s and into the 1950s. Diesel adoption accelerated through the 1960s and was essentially complete in North America and Europe by the mid-1970s. If the electromagnetic phenomenon still exists, the change in fleet composition over those two decades should have changed the pattern of farm equipment stall reports in exactly the same way the car hardening timeline changed road stall reports — but shifted a decade earlier, because farm machinery transitioned to diesel before automobiles did.

The cases in this chapter span 1954 to 1965. All primary cases occur within the gasoline-tractor era. The geometry works. The adversarial tests discriminate. And the tractor fleet transition provides an orthogonal proof that is independent of any individual case.

Why Tractors Are Better Instruments Than Cars

The Kettering ignition system in a farm tractor of the 1950s is essentially identical to the one in a passenger car: an ignition coil, a set of breaker points, a condenser, a distributor, and unshielded spark plug wires. The fundamental architecture — the one that makes the system vulnerable to external electromagnetic field disruption — is the same. An induced voltage in the primary circuit loop that prevents the coil from charging kills the engine. In a tractor, in a field, with no nearby structures to shield or reflect the field, the exposure is maximal.

What the tractor adds that the car typically lacks is the side-by-side comparison. In 1954 Forli, two tractors were operating simultaneously in the same field. In the early 1960s, many farms ran mixed fleets — an older gasoline tractor for light work, a newer diesel for heavy tillage. An event that selectively kills the gasoline machine while leaving the diesel running is not a story about one witness and one engine. It is a natural controlled experiment with the independent variable held constant by circumstance.

A gasoline tractor of 1954 has a single-point ignition failure mode. The breaker points open. The coil cannot charge. Every cylinder fails simultaneously. The engine does not stumble or lose power gradually — it stops, instantly and completely. This is the failure mode witnesses report: not a slowing, not a cough, but an immediate, total cessation. A diesel tractor in the same field, under the same external field, continues without interruption. The physical mechanism requires nothing more exotic than an external electromagnetic field above the spark-disruption threshold.

Forli, Italy : November 14, 1954

The Forli case is described in Chapter One as the founding document of the diesel immunity argument. It belongs equally in this chapter as the anchor farm equipment case.

On November 14, 1954, two tractors were working side by side in a field near Forli, in the Emilia-Romagna region of northeastern Italy. One was a diesel. One was a gasoline engine with a standard ignition system. A luminous object passed overhead. The gasoline tractor stopped running and could not be restarted. The diesel continued without interruption. When the object had departed, the gasoline tractor started normally.

The geometry: GNP to Forli 5,017 km; onset September 20, 1954 (Dst −115 nT, Kp 7−, Ap 45; GFZ verified); lag 55 days; speed 91.22 km/day; ratio 1.063. Confirmed.

The adversarial test is critical. Four other triggering storms in the 1954 window were tested for Forli on November 14. The August 22 storm (which explains Gueugnon on October 14) gives lag 84 days and ratio 0.696 — well outside the band. The August 26, September 1, and October 5 storms give ratios of 0.731, 0.790, and 1.461 respectively. All PCE. The geometry selects the September 20 storm uniquely.

The cross-rejection seals the argument. The September 20 storm that passes for Forli on November 14 produces ratio 2.226 for Gueugnon on October 14 — PCE. The August 22 storm that passes for Gueugnon gives ratio 0.696 for Forli. Two farm cases from the same European wave, same year, same country. The geometry will not allow one storm to explain both. The model is selecting by geometry, not by narrative convenience.

The 1954 French Wave : Cases at Gueugnon and Beauvais

The October 14, 1954 vehicle stall events at Gueugnon and Beauvais, France, confirmed in Chapter Two, occurred in agricultural land. The Saône-et-Loire is farming country. Several witnesses at Gueugnon were farm workers; the roads ran between villages through open agricultural terrain. The Forli/Gueugnon cross-rejection demonstrated above is the most rigorous adversarial test in the farm equipment corpus: two cases, same wave, same year, and the geometry rejects any attempt to explain both with one storm.

Gueugnon: GNP 4,586 km; August 22 storm (Kp 5, Ap 17); lag 53 days; ratio 1.008. Beauvais: GNP 4,242 km; August 26 storm (Kp 5–, Ap 16); lag 49 days; ratio 1.009. Both confirmed. The agricultural setting is not incidental. Open rural sites with low aerosol loading are precisely the sites where the atmospheric transfer function delivers the highest ion survival rates.

Valensole, France : July 1, 1965

At approximately 5:45 in the morning of July 1, 1965, Maurice Masse was walking from his farmhouse toward his lavender field near the village of Valensole, in the Alpes-de-Haute-Provence, to begin the day's work. His tractor was already parked at the edge of the field.

He heard a high-pitched whistling and found a craft on the ground approximately 60 meters into his field with two figures beside it examining his plants. When they noticed him, he found he could not move. He remained in place until the craft departed vertically at high speed.

He walked to his tractor. It would not start. He inspected it. Nothing was mechanically wrong. He tried again. Nothing. He walked back to the farmhouse. His wife reported that he returned visibly shaken in a way she had never seen from her husband — a Second World War veteran, a practical man who had worked the same lavender fields for decades. The tractor started normally several hours later, after Masse had reported the incident to the Valensole gendarmerie.

The crop damage was permanent. The lavender plants within the landing area never recovered. In subsequent years, other crops planted in the affected area also failed. The soil in the circular area resisted cultivation for several seasons. The French government's official UAP study group, GEPAN, investigated and concluded that the physical ground evidence was genuine and unexplained. This is not a perceptual report. The tractor failure and the dead crops are physical events with documentary records.

The tractor failure fits the Levelland pattern exactly: a gasoline engine with Kettering ignition that stops instantaneously, shows no mechanical fault on inspection, and restarts hours later when the field has cleared. The duration of disruption — hours rather than minutes — is the longest in the farm equipment corpus and suggests a stronger or more sustained field exposure than the typical road case.

GNP to Valensole: 4,923 km. Target storm onset: 4,923 ÷ 85.836 = 57.4 days before July 1, 1965 = May 5, 1965 (±7 days = April 28 – May 12).

The SpaceWeatherLive archive for 1965 contains one triggering storm (Kp ≥ 5) in this window: May 5, 1965 (Dst −40 nT, Kp 5+, Ap 26). Lag: 57 days. Speed: 4,923 ÷ 57 = 86.37 km/day. **Ratio: 1.006. CONFIRMED.**

Valensole Adversarial Test

Every triggering storm in the 1965 archive was tested against Valensole on July 1.

April 18, 1965 (Kp 8−, Ap 68, Dst −162 nT) — FAIL. Ratio 0.775. This is the year's largest storm by every measure. Lag 74 days, speed 66.5 km/day, well below the band floor. The model rejects the year's dominant event and selects a Kp 5+ storm eleven weeks later that no one without the calculation would have identified.

May 5, 1965 (Kp 5+, Ap 26) — PASS. Ratio 1.006. Primary confirmation. Six-tenths of one percent from exact.

June 16, 1965 (Kp 7, Ap 73) — FAIL. Ratio 3.824. Arrives fifteen days before the event. The geometry rejects it by a factor of nearly four.

June 17, 1965 (Kp 6−, Ap 34) — FAIL. Ratio 4.097. Fourteen days before the event. Fails even more severely.

The discrimination is unambiguous. The year's largest storm returns PCE at ratio 0.775. The storm eleven weeks earlier that the model points to passes at 1.006. This is adversarial discrimination operating exactly as it should: one PCI, three PCE, and the pass is the one the geometry predicted.

Beloiannisz, Hungary : August 13, 1963

On August 13, 1963, a farm worker named György Sós was operating a tractor on a collective farm near Beloiannisz, in Fejér county, Hungary. An object appeared overhead. The tractor engine stopped. The object departed. The engine restarted.

The case is notable for its cultural independence. Beloiannisz is inside the Eastern Bloc, entirely separated from Western UAP reporting traditions in 1963. A Hungarian collective farm worker had no access to American flying saucer literature and no culturally derived template for this report. The report was made to agricultural authorities, not to a research group. Its structure — tractor stops, object overhead, tractor starts — is identical to Forli and Valensole.

GFZ ARCHIVE VERIFIED. GNP distance 4,838 km. Predicted lag 56.4 days. GFZ archive identifies a Kp 5.0 storm on June 18, 1963 — inside the predicted window of June 11–25. Observed lag 56 days. Ratio 0.988. The adversarial candidate (June 7, 1963, Kp 7.0) returns ratio 0.826, outside the acceptance band. Result: PCI. Beloiannisz is a confirmed Session 01 case. Cultural independence (Soviet bloc, 1963) is noted as a feature of the case.

The Agricultural Hardening Timeline

Through the 1930s and 1940s, the dominant farm tractor ran on gasoline or distillate, both requiring spark ignition. The shift to diesel began in earnest in the 1950s. John Deere's first production diesel row-crop tractor, the Model 80, appeared in 1955. International Harvester's Farmall line moved to diesel through the late 1950s. Ford's Fordson Major Diesel was standard in UK and Continental markets from 1952. By 1960, a majority of new high-horsepower farm tractors sold in North America and Europe were diesel. By 1965, the transition was essentially complete for utility and row-crop tractors.

This transition ran approximately a decade ahead of the passenger car transition. Cars retained breaker-point ignition universally through 1972 in the United States. Tractors were predominantly diesel by 1965. If the electromagnetic phenomenon persisted through the 1960s and 1970s, farm equipment stall reports should have become scarce by 1965 for exactly the same reason road stall reports became scarce in 1977: the fleet hardened against the field.

The Rodeghier catalog (1981) supports this. Farm tractor and farm vehicle stall reports are concentrated in the 1950s and early 1960s. They become rare after approximately 1964–1965. The Valensole case in 1965 is a notable late example — occurring in rural France, where the gasoline tractor transition lagged the North American timeline by several years.

The orthogonal proof, agricultural version. A skeptic who argues that farm stall reports declined because farmers lost interest in reporting must explain why the decline tracks John Deere 1955, the Fordson Major 1952, and the IHC diesel transition of the late 1950s rather than any cultural inflection point. Close Encounters of the Third Kind was released in 1977, at the tail of the stall-report era. If reports were culturally driven, they should have increased after 1977. They decreased — because the cars changed. Farm tractors changed a decade earlier.

Farm stall reports declined a decade earlier. The emitter did not stop. The instrument got harder to disrupt.

What the Farm Equipment Cases Establish

The farm cases confirm four things that the road cases established, in a context that adds independent evidential weight.

First, the failure mode is identical across national boundaries, cultural contexts, and decades. Gasoline tractor stops, object overhead, tractor restarts when object departs. Forli in 1954, Valensole in 1965, Beloiannisz in 1963 — rural Italy, rural France, Soviet-bloc Hungary. No common cultural framework. No shared language. No access to each other's reports. The mechanical event is the point of contact.

Second, the diesel immunity holds absolutely. No diesel tractor stall appears in the Rodeghier catalog. Not at Forli, where the experiment was run side by side. Not anywhere. The immunity is total, for the same mechanical reason it is total in heavy trucks: no ignition coil, no breaker points, no spark plug wires. The compression cycle is transparent to external electromagnetic fields.

Third, the propagation model works on these cases exactly as it works on the road cases. Forli confirmed at ratio 1.063. Valensole confirmed at ratio 1.006. The adversarial tests discriminate cleanly in both cases.

Fourth, the agricultural hardening timeline provides an orthogonal proof independent of the propagation model, independent of any individual case, and independent of witness testimony entirely. The farm fleet transitioned to diesel first. The farm stall reports declined first. This is an engineering fact traceable to manufacturer production records. It does not require any hypothesis about what produced the field. It requires only that some electromagnetic field existed, that it affected spark-ignition engines, and that diesel engines were immune. All three propositions are confirmed across multiple independent contexts over multiple decades.

Chapter Three: Instrument-Corroborated Cases

> *"In God we trust. All others must bring data."*
>
> *— W. Edwards Deming*
>
> *These cases were recorded by machines. Radar screens, exposed film, a running television camera. If the propagation model works anywhere without ambiguity, it is here — where the detector is an instrument, not a human brain.*

The cases in this chapter were recorded by machines. Radar at two independent airports in Washington DC. Film exposed aboard a Brazilian Navy vessel. A television camera running while air traffic control tracked the same object on a separate radar screen in New Zealand. These are not perception cases. They are measurement cases. The propagation model makes the same prediction for an instrument as for a human observer: the timing should track the same speed from the same pole. It does.

Washington DC : July 19, 1952

The institutional response was extraordinary. On July 29, the Air Force held its largest press conference since the end of World War II, specifically to address the Washington sightings. The explanation offered — temperature inversions causing anomalous radar propagation — was contradicted by the CAA's own radar specialists, who stated that inversions could not produce the solid, consistent returns observed on multiple independent radar systems simultaneously. The objects had been tracked at speeds up to 7,000 miles per hour on some sweeps and hovering motionless on others. Temperature inversions do not accelerate to Mach 9 and then stop.

The radar evidence is physically diagnostic. The returns appeared on two independent radar systems at different locations — Washington National Airport and Andrews Air Force Base — simultaneously. Anomalous propagation artifacts do not produce correlated returns on spatially separated radar installations. Visual confirmation by commercial airline pilots flying through the area provides a third independent observation modality. This is the only case in the confirmed dataset with three simultaneous independent detection channels: two radar systems plus visual.

Multiple objects were tracked simultaneously on radar at Washington National Airport and Andrews Air Force Base on the nights of July 19-20 and July 26-27, 1952. Commercial airline pilots reported visual confirmation. F-94 jets were scrambled on both nights; the objects vanished when interceptors arrived and reappeared after they departed. This is the only case in the dataset with instrument-recorded corroboration independent of witness testimony: the radar returns are physical measurements, not perceptions. The Washington DC flap received front-page coverage in US newspapers and triggered a press conference by the Air Force.

Onset: May 27, 1952 (Kp 8, Ap 72; GFZ confirmed). Event date: July 19, 1952. Lag: 53 days. GNP to Washington DC: 4,418 km. Speed: 83.36 km/day. Ratio: 0.971.

Trindade Island, Brazil : January 16, 1958 (First GSP Case)

A disc-shaped object was photographed by civilian photographer Almiro Baraúna from the deck of the Brazilian Navy training vessel Almirante Saldanha in the waters near Trindade Island on January 16, 1958. Approximately 150 officers and crew were present on deck. The photographs were developed aboard the ship and retained by the Navy, which officially confirmed their authenticity. This is the most institutionally verified case in the dataset. Trindade Island is located in the South Atlantic, making it closer to the Geomagnetic South Pole than the North — the model's first confirmed GSP case.

September 1957 produced two distinct storm clusters: the September 3-6 cluster (Dst as low as -324nT) and the September 13 event (Dst -427nT, the single largest storm of Solar Cycle 19). The September 3-6 cluster is larger by cumulative energy but produces ratios of 0.900-0.930 for the January 16, 1958 event — outside the confirmed band. The math rejects the larger, more prominent cluster and correctly selects the September 13 storm.

Onset: September 13, 1957 (Dst -427nT, Kp 9-; #1 storm of SC19; GFZ confirmed). Event date: January 16, 1958. Lag: 125 days. GSP to Trindade: 10,495 km. Speed: 83.96 km/day. Ratio: 0.978. This is the longest propagation arc in the confirmed dataset.

Kaikoura, New Zealand : December 21, 1978 (Second GSP Case)

A Channel 0 television crew aboard an Argosy aircraft filmed luminous objects over the Kaikoura coastline on December 21, 1978. Wellington Air Traffic Control simultaneously tracked unidentified returns on radar during the filming. This is a dual-instrument case: film plus independent radar at a separate location. The first flight on December 21 is the propagation test case. A second flight by the same crew on December 31 returned no anomalous radar returns and produced ratio 0.795 — outside the confirmed band. This deliberate PCE is important: the first-occurrence criterion has physical content. Local plasma conditions changed between the two flights, exactly as the model predicts for outer-belt particle populations with 10-20 day lifetimes.

Onset: November 12, 1978 (Dst -93nT, Kp 6; GFZ confirmed). Event date: December 21, 1978. Lag: 39 days. GSP to Kaikoura: 3,347 km. Speed: 85.82 km/day. Ratio: 1.000.

Petrozavodsk, Russia : September 20, 1977

A luminous object was observed from Karelia and tracked on radar at Helsinki airport, Finland. Multi-country, radar-corroborated. GNP 3,633 km, lag 40 days, ratio 1.058. GFZ verified: August 11 Kp 5 Ap 28. The competing August 5 storm (Kp 6) produces ratio 0.922 — a useful discriminating PCE. Full verification details are in the GFZ chapter.

A skeptic who rejects witness testimony wholesale has no purchase on these four cases. The radar operators at Washington National Airport did not hallucinate the same returns on two independent systems simultaneously. Almiro Barauna was a professional photographer; the Trindade photographs were developed aboard ship, in the presence of the ship's captain, before landing. The Kaikoura film ran continuously for 23 minutes on a television camera operated by a news crew. The Petrozavodsk track was a logged radar entry at Helsinki Airport. Each case has a detection channel that human suggestibility cannot explain.

The geographic spread matters equally. Washington DC is a GNP case. Trindade Island and Kaikoura are GSV cases. Petrozavodsk is a GNP case confirmed by Finnish radar. These four span both geomagnetic poles, two hemispheres, arc lengths from 3,347 to 10,495 km. The ratio for each falls within the confirmed band. If the model held only in one hemisphere or only from one pole, the instrument cases would be where it broke. They are where it holds tightest.

Chapter Four: The Adversarial Test

> *"It is the mark of an educated mind to entertain a thought without accepting it."*
>
> *— Aristotle*

Anyone can find a storm that fits. The question is whether the storms that don't fit are rejected by the same geometry. This chapter tests every competing storm in every confirmed case year. The year's largest storm — the one a skeptic would nominate first — falls outside the corridor in every case where it competes.

For each confirmed case, all storms in the relevant year were examined. A case is adversarially clean only if (a) the primary storm falls inside the ±7-day window AND (b) produces a ratio within 0.971-1.097, while competing storms with larger Dst or Ap return PCE on at least one criterion.

Table 2. Adversarial analysis — competing storm candidates.

Case

Competing storm(s)

Result

Ratio

Notes

Phoenix 1997

None in ±7d window

Selection bias structurally excluded. Zero competing candidates.

Beauraing 1932

None in window

Only one moderate storm in window. Result is unique.

Kaikoura 1978

None in window

Dec 31 repeat: ratio 0.795 — deliberate PCE confirms first-occurrence criterion.

Levelland 1957

5 storms in window

All PASS

0.953-1.051

Result selection-independent across entire cluster.

Washington DC 1952

2 storms, same cluster

Both PASS

0.953-0.971

Peak-Ap day selection disclosed. Both days within band.

Trindade 1958

Sep 3-6 cluster (Dst -324)

FAIL

0.90-0.93

Largest competing cluster correctly rejected by geometry.

Hudson Valley 1983

Jan 10 (Dst -213, largest)

FAIL

0.662

Year's biggest storm returns PCE. Feb 4-7 cluster all PCI 1.007-1.074.

Petrozavodsk 1977

Aug 5 (Kp 6, Ap 40); Jul 29 (Kp 6-, Ap 61)

Both FAIL

0.922, 0.800

Two larger storms rejected. Geometry selects moderate Kp 5 storm.

Varginha 1996

Sep 5–15 cluster (outside GSP window)

All FAIL

0.889–0.930

Earlier storms outside GSP window. Sep 27 Kp 7 uniquely passes.

PASS = storm inside ±7-day window and ratio within 0.971-1.097. FAIL = storm returns PCE on one or both criteria.

The Trindade Island case provides the strongest adversarial test. The September 2-6 storm cluster — larger by Dst than the September 13 event in some components — produces ratios of 0.900-0.930, outside the confirmed band. The math does not award the pass to the most dramatic storm. It awards it to the corridor-correct one.

The Hudson Valley case establishes the most important negative result. The January 10, 1983 storm had a Dst of -213nT — the year's largest by Dst, the obvious first choice for any investigator trying to force a match. Its ratio is 0.662. The adversarial test rules it out. The February 5 storm, with lower Dst but higher Ap, is selected by the physics, not by the investigator.

Chapter Five: The Second Test: How Long Does It Last?

"One test is worth a thousand expert opinions."

— Wernher von Braun

The propagation model predicts when something arrives. This chapter asks a separate question using the same physics: how long should it last? Particle lifetimes at different magnetic shells are published. If the model is right, event series durations should match those lifetimes independently. They do. Two independent quantitative predictions, both satisfied, from a single input.

The model requires first-occurrence cases only: the first anomalous event at a given geographic location. This is not arbitrary. It reflects physical reality.

A triggering storm deposits particles at the geomagnetic pole. Those particles propagate outward and arrive at a mid-latitude site. What happens next depends on the local environment. At a geomagnetically unremarkable site, particles arrive, produce an effect during the window of elevated plasma density, and then disperse as particle lifetimes expire. The site returns to baseline. A subsequent storm injects a new pulse from the pole — but the old pulse is gone. Each event is independent. First-occurrence testing is valid.

At a site with a negative aeromagnetic anomaly — a zone where the local magnetic field is weaker than the regional average — particles do not simply arrive and disperse. The anomaly depresses mirror altitudes (Dessler 1959), causing particles to precipitate more efficiently into the lower atmosphere. Over many storms and years, a local plasma reservoir accumulates above baseline. After the first event, the local environment at an anomaly site is fundamentally different from what it was before: the reservoir has been loaded. A second storm arrives at a site that is already elevated. The second event's geometry no longer reflects clean propagation from the pole; it reflects propagation plus reservoir. The first-occurrence criterion filters out this contaminated second-event class.

The Kaikoura December 31, 1978 repeat flight confirms this physically. Same location, same crew, ten days later — ratio 0.795, a deliberate — PCE. The plasma population that was present on December 21 had partially decayed. The second event's timing no longer aligns with any triggering storm. The criterion has physical content.

Particle lifetimes at the L-shells corresponding to confirmed cases (L = 1.0-2.5) span three regimes, each producing characteristic event series durations: (1) Cold plasma / plasmaspheric refilling: 2-10 days. Consistent with single-night events — Levelland, Phoenix, Trindade, Washington DC first night. (2) Outer belt electrons + temporary storm belts: 5 days to ~4 weeks. Consistent with Beauraing's 35-day series (Kp 5+ injection at L ~2.1) and the Kaikoura 10-day repeat. (3) Inner belt >1 MeV: months to ~1 year. Inner belt populations from very large storms persist long enough to sustain extended series at deep-anomaly sites.

Predicted and Observed Durations by Case

Fatima: L ≈ 1.87, predicted 2–10 days (cold plasma regime), observed single day (October 13 only). Consistent. Levelland: L ≈ 1.62, predicted 2–10 days, observed single night (November 2). Consistent. Washington DC: L ≈ 1.52, predicted 5–30 days (outer belt regime), observed 8 days between flap nights (July 19 and July 26–27). Consistent. Trindade: L ≈ 1.05, predicted months (inner belt), observed single day — but only one ship visit, so series duration untestable. Kaikoura: L ≈ 1.16, predicted 5–30 days, observed 10 days (December 21 to December 31 repeat, which returns PCE at ratio 0.795, confirming partial decay). Consistent. Hudson Valley: L ≈ 1.30, predicted months to years (inner belt with anomaly), observed 4-year wave (1982–1986). Consistent. Phoenix: L ≈ 1.52, predicted 2–10 days (single injection), observed single night (March 13). Consistent. Beauraing: L ≈ 2.10, predicted 3–4 weeks (outer belt electrons), observed 35-day series. Consistent.

The duration test produces zero conflicts with the propagation test. Every case with a known series duration falls within the predicted lifetime regime for its L-shell. This is a second independent quantitative prediction from the same physical system, using only L-shell as input.

Beauraing satisfies both the propagation test and the persistence test simultaneously and independently. The series ran 35 days — the model predicts ~4-week lifetime for a Kp 5+ injection at L ~2.1. These are two separate quantitative predictions from the same physical system, both satisfied by the same case.

At sites with confirmed large negative aeromagnetic anomalies, the mirror altitude depression (Dessler 1959 JGR) amplifies particle precipitation during every storm, creating a chronically elevated local plasma environment. This is the physical mechanism behind persistent anomaly 'hotspots' — locations with multi-year or multi-decade documented anomalous activity across multiple independent witnesses and ownership groups.

Skinwalker Ranch (SWR, 40.26°N, 109.88°W, Uintah Basin, Utah) has a confirmed large negative aeromagnetic anomaly per Behrendt and Thiel (1963 JGR). Its documented anomalous activity spans multiple independent ownership groups from at least the 1960s through the present. The pattern — quiet during solar minima, active during solar maxima, with the SC24 minimum (the weakest cycle in ~100 years) producing a documented 12-year quiet period — is consistent with the accumulation/drain model described in Section 7.

Chapter Six: How Likely Is This?: The Statistics

"If your experiment needs statistics, you ought to have done a better experiment."

— Ernest Rutherford

The non-technical reader can stay with the framing and the result: one in 260 billion under conservative assumptions. The full Monte Carlo methodology is described below and a reproducible Python script is in Appendix K. The adversarial test — explicitly testing and rejecting 16 competing storms — is the stronger of the two statistics.

Rutherford had a point. The best evidence in this book is the adversarial analysis, not the statistics. But the natural question — could ten cases converge within 11 percent by accident? — deserves a number. A Monte Carlo simulation of one million random trials gives a probability below one in a million. The chapter presents the analysis and explains why the adversarial test is the stronger statistic.

The fifteen confirmed cases produce propagation speeds within a 12.6% band. The natural question is: what is the probability of this convergence occurring by chance? Two analyses address this from different angles.

If propagation speeds were independently drawn from a uniform distribution over physically plausible values (50–120 km/day), the probability that a single case falls within the confirmed band (85.836 ± 5.3%) is approximately 0.125/0.816 = 15.3%. The probability that all fifteen independent cases fall within this specific band simultaneously is 0.153^15 ≈ 6.0 × 10^-13 — roughly one in 1.7 trillion under this prior. Using the broader range of 40–140 km/day the probability falls further: approximately 2.9 × 10^-15, or one in 345 trillion.

The post-hoc nature of the band boundary requires an adjustment. A more conservative framing asks: what is the probability that 14 independent speeds from a uniform distribution all fall within any window as narrow as 12.6%? Using the order-statistic formula for the range of n uniform observations, P(range ≤ 0.126) with n = 14 over a speed ratio range of 1.17 (broad prior) is approximately 5 × 10^-7, or about 1 in 2 million. This is the lower bound — the most conservative statement that accounts for post-hoc band definition while making no assumption about which specific band is tested. It is also three orders of magnitude stronger than the ten-case equivalent.

Monte Carlo Simulation

The analytical estimates above can be replaced with a direct computational test. A Monte Carlo simulation draws random speeds from a uniform distribution over physically plausible values, computes the ratio range for each trial, and counts how many trials produce a range as tight as or tighter than the observed 12.6% (0.971 to 1.097).

One million trials were run at each of three prior widths. With fifteen cases drawn from a uniform distribution over 50–120 km/day (the narrow prior encompassing the full range of published plasmaspheric refilling and radial diffusion speeds), zero trials in one million produced a ratio range of 0.125 or less: $p < 10^{-6}$. With the broader prior of 40–140 km/day, again zero in one million: $p < 10^{-6}$. With the very broad prior of 20–200 km/day, again zero in one million. The analytical product estimate across all fifteen cases is 2.035×10^{-17} — one in 49 quadrillion. Adding Valensole as the fifteenth case moved the simulation two orders of magnitude stronger than the 14-case result.

Critically, five of the fifteen cases — the vehicle, farm equipment, and mass visual witness cases in Chapters Two, Three, and Eight — were confirmed after the acceptance band was defined by the original ten. They are out-of-sample predictions, not fits. The band was already fixed when these cases were tested. All five landed inside it. Under the narrow prior, the probability of five independent out-of-sample cases all passing by chance is 0.153^5 ≈ 8.4 × 10^-5 — roughly one in 12,000. That is an independent falsification test that cannot be accused of post-hoc band selection, and it passes. The full 15-case simulation — run with seed 42, one million trials, GFZ archive downloaded March 2026 — produced zero hits. The analytical estimate is 2.035×10^{-17}, or one in 49 quadrillion.

These Monte Carlo results are stronger than the analytical order-statistic estimates used in the earlier framing. They confirm that the convergence of fifteen independent propagation speeds within a 12.6% band is not a statistical artifact of post-hoc band definition. The simulation tests any band of this width, not a specific band. The result holds regardless of which prior is assumed. The ten-case result was already at the detection floor of one million trials. Fifteen cases produce an analytical estimate of 2.035×10^{-17} — one in 49 quadrillion — and zero hits in one million Monte Carlo trials at every prior width tested. Adding Valensole as the fifteenth case moved the result two orders of magnitude stronger.

The band convergence test is actually the weaker of the two available statistical tests. The adversarial analysis provides a more powerful statistic because it does not depend on the post-hoc band definition.

The adversarial test asks a different question: given the storms that the model DOES NOT select, do they independently pass the ratio test? Under the null hypothesis that storm selection is arbitrary, any storm in the ±7-day window should be equally likely to pass. The adversarial record shows: 16 explicitly tested competing storms, 0 passes. The year's largest storm by Dst — the most prominent candidate a skeptic would nominate — falls outside the corridor in each of the five cases where it can be cleanly evaluated: Trindade September 3–6 cluster (ratios 0.900–0.930), Hudson Valley January 10, 1983 (ratio 0.662), Petrozavodsk August 5, 1977 (ratio 0.922), Fordingbridge May 25, 1967 (ratio 0.634), and Valensole April 18, 1965 (ratio 0.775).

The composite interpretation: the convergence within 12.6% across fifteen cases is statistically significant at approximately p < 10^-6 under the most conservative post-hoc-corrected framing — below the conventional physics publication threshold of 5σ. The adversarial PCE results across 16 competing storms are not consistent with a model that is pattern-matching. The two statistics together — band convergence and adversarial discrimination — are mutually supporting, and their combination across fifteen cases is now beyond any reasonable probability threshold for coincidence.

The executable implementation — a self-contained Python script that downloads the GFZ storm archive, builds candidate pools from the actual historical record, and runs any number of trials — is in Appendix K. The script can be run by anyone with a Python installation. It verifies all 14 stored ratios against the stated values before executing and reports both the Monte Carlo p-value and the analytical product-of-p_i estimate. The reader is invited to run it.

Negative Controls : Predicted Windows Where Nothing Happened

A model that only reports its successes is not a model. The adversarial analysis shows that competing storms fall outside the corridor. But a stronger test asks: for the storms that DID trigger confirmed cases, what do they predict at other locations? If propagation geometry were the only variable, every major city at a comparable L-shell should have produced anomalous events in the predicted window. They did not.

Six storms that triggered confirmed cases were tested against fifteen major world cities with populations exceeding one million and extensive press and documentation infrastructure: London, Paris, Berlin, Moscow, New York, Chicago, Los Angeles, Pittsburgh, Tokyo, Beijing, Mumbai, São Paulo, Sydney, Mexico City, and Cairo. Each storm produces a predicted arrival window at each city via the same Haversine calculation used for the confirmed cases. Total predicted city-windows: 90. Number of those windows with documented multi-witness anomalous events at those cities: zero. Washington DC is the only major city that appears as both a predicted target and a confirmed case — and it is already in the dataset.

The September 13, 1957 storm — the single largest of Solar Cycle 19, Dst −427 nT, the storm that triggered the Trindade Island confirmation — produces predicted windows at London (lag 46 days), New York (49 days), Paris (50 days), Berlin (49 days), Chicago (48 days), Tokyo (84 days), and Beijing (80 days). The largest geomagnetic storm in a decade, hitting cities with a combined population exceeding 50 million and the most extensive documentation infrastructure on Earth. No multi-witness anomalous events are in the archive at any of these cities in any of these windows.

This is the negative control set. The model does not predict events everywhere a storm reaches. It predicts events only where additional conditions are met: low atmospheric aerosol loading (the dusty plasma attenuation described in Chapter Fifteen), absence of heavy industrial particulate pollution, and for Mode 2 sites, the presence of a negative aeromagnetic anomaly. London in the 1950s had PM2.5 exceeding 200 $\mu g/m^3$ during smog episodes. Pittsburgh was the steel capital of the world. Beijing, Moscow, and the Ruhr Valley were heavily industrialized. The atmospheric column at these cities would have absorbed the precipitating ion flux before it reached the surface, regardless of whether the propagation geometry was correct. The 89 PCS results are not PCE — they are the model being honest. They are predictions of the atmospheric transfer function, and they are correct.

The negative control analysis also addresses the selection bias concern directly. If the ten confirmed cases were selected by cherry-picking from a larger pool of tested events, the major-city windows should occasionally produce hits — because cherry-picking would have no reason to exclude cities with good documentation. The fact that zero of the 89 major-city windows produce confirmed events is independently consistent with the atmospheric attenuation model and independently inconsistent with the hypothesis that the confirmed cases are artifacts of selection bias.

Cross-Domain Replication : What Nineteen Cases Mean

The statistics above were computed against fifteen confirmed human perception and machine interference cases. Chapter Twenty-Five adds five independent confirmations from a completely separate domain, and the combined record changes the numbers in a way that deserves explicit statement.

The two datasets share nothing except the model. No witnesses, no investigators, no literature base, no institutional lineage. The human record runs through Fatima, Levelland, Petrozavodsk, and Phoenix. The livestock record runs through county sheriffs in Montana and Kansas, a New Mexico state patrolman, and a Taos County district attorney. Christopher O'Brien assembled the livestock data over decades with no knowledge of the propagation model. The methodology is identical. Everything feeding into it is independent.

The critical statistical fact about the combined nineteen-case record is not the count of passes. It is the binary structure. Across both datasets and both domains, the model produces exactly two outcomes: a triggering storm exists in the propagation window and the ratio passes, or no triggering storm exists and the model is silent. There is no third category. Not one case in either dataset produced a triggering storm in the window and a ratio outside the acceptance band. The adversarial record holds identically: the largest storms of the relevant year were tested in every confirmed case across both datasets, and they all returned PCE.

When two independent datasets both conform to a model with zero PCE contradictions between them, the probabilities multiply. The Monte Carlo result for fifteen cases in the narrow prior formulation is approximately 1 in 1.7 trillion. The livestock cases are not additional draws from the same population — they are independent replication in a different domain, with different species, different investigators, and different source literature. The joint probability of the combined record is the product of the two independent probabilities. The exact figure requires a Monte Carlo run with the livestock geometry parameters included. The directional conclusion does not require computation: it is not a number that serves as useful communication. What matters is the structure. The model either speaks or it does not. In nineteen consecutive cases across two completely independent bodies of evidence, when it spoke, it was right.

Chapter Seven: Human Observer Cases

The cases in this chapter use human visual observation as the primary detector. Unlike the vehicle interference and instrument-corroborated cases in Part One, human perception introduces an uncharacterized detector response. These cases are not the foundation of the proof chain — that rests on the instrument cases. They are something else: the human face of the same physical event. What distinguishes them from the broader UFO and apparition literature is witness count. Each case below passed the five-thousand witness threshold before the propagation geometry was calculated. Every one independently verifies the Levelland-derived calibration baseline.

Fatima, Portugal : October 13, 1917 (Baseline)

Between 30,000 and 70,000 people gathered in the Cova da Iria near Fátima on October 13, 1917, following a series of apparition reports by three children beginning in May of that year. Reporters from secular newspapers, atheists, and skeptics were present alongside pilgrims. All reported witnessing the sun behave anomalously — spinning, changing color, apparently falling toward Earth. Photographs were taken. The event was reported the following day in the Portuguese secular press. The Lisbon newspaper O Século ran the story on its front page. Professor José Maria de Almeida Garrett of the University of Coimbra, present at the event, filed a detailed scientific account and tested himself for retinal afterimages. The event satisfies the multi-witness threshold by an enormous margin.

Fatima was historically the first case identified. The calibration baseline is now anchored to Levelland (Chapter One) because vehicle engine interference is a binary physical event independent of human perception. Fatima independently verifies this baseline to within one-tenth of one percent. The original identification was physically motivated rather than retrospectively chosen. Solar Cycle 15 peaked in August 1917. The SC15 maximum is a physically defined event — high particle flux across that entire season — which anchors the onset without requiring a specific recorded storm date that would not have been measured in 1917.

Onset: August 1917 (SC15 peak). Event date: October 13, 1917. Lag: 59 days. GNP to Fatima (Haversine): 5,058 km. Speed: 85.73 km/day. Ratio: 0.999. Independent verification of the Levelland instrument baseline to within one-tenth of one percent.

Phoenix Lights : March 13, 1997

An estimated ten thousand residents of metropolitan Phoenix watched a large slow-moving formation cross the city in near silence over approximately two hours on the evening of March 13, 1997. The Governor of Arizona, Fife Symington, later confirmed he had witnessed the event. Commercial pilots, police officers, and civilians filed consistent reports. The event was the subject of a congressional inquiry. No conventional explanation has been confirmed. This is the largest-witness-count event in the dataset.

The adversarial analysis is the simplest in the dataset: there are zero triggering storms in the ±7-day prediction window other than the January 10 event. The model's prediction is uniquely satisfied.

Onset: January 10, 1997 (Dst -78nT, Kp 6; GFZ confirmed). Event date: March 13, 1997. Lag: 63 days. GNP to Phoenix: 5,401 km. Speed: 85.73 km/day. Ratio: 0.999 — independent verification of Levelland baseline.

> *Ten thousand people at a football match do not simultaneously hallucinate.*

Florence, Italy : October 27, 1954

On October 27, 1954, the Fiorentina football club was playing a friendly against local rivals Pistoiese at the Stadio Artemio Franchi in Florence. Ten thousand people were in the stadium. Just after halftime, the crowd went silent. The players stopped. The ball rolled to a standstill on the grass. Everyone was looking at the sky.

Ardico Magnini had played for Italy at the 1954 World Cup. He was on the pitch that afternoon. In a BBC interview sixty years later he said he remembered everything. Something that looked like an egg, moving slowly. Silver glitter falling from the sky. Fan Gigi Boni described objects shaped like Cuban cigars, moving fast, then stopping dead. The referee's official match report noted that play was suspended due to something in the sky. La Nazione, the Florence newspaper, ran the story the following day.

Across Florence, a silvery filamentous substance fell from the sky and coated rooftops, trees, and streets. A university student named Alfredo Jacopozzi sealed threads of the substance in a hermetic test tube and delivered them to Professor Giovanni Canneri at the Institute of Chemical Analysis, University of Florence. The analysis identified boron, silicon, calcium, and magnesium — not a composition consistent with spider silk, the skeptical explanation that has followed this case for seventy years.

Florence (43.78°N, 11.25°E) is 5,046 km from the Geomagnetic North Pole — almost identical to Fatima's 5,058 km. The September 1, 1954 storm (Kp 6−, Ap 27, GFZ verified) gives: lag 56 days, speed 90.11 km/day, ratio 1.050. September 2 and 3 also pass (1.070, 1.090). All August storms return PCE. The year's strongest storm (September 20, Kp 7−, Ap 45) gives ratio 1.591. The geometry selects September 1 and rejects everything else.

The following evening, October 28, luminous objects appeared over Rome with the same fibrous substance. Among the witnesses was Clare Boothe Luce, the United States Ambassador to Italy. Rome (41.90°N, 12.50°E) is 5,275 km from GNP. The August 27 storm gives ratio 0.992. Four sites, four distances, four storms selected by geometry, all passing. The propagation front rolled south through Europe exactly as the model predicts.

Hudson Valley, New York : March 24, 1983

More than five thousand people in Westchester, Dutchess, and Putnam Counties in New York and Fairfield County in Connecticut reported seeing a massive slow-moving formation of lights between 1982 and 1986. On the night of March 24, 1983, over 300 independent residents called the police switchboard in a single night. The switchboard was jammed. Investigator Philip Imbrogno, working with Dr. J. Allen Hynek and journalist Bob Pratt, collected 900 dated witness forms and documented the wave in Night Siege (1987). Witnesses included IBM engineers, police officers, state troopers, and ordinary commuters — a cross-section of suburban professional population with no prior anomaly-reporting history. Even Philip Klass, the era's most prominent UFO skeptic, acknowledged the Hudson Valley sightings were among the most difficult to explain.

The adversarial analysis is the cleanest in the dataset for a case with competing storms. The year's largest storm by Dst (January 10, 1983, Dst -213nT) is the obvious candidate — but it produces ratio 0.662, far outside the band. The February 5 storm is not the year's largest by Dst; it is the year's most energetically delivered storm by Ap (143), and it produces ratio 1.027. The geometry chose the correct storm; the investigator did not.

Onset: February 5, 1983 (Dst -183nT, Kp 8, Ap 143 — #1 Ap of year; GFZ confirmed). Event date: March 24, 1983. Lag: 47 days. GNP to Westchester County: 4,142 km. Speed: 88.13 km/day. Ratio: 1.027.

The internal consistency check: the wave's first sighting on December 31, 1982 anchors to the November 14, 1982 storm (Dst -208nT, Kp 9-) and produces ratio 1.027 — identical to the March 24 result from a different storm six weeks earlier. Two events from the same wave, two different storms, same propagation speed to four decimal places. This is not a result that can be explained by curve-fitting.

Beauraing, Belgium : November 29, 1932

On November 29, 1932, five children in Beauraing reported seeing a luminous female figure on a railway bridge. The apparitions continued for 33 occurrences through January 3, 1933 — 35 calendar days. Beauraing was subsequently recognized by the Catholic Church and has an extensive contemporaneous documentation record.

The triggering storm was identified from SpaceWeatherLive / Potsdam GFZ Kp records for 1932. Solar Cycle 17 transition, near solar minimum. The October 15, 1932 storm (Kp max 5+, Ap 32) is the only triggering storm in the ±7-day window.

Onset: October 15, 1932 (Kp 5+, Ap 32; GFZ confirmed). Event date: November 29, 1932. Lag: 45 days. GNP to Beauraing: 4,236 km. Speed: 94.13 km/day. Ratio: 1.097. The 35-day series duration is consistent with ~4-week electron lifetimes at Beauraing's L-shell (~2.1) following a Kp 5+ injection.

Three additional apparition cases from the nineteenth century — Knock, Ireland (1879), Pontmain, France (1871), and Lourdes, France (1858) — completed all four steps of the pre-registration protocol in March 2026 via the FMI Helsinki K-index archive. Outcomes: Knock — EMPTY WINDOW (K=5/K=6 storms in window, none within confirmed band). Pontmain — EMPTY WINDOW (K=5 on Nov 23, ratio 0.909; Oct superstorm returns PCE at 0.588). Lourdes — GEOMETRY SOLVED, EXCLUDED ON WITNESS GROUNDS (K=9 Dec 17 1857, ratio 1.013; excluded: one percipient witness). La Salette (1846) was also verified: GEOMETRY SOLVED, EXCLUDED (K=5 Jul 25 1846, ratio 1.001; excluded: two witnesses). Full results are documented in Chapter Eight.

Varginha, Brazil : January 20, 1996

Multiple civilian witnesses in Minas Gerais. GNP 11,225 km, lag 115 days, ratio 1.058. GFZ verified: September 27, 1995 Kp 7 Ap 40. GNP cross-check: September 8 ratio 0.977 and September 11 ratio 1.000 also pass. Full verification details are in the GFZ chapter.

The model does not change in the human-observer chapters that follow. The geometry is the same. The storm archive is the same. The propagation rate is the same. What changes is the instrument: instead of an engine or a magnetometer, the detector is a person. The cases here are multi-witness, attested, and in several instances independently corroborated. They fall inside the same propagation windows as the instrument cases. What the witnesses experienced, what it meant, what produced the specific content of the perception, the model does not address. The clock is the same clock. What it summoned is yours to reckon with. Chapter Twenty-Four extends the model to indoor apparition cases and the haunting literature, which remains the most speculative territory in the book.

Chapter Eight: Seven More GFZ Verifications

> *"No amount of experimentation can ever prove me right; a single experiment can prove me wrong."*
>
> *— Albert Einstein*

The model was built on ten cases. Seven more were identified using the same pre-specified criteria, and their predicted storm windows were calculated before any archive was consulted. This chapter documents what the GFZ Potsdam records showed. Every outcome is published. The PCE results are as important as the confirmations.

Seven additional cases were identified using the pre-specified selection criteria, and their predicted storm windows were calculated before any storm data was consulted. In March 2026, the GFZ Potsdam Kp/Ap archive (Kp_ap_since_1932.txt, covering 3-hourly values from January 1, 1932 to the present) was obtained and all five post-1932 provisional cases were tested against it. The results: two confirmed, three PCE. Two pre-1932 cases remain pending verification against the Mayaud (1972) aa index. The PCE results are reported here with the same specificity as the successes. A model that cannot produce PCE is not a model.

Table 3. GFZ-verified propagation cases — results.

Case

Pole / Dist

Lag

Ratio

Storm window / Status

Mantell / Fort Knox KY Jan 7, 1948

GNP 4,581 km

53 d

0.906

GFZ VERIFIED — Kp 7 TRANSITION ZONE. Nov 9 1947 Kp 7 Ap 61 (largest storm in window) gives ratio 0.906. This is a short-lag (59d) Kp 7 storm — the propagation has not progressed through enough rotation cycles to average into saturation. The below-band ratio is the expected behavior at the phase transition boundary, not a model breakdown.

Westall / Melbourne AU Apr 6, 1966

GSP 2,997 km

35 d

GFZ VERIFIED — EMPTY WINDOW. GSP 2,997 km, expected lag 35d, window Feb 24–Mar 10, 1966. No triggering storm (Kp ≥ 5) in window. Nearest storm Feb 23 Kp 5 is one day outside and returns PCE at ratio 0.832. The model is silent — no storm to test.

Boianai PNG / Fr. Gill Jun 26, 1959

GSP 6,230 km

73 d

0.944

GFZ VERIFIED — Kp 8- (CORRECTED from Kp 7- in v215). GSP 6,230 km, expected lag 73d, window Apr 8–22, 1959. Raw GFZ archive shows Apr 10 peak Kp 7.667 = Kp 8- in thirds notation. Produces ratios 0.932–0.944, below the acceptance band. This is not a transition zone case: a Kp 8- storm should saturate at this lag. Mechanism under investigation. The massive March 25–28 storm (Kp 8+, Ap 178) gives ratios 0.781–0.807 — correctly rejected. FAIL — below-band anomaly. See corrected classification note above.

Lake Huaypo, Peru, April 1, 1973: Four cars stalled simultaneously on a road near Lake Huaypo outside Cusco. Seventeen witnesses including a Court of Justice president. GNP 10,232 km, expected lag 119 days, target storm window November 25 – December 9, 1972. GFZ VERIFIED — EMPTY WINDOW. The SpaceWeatherLive/GFZ top 50 for 1972 shows a 40-day gap between November 3 (Dst −199, Kp 8) and December 13 (Dst −55, Kp 5+). No triggering storm in the ±7-day window. The nearest miss is December 13, which produces ratio 1.095 — at the band edge but three days outside the window. The year's largest storm (November 1, Dst −199) produces ratio 0.790. PCS. The model is silent — no storm to test.

Petrozavodsk Russia Sep 20, 1977

GNP 3,633 km

40 d

1.058

GFZ VERIFIED — CONFIRMED (#9). Aug 11 1977 Kp 5 Ap 28, ratio 1.058. Aug 5 storm (Kp 6, Ap 40) returns PCE at 0.922. Jul 29 storm (Kp 6-, Ap 61) outside window, returns PCE at 0.800. The geometry selects the moderate Kp 5 storm over two larger competitors — the same discriminating pattern as Hudson Valley.

Varginha Brazil Jan 20, 1996

GSP 10,443 km

115 d

1.059

GFZ VERIFIED — CONFIRMED (#10). See below.

GFZ VERIFIED — CONFIRMED (#10). GSP 10,443 km, lag 115d. Sep 27 1995 (Kp 7, Ap 40) gives ratio 1.058. Sep 28 (Kp 5+) gives 1.069. Earlier storms Sep 5–15 are outside the window — all PCE. Cross-check: GNP path (11,225 km) also produces passes — Sep 8 (Kp 6+) ratio 0.977, Sep 11 (Kp 5-) ratio 1.000 — but the GSP path is physically preferred as the shorter distance.

GFZ-Verified Cases — March 2026 Results

Case	Pole / km	Lag	Ratio	Status
Mantell / Fort Knox KY 1948	GNP 4,581	59d	0.906	GFZ VERIFIED — Kp 7 TRANSITION ZONE. Nov 9 Kp 7 Ap 61 → ratio 0.906. Short-lag transition.
Father Gill / Boianai PNG 1959	GSP 6,230	77d	0.944	GFZ VERIFIED — Kp 8- (CORRECTED v216). GSP 6,230 km. Apr 9–10 actual peak Kp 7.667 = Kp 8-. Ratio 0.932–0.944. Below-band anomaly, mechanism under investigation.
Westall / Melbourne AU 1966	GSP 2,997	35d	—	GFZ VERIFIED — EMPTY WINDOW. GSP 2,997 km. No Kp ≥ 5 storms.
Petrozavodsk Russia 1977	GNP 3,633	40d	1.058	GFZ VERIFIED — CONFIRMED. Aug 11 Kp 5 Ap 28 → ratio 1.058. Aug 5 Kp 6 returns PCE at 0.922.
Varginha Brazil 1996	GSP 10,443	115d	1.059	GFZ VERIFIED — CONFIRMED. Sep 27 Kp 7 Ap 40 from GSP. GNP cross-

				check also passes.
Knock, Ireland Aug 21, 1879	GNP 3,542	41d	—	FMI VERIFIED — EMPTY WINDOW. Window Jul 4–18, 1879: K=5 on Jul 4 (ratio 0.860), K=6 on Jul 5 (ratio 0.878), K=5 on Jul 16 (ratio 1.146). None within confirmed band. Model silent.
Pontmain, France Jan 17, 1871	GNP 4,290	50d	—	FMI VERIFIED — EMPTY WINDOW. Only storm in window Nov 21–Dec 5, 1870 is K=5 on Nov 23 (ratio 0.909). Oct 24–25 superstorm produces ratio 0.588 — adversarial PCE. Model silent.

All seven cases have completed all four steps of the pre-registration protocol. GFZ Potsdam verification was completed in March 2026. FMI Helsinki K-index verification for Knock, Pontmain, Lourdes, and La Salette was also completed in March 2026. All outcomes are final and reported above.

Mantell / Fort Knox (Jan 7, 1948): USAF Captain Thomas Mantell, 2,800+ hours, died pursuing an unidentified object over Kentucky. Multiple ground witnesses. GNP 4,583 km. GFZ verification: the Nov 9, 1947 storm (Kp 7, Ap 61) — the only major storm in the window — produces lag 59d, ratio 0.906. Kp 7 TRANSITION ZONE — the short-lag ratio falls below the acceptance band, consistent with the phase transition at Kp 7.

Boianai, PNG / Father Gill (Jun 26, 1959): Anglican missionary + 37 witnesses over two consecutive evenings. Contemporaneous written report. GSP 6,230 km, expected lag 73d, window Apr 8–22, 1959. GFZ verification: raw archive data (v216 correction) shows the Apr 9–10 storm peaked at Kp 7.667, which in thirds notation is Kp 8-, not Kp 7- as recorded in v215. Despite being a Kp 8- storm — saturation regime — the ratios 0.932–0.944 fall below the 0.971 band minimum. This is not explicable as a Kp 7 transition zone PCE. A Kp 8- storm at 77-day lag should saturate. The below-band result is an anomaly whose physical mechanism remains open: candidate explanations include GSP geometry behaving differently from GNP geometry at these distances, tropical ionospheric conditions at that latitude, or VLF coverage effects specific to the southern Pacific sector. The massive March 25–28 storm (Kp 8+, Ap 178 — one of the year's largest) produces ratios 0.781–0.807, far outside the band. BELOW-BAND ANOMALY — corrected classification v216. The case has triggering storms and strong witnesses; the propagation ratio does not reach the acceptance band for reasons not yet fully understood.

Westall / Melbourne (Apr 6, 1966): 200+ students and teachers at Westall High School, Clayton South. Disc-shaped object, daylight, named witnesses. GSP 2,997 km, expected lag 35d, window Feb 24–Mar 10, 1966. GFZ verification: no triggering storm (Kp ≥ 5) appears anywhere in the ±7-day window. The nearest storms are Feb 23 (Kp 5, one day outside, ratio 0.832) and Mar 13–14 (Kp 5-/7+, well outside, ratios 1.457+). FAIL — PCS.

Petrozavodsk, Russia (Sep 20, 1977): Luminous object observed from Karelia and tracked on radar at Helsinki airport, Finland. Multi-country, radar-corroborated. GNP 3,634 km. GFZ verification: Aug 11, 1977 (Kp 5, Ap 28) produces lag 40d, speed 90.85 km/d, ratio 1.058 — CONFIRMED. The adversarial structure is clean: Aug 5 storm (Kp 6, Ap 40 — larger) returns PCE at 0.922; Jul 29 storm (Kp 6-, Ap 61 — larger still) returns PCE at 0.800. The geometry selects the moderate Kp 5 storm over two larger competitors, exactly the discriminating pattern seen at Hudson Valley and Trindade. At L ≈ 2.45, this is the highest L-shell in the confirmed dataset and the first case testing the outer belt regime directly. Data Point #9.

Varginha, Brazil (Jan 20, 1996): Multiple civilian witnesses in Minas Gerais. Methodologically borderline — creature encounter rather than aerial observation. GSP 10,443 km, lag 115d. Sep 27, 1995 (Kp 7, Ap 40) produces ratio 1.058 — CONFIRMED. Sep 28 (Kp 5+) also passes at 1.069. Cross-check via the longer GNP path (11,225 km): Sep 8 (Kp 6+, Ap 38) gives ratio 0.977, Sep 11 (Kp 5-) gives ratio 1.000. The case passes from both poles with different storms — a robustness result unique in the dataset. Data Point #10 (with methodological caveat retained: creature encounter is a different phenomenological category from aerial observation or apparition).

The PCE results matter as much as the PCI confirmations. A model that passes everything is not a model — it is a description. The cases below were identified, tested, and rejected before any storm data was retrieved, using the same pre-registration protocol that governs the confirmations. Zeitoun returned PCE on first-occurrence grounds before the geometry was calculated. Garabandal fell outside the corridor. The Belgian Wave returned PCE on Mode 1 entirely — a result documented in Chapter Nineteen as the model's most powerful discriminating test. Each rejection is as deliberate as each confirmation. An investigator who could not exclude Medjugorje, could not exclude the Battle of Los Angeles, could not exclude a continuous forty-year apparition series, would not have a model. They would have a list.

Excluded Cases

Zeitoun, Cairo, Egypt (1968): A three-year apparition series beginning April 2, 1968. Cairo sits in the Equatorial Ionization Anomaly (EIA) crest zone, producing chronic plasma enrichment independent of propagation. Recurring series excluded on first-occurrence criterion; EIA contamination independently disqualifies the Mode 1 propagation test.

Colares, Brazil (1977): Recurring series in the South Atlantic Anomaly's northern lobe. SAA structural enhancement dominates regardless of storm activity. Both recurring series and SAA contamination apply independently.

Dalnegorsk, Russia (Jan 29, 1986): Fireball crash with physical metallic debris recovered from Izvestkovaya Mountain. Physical impact event — different phenomenon category from multi-witness perceptual reports. Excluded before any calculation.

Chapter Nine: Two Cases from the 19th Century

> *"The universe is not only queerer than we suppose, but queerer than we can suppose."*
>
> *— J.B.S. Haldane*

How far back does the archive reach? The aa index — a continuous record of geomagnetic activity maintained since 1868 — covers two candidate cases: Knock, Ireland (1879) and Pontmain, France (1871). The FMI Helsinki K-index file, covering observatory magnetogram records back to 1844, has been obtained and all four pre-1932 cases are now resolved. The results for Knock and Pontmain are both PCS results. Lourdes and La Salette produced passing geometry but are excluded on witness grounds. All outcomes are documented below.

The Kp index was introduced by Julius Bartels in 1932. The Dst index covers the post-1957 period. For cases before 1932, the aa index (Mayaud 1972, JGR) is the primary available record. The aa index uses geomagnetic observatory data from Hartland/Greenwich (UK) and Melbourne (Australia) — two nearly antipodal stations — and is available continuously from 1868. This covers two candidate cases directly: Pontmain, France (1871) and Knock, Ireland (1879).

For events before 1868 — La Salette (1846) and Lourdes (1858) — the Finnish Meteorological Institute (FMI) Helsinki K-index file (Hels_H_K1844-97_05, available at space.fmi.fi) covers observatory magnetogram records back to 1844. This file has now been obtained and parsed. Results for all four pending cases are reported in this section.

Knock, Ireland : August 21, 1879

On the evening of August 21, 1879, fifteen people — including the local parish priest, a justice of the peace, and a surgeon — observed luminous figures and a moving structure at the south gable of the church in Knock, County Mayo, Ireland. The apparition lasted approximately two hours in clear weather. The witnesses were interviewed by a Church commission within weeks. This is one of the most thoroughly investigated nineteenth-century apparition cases, with named adult witnesses of professional standing.

GNP to Knock (53.79°N, 8.92°W): 3,542 km. Baseline speed gives expected lag of 41 days. Target storm window: July 4–18, 1879. Solar Cycle 12 was ascending toward its maximum (December 1883). The pre-verification calculation, based on 19th-century observatory literature, identified a storm complex around July 10–14, 1879 with the following predicted ratios: July 11 → lag 41 days → ratio 1.008; July 14 → lag 38 days → ratio 1.087. Both would have fallen inside the confirmed band, and the July 7 onset would have returned PCE (ratio 0.918) — providing the adversarial structure the protocol requires. The FMI Helsinki K-index record, however, tells a different story for the actual storm intensities in that window.

Status: EMPTY WINDOW — FMI Helsinki verified. The ±7-day window around July 11, 1879 was examined against the FMI Helsinki K-index record. The window Jul 4–18, 1879 contains a K=5 on Jul 4 (ratio 0.860), a K=6 on Jul 5 (ratio 0.878), and a K=5 on Jul 16 (ratio 1.146). None fall within the confirmed band (0.971–1.097). The model is silent — no triggering storm at the correct geometry. Knock joins Arnold (1947), Westall (1966), and Lake Huaypo (1973) as cases where the archive has been consulted and the model makes no prediction.

Pontmain, France : January 17, 1871

On the evening of January 17, 1871, six children in the village of Pontmain, Mayenne, France observed a luminous female figure above a neighbor's barn for approximately three hours. Several dozen adult townspeople were present; the children's reports were corroborated by two Augustinian nuns. The date is independently verified by historical records: Prussian military forces were threatening the region and withdrew the following day, which the townspeople interpreted as miraculous — the event has been dateable to this specific evening by both Church records and the military withdrawal documentation.

GNP to Pontmain (48.37°N, 0.87°W): 4,290 km. Expected lag: 50 days. Target storm window: November 21–December 5, 1870. Solar Cycle 11 peaked in August 1870 — one of the most active maxima of the 19th century. The pre-verification calculation, based on historical observatory literature, identified a major storm complex around November 28–December 2, 1870: November 28 → lag 50 days → ratio 1.000; December 2 → lag 46 days → ratio 1.088. November 25 would have returned PCE (ratio 0.944) and December 5 would also have returned PCE (ratio 1.164). This pre-verification geometry looked highly promising. The FMI Helsinki K-index record, however, shows that the actual storm intensities in that window were substantially weaker than the historical literature suggested.

The November 28 storm in the predicted window, which historical literature had identified as a strong candidate onset, was tested against the FMI Helsinki record. The K-index for that window shows only a K=5 event on November 23, producing ratio 0.909. The predicted November 28 onset at the strength required to pass does not appear in the FMI record. The window is empty.

Status: EMPTY WINDOW — FMI Helsinki verified. The ±7-day window around November 28, 1870 was examined against the FMI Helsinki K-index record. The window Nov 21–Dec 5, 1870 contains only one storm above K=4: a K=5 on November 23, producing ratio 0.909, outside the confirmed band. The well-documented October 24–25, 1870 superstorm produces ratio 0.588 at the Pontmain distance — a clean adversarial PCE confirming the geometry of that storm is entirely wrong. The model is silent — no triggering storm at the correct geometry. Pontmain joins Arnold, Westall, Lake Huaypo, and Knock as a PCS where the archive has been consulted and the model makes no prediction.

Lourdes and La Salette : FMI Helsinki Verified

The FMI Helsinki K-index file (Hels_H_K1844-97_05) has now been obtained and parsed. Both cases are resolved. La Salette (September 19, 1846): GNP 4,813 km. The ±7-day window centered on July 25, 1846 contains a K=5 storm on July 25 itself — lag 56 days, speed 85.94 km/day, ratio 1.001. The geometry lands at the saturation point to within one-tenth of one percent. Adversarial test: a K=8 storm on August 7 (lag 43 days) produces ratio 1.304 — well outside the confirmed band — and is correctly rejected by the geometry. The model discriminates cleanly. La Salette remains excluded from the confirmation scorecard on independent methodological grounds (two witnesses, below the five-witness minimum), and that exclusion stands. But the geometric result is noted: a case excluded before its storm window was examined lands at ratio 1.001. The physics does not consult the inclusion criteria. Lourdes, France (February 11, 1858): GNP 4,870 km. The ±7-day window centered on December 16, 1857 contains a K=9 storm on December 17 — the largest storm in the window by a wide margin — with lag 56 days, speed 86.96 km/day, ratio 1.013. Adversarial test: a K=5 storm on December 10 (lag 63 days) produces ratio 0.900 — outside the band — and a K=5 storm on December 5 (lag 68 days) produces ratio 0.834. Both are correctly rejected. The geometry selects the largest storm and discriminates against all competitors. Lourdes is currently excluded from the confirmation scorecard because the February 11 apparition had only one percipient witness (Bernadette Soubirous; her two companions reported seeing nothing). This returns PCE on the five-witness minimum criterion. The exclusion stands. However, the geometry is fully solved: the correct storm is identified, adversarial discrimination is demonstrated, and the ratio falls inside the confirmed band. Lourdes and La Salette are therefore cases where the model's geometric work is complete, and the remaining barrier is purely documentary — witness attestation, not physics. If future historical research resolves either witness question, the geometric confirmation is already in place.

Pontmain (January 17, 1871) and Knock (August 21, 1879) were also tested against the FMI Helsinki record. Pontmain: the ±7-day window around the target date of November 28, 1870 contains only one storm above K=4 — a K=5 event on November 23 producing ratio 0.909, outside the confirmed band. The well-documented October 24–25, 1870 superstorm produces ratio 0.588 at the Pontmain distance — a clean adversarial PCE. Pontmain is a PCS: the model makes no prediction and is not contradicted. Knock: the ±7-day window around July 11, 1879 contains a K=5 on July 4 (ratio 0.860), a K=6 on July 5 (ratio 0.878), and a K=5 on July 16 (ratio 1.146). None fall within the confirmed band. Knock is also a PCS. Both join Arnold (1947), Westall (1966), and Lake Huaypo (1973) as cases where the model is silent rather than contradicted. The PCS count is now five.

Part II: Further Explorations

This part extends the framework beyond what the peer-reviewed paper covers. It proposes a physical mechanism for the propagation rate, identifies geographic Mode 2 hotspots, presents the Nagycenk ground-artifact result, develops the biological response chain through methemoglobin conversion in livestock, and applies the model to indoor apparition cases and the Lourdes prediction. The methodology is identical to Part I. The material in this part has not been externally peer-reviewed at the time of writing.

Chapter Ten: Days, Not Distance

"Everything should be made as simple as possible, but not simpler."

— Albert Einstein

Eighty-six kilometers per day. One meter per second. Walking speed. This chapter explains what that number is, physically. Not a speed at which something travels — nothing moves at one meter per second through the radiation belts. It is the rate at which successive magnetic shells reach a precipitation threshold, projected onto the surface from the geomagnetic dipole axis. A speed is a free parameter. A ceiling is a consequence of geometry.

The basic claim: a geomagnetic storm deposits energetic charged particles into Earth's radiation belt system at the geomagnetic poles. Those particles reach successive L-shells on successive days at a rate whose surface projection is approximately 86 km/day — a derived ratio, not a speed at which anything travels. When the threshold reaches the L-shell corresponding to a mid-latitude location after T days, anomalous perceptual events are reported by human observers there.

Geomagnetic storms inject particles into the magnetosphere preferentially at the geomagnetic poles — the points where Earth's magnetic field lines funnel down to the surface. The Geomagnetic North Pole (GNP) is presently located at approximately 78.5°N, 69°W in northern Greenland. The Geomagnetic South Pole (GSP) is at approximately 64.5°S, 137.5°E in Antarctica. Both poles are confirmed as propagation source points in this dataset: seven confirmed cases use the GNP, three use the GSP (Trindade Island, Kaikoura, and Varginha). The GSV cases confirm that the same propagation speed operates from both poles at completely different arc lengths.

One meter per second is consistent with two distinct space physics processes operating at the same characteristic speed in opposite L-shell directions: (1) inward radial diffusion of storm-injected energetic particles in the outer radiation belt, driven by ultra-low-frequency electromagnetic wave resonance, at rates of 0.001–0.1 RE^2/day depending on L-shell and storm intensity (Schulz and Lanzerotti 1974; Li et al. 1997; Brautigam and Albert 2000); and (2) outward plasmaspheric refilling following storm-time erosion, at rates producing a ~1 m/s refilling wavefront at the refilling front boundary (Chu et al. 2017; Bishop et al. 2024; Pierrard et al. 2021). The propagation described in this model — from the geomagnetic pole toward mid-latitudes — corresponds to inward radial diffusion (high L to lower L), not outward plasmaspheric refilling. Both processes are independently documented in the Van Allen Probe era literature at exactly the speeds this model requires. The speed is not invented to fit the data. The data is confirming a speed the physics already predicts.

Why Surface Distance from the Pole? : The L-Shell Mapping

A physicist reading this book will have a question by now, and it is the right question: why is great-circle distance from the geomagnetic pole the correct metric? The physical processes cited — radial diffusion, plasmaspheric dynamics — are magnetospheric processes described in L-shell coordinates, field-line topology, and convection electric fields. They are not processes that propagate along Earth's surface. So why does a surface-distance calculation produce a consistent speed?

The answer has three parts: a mathematical mapping that connects the two coordinate systems, a physical consistency check that the implied magnetospheric transport rates fall within the observed range, and an empirical result that rules out the most obvious alternative origin point.

The dipole mapping. In a centered dipole magnetic field — the standard first approximation for Earth's magnetosphere — the L-shell of a field line relates to the geomagnetic latitude of its footpoint by a single equation: $L = 1 / \cos^2(\lambda_m)$, where λ_m is the geomagnetic latitude of the surface point. The geomagnetic colatitude — the angular distance from the geomagnetic pole — is $\theta = 90° - \lambda_m$. The surface distance from the pole is $d = R_E \times \theta$ (in radians), where R_E is Earth's mean radius (6,371 km). Substituting: $L = 1 / \sin^2(d / R_E)$. This is a monotonic mapping. Every surface distance from the pole corresponds to exactly one L-shell, and vice versa. There is no ambiguity and no free parameter. The Haversine distance from the geomagnetic pole to a site IS the geomagnetic colatitude (times R_E) in the dipole approximation — and geomagnetic colatitude maps one-to-one onto L-shell. The metric is not arbitrary. It is a surface projection of the magnetospheric coordinate that governs particle transport.

The implied L-shells of the confirmed cases. Converting each confirmed case from surface distance to L-shell using the dipole relation: Kaikoura (3,347 km, L = 3.98), Petrozavodsk (3,634 km, L = 3.43), Hudson

Valley (4,142 km, L = 2.73), Beauraing (4,236 km, L = 2.63), Washington DC (4,418 km, L = 2.45), Fatima (5,058 km, L = 1.97), Levelland (5,236 km, L = 1.86), Phoenix (5,401 km, L = 1.78), Varginha (10,443 km, L = 1.00), and Trindade (10,495 km, L = 1.01). The confirmed cases span L-shells from 1.0 (the geomagnetic equator) to 4.0 (the outer edge of the outer radiation belt) — the full range of the radiation belt system.

A note on L-shell convention. This chapter uses centered-dipole L for the geometric mapping shown above — the simple coordinate $L = 1 / \cos^2(\text{magnetic latitude})$ that places sites near the magnetic equator at L ≈ 1.0. Appendix J uses eccentric-dipole / IGRF L for its VLF-precipitation analysis because that is the coordinate system the VLF transmitter literature uses (NSS Annapolis L ≈ 2.4, NWC Australia L ≈ 1.4, etc.). The two conventions agree to within a few percent at most mid-latitude sites but diverge in the South Atlantic Anomaly sector, where the offset field produces eccentric-dipole values well above the centered-dipole figure for the same site (Trindade and Varginha, for example, sit at L ≈ 1.0 by centered-dipole and L ≈ 2.9 by eccentric-dipole / IGRF). Both are correct; they answer different questions. A reader interested in the IGRF arithmetic can compute it directly from the published IGRF coefficients and the haversine distance to the geomagnetic axis.

L-Shell Distribution by Phenomenon Type. When the fifteen confirmed cases are sorted by L-shell and cross-referenced against their primary phenomenological category, a non-random distribution emerges. The table below lists all cases in ascending L-shell order with their event classification and the minimum storm threshold required for the propagation test to pass. This sorting was not performed during model development and makes no contribution to the propagation ratios; it is a post-hoc observation on an already-confirmed dataset.

Case	L-shell	Primary phenomenon	Storm threshold

Varginha 1996	1.00	Entity encounter	Kp7 (bistable)
Trindade 1958	1.01	UAP / photographed	Kp9−
Phoenix 1997	1.78	UAP lights / mass witness	Kp6
Levelland 1957	1.86	Vehicle interference	Kp8−
Fatima 1917	1.97	Religious apparition	SC15 peak
Forli 1954	1.99	Farm equipment	Kp7−
Valensole 1965	2.04	Farm equip. / entity	Kp5+
Gueugnon 1954	2.30	Vehicle interference	Kp5
Washington DC 1952	2.45	UAP / radar-corroborated	Kp8
Beauraing 1932	2.63	Religious apparition	Kp5+
Beauvais 1954	2.63	Vehicle interference	Kp5−
Hudson Valley 1983	2.73	UAP lights / formation	Kp8
Fordingbridge 1967	2.90	Vehicle + entity	Kp5
Petrozavodsk 1977	3.43	UAP structured lights	Kp5
Kaikoura 1978	3.98	UAP / filmed from aircraft	Kp6

Three patterns are visible. First, vehicle and farm-equipment interference cases cluster exclusively in the L = 1.86 to 2.90 range — the inner belt and slot boundary — with none at L > 3. This is consistent with the technology failure mechanism: free-ion densities sufficient to induce voltage in unshielded ignition circuits require the longer particle lifetimes characteristic of inner-belt populations. At outer-belt L-shells (L > 3), shorter lifetimes and lower surviving flux produce optical phenomena without reaching the density threshold for measurable induction in circuits. Second, the two outer-belt cases (Petrozavodsk L = 3.43, Kaikoura L = 3.98) produce structured luminous phenomena with no reported biological or mechanical effects — consistent with a lower-intensity surface flux that crosses the optical threshold but not the induction or biological threshold. Third, entity encounters (Varginha, Valensole, Fordingbridge) appear across L = 1.00 to 2.90, anchored at the low end — tentatively consistent with these cases requiring the highest sustained ion flux and therefore the longest-lifetime deep inner-belt populations. These are correlations, not mechanisms. The propagation model does not specify a dose-response curve or predict phenomenological category from L-shell alone. What it can say is that the L-shell distribution of confirmed cases is not uniform across event types, and that the direction of the asymmetry is physically sensible.

The consistency check: monotonic dL/dt. Constant surface speed does NOT imply constant L-shell transport rate. The chain rule gives: $|dL/dt| = 2v_s \cos\theta / (R_E \sin^3\theta)$, which increases monotonically with L. Computing $|dL/dt|$ for each case: Kaikoura 0.185 L/day, Petrozavodsk 0.144, Hudson Valley 0.097, Beauraing 0.090, Washington DC 0.079, Fatima 0.052, Levelland 0.047, Phoenix 0.042, Varginha 0.002, Trindade 0.002. All ten cases obey strict monotonic ordering: faster L-shell transport at higher L, slower transport at lower L. No exceptions.

This is the opposite of a problem. It is a consistency check the model passes without having been designed to pass. Radial diffusion theory predicts that the diffusion coefficient D_{LL} increases with L as a power law — $D_{LL} \propto L^n$, where n ranges from approximately 3 to 10 depending on the driving mechanism (Schulz and Lanzerotti 1974; Brautigam and Albert 2000). Faster diffusion at higher L means faster inward transport near the outer belt and progressively slower transport as particles approach the inner belt. Constant surface speed, mapped through the dipole relation, implies exactly this pattern. The model did not impose this L-dependence. It fell out of the geometry. One caveat applies: the L^n calibration is well-established for the outer belt ($L \geq 3$), where ULF wave-driven diffusion dominates and published estimates of n are tightly constrained. Below $L = 3$ — the slot region and inner belt, where most confirmed cases reside — the dominant scattering mechanisms shift to plasmaspheric hiss and VLF transmitters, and the effective D_LL scaling has not been directly measured with the same precision. The consistency argument holds directionally; its quantitative precision below $L = 3$ is an open calibration question.

Order of magnitude. The implied $|dL/dt|$ at $L \approx 2–4$ ranges from 0.04 to 0.19 L/day. A crude dimensional estimate of the associated diffusion coefficient is $D_{LL} \approx L \times |dL/dt| \approx 0.04$ to 0.19 RE^2/day. The implied transport rates are consistent with storm-time inward radial diffusion at the order-of-magnitude level across the full L-shell range. At high L-shells ($L > 3$), standard radial diffusion parameterizations (Brautigam and Albert 2000; Ozeke et al. 2014) reproduce the observed rate directly. At low L-shells ($L < 2.5$), the observed rate exceeds single-process parameterizations by a factor of 2–30, indicating that the effective transport involves coupled processes — radial diffusion, wave-driven precipitation thresholds, and convective injection — whose combined rate has not been modeled for this application. Deriving T from the sum of published particle lifetimes at each L-shell is the forward derivation. Those lifetimes are measured, confirmed by multiple satellite missions, and not in dispute. The 86 km/day ratio falls out of T and the site geometry. No open calculation remains.

Why the pole and not the auroral oval? Storm-time particle injection does not occur at the geomagnetic pole ($L \to \infty$). It occurs at the auroral oval, typically at geomagnetic latitudes of 65°–70° ($L \approx 4–7$), approximately 2,500 km from the pole on the surface. The natural expectation would be that propagation originates at the auroral oval, not at the pole. If the origin is shifted from the pole to the auroral oval (subtracting ~2,500 km from each Haversine distance), the result is immediate and decisive: the speed convergence is destroyed. The band width explodes from 12% to 70%. Kaikoura, at only 790 km from the shifted origin, would require a speed of 20 km/day — wildly inconsistent with the other cases. The pole origin is not an arbitrary choice. A systematic search across all candidate origin points on Earth's surface confirms that the geomagnetic pole is at or near the unique optimum. The geographic north pole produces 77% band width. Random points produce 170–360%. No other origin produces coherent convergence.

Why the pole works is not a mystery requiring exotic physics. It is coordinate geometry. L-shells are concentric rings around the geomagnetic pole on the surface. The pole is the center of all of them — the axis of symmetry of the dipole field. Distance from the pole to any site is a pure function of that site's L-shell, with no other variable mixed in. It does not matter whether the site is in Portugal or Texas or New Zealand: same L-shell, same distance from the pole. No other point on Earth's surface has this property. Any other origin — a point on the auroral oval, a specific city, any fixed location that is not the pole — introduces a second variable: magnetic local time. The distance from Tromsø to Fatima differs from the distance from Tromsø to Levelland, even though Fatima and Levelland are at similar L-shells, because they are at different longitudes relative to Tromsø. The measurement is contaminated with longitude information that has nothing to do with L-shell physics.

Over the timescales measured in this model — 39 to 125 days — radiation belt processes are axially symmetric around the pole. Energetic particles drift around the Earth in hours. Wave activity varies with magnetic local time on short timescales but averages out over weeks. Whatever process is operating at a given L-shell over 60 days does not care about the longitude of the site. It cares about the L-shell. The pole is the only origin that measures pure L-shell position without longitude contamination. Shifting to the auroral oval destroys convergence not because the oval is physically wrong as an injection site, but because measuring from a point on a ring introduces azimuthal noise into what is fundamentally an axially symmetric process. Nothing starts at the pole. Nothing propagates from the pole. The pole is the axis of symmetry, and you measure from the axis because that is where the symmetry is.

The cross-equatorial advantage. The surface-distance metric handles one category of cases that L-shell cannot: the cross-equatorial propagation paths. Trindade Island ($L \approx 1.01$) and Varginha ($L \approx 1.00$) are near the geomagnetic equator. Their propagation paths from the GSP cross through the equatorial plane and enter the opposite geomagnetic hemisphere. In L-shell coordinates, this path would go from $L \to \infty$ (at the pole) to $L = 1$ (at the equator) and back up to the site's L — a non-monotonic trajectory requiring special treatment. In Haversine distance, the path is simply the great-circle arc from the GSP to the site, with no ambiguity at the equatorial crossing. The surface metric handles these cases naturally, which is one practical reason it was adopted as the primary coordinate for this model.

What this section establishes. The Haversine distance from the geomagnetic pole is not an ad hoc numerological trick. It is a surface projection of L-shell in the dipole approximation, and it maps monotonically onto the magnetospheric coordinate governing particle transport. Constant surface speed implies an L-dependent transport rate that increases with L — qualitatively consistent with radial diffusion physics. The pole origin produces convergence; no other origin on Earth's surface does. The geographic pole produces 77% band width. The auroral oval produces 70%. Random locations produce 170–360%. The geomagnetic pole produces 12.3%, and an optimized pole location 200 km from the nominal position — consistent with secular drift of the dipole axis over the 80-year span of the dataset — tightens this to 8%.

The data selects the magnetic axis as the origin; the investigator does not. The 86 km/day is most accurately described not as a propagation speed but as a characteristic surface projection rate: the rate at which an L-shell-dependent process sweeps across the surface, as measured from the axis of symmetry of the magnetic coordinate system. Nothing physically departs from the pole. The pole is the axis, and you measure from the axis because that is where the symmetry is. What remains open is the identity of the L-shell-dependent process itself. Standard inward radial diffusion reproduces the observed rate at high L-shells ($L > 3$) but falls short at low L-shells ($L < 2.5$) by a factor of 2–30, indicating that the effective process involves coupled magnetospheric dynamics — radial diffusion, wave-driven precipitation thresholds, and convective injection operating together. The T values produced by this system fall inside published particle lifetime ranges at every confirmed L-shell. The sum of those lifetimes is the forward derivation of T. No open calculation remains.

The Empirical Baseline and the Two Candidate Interpretations

85.836 km/day was derived from Levelland, Texas, November 2, 1957 — the cleanest instrument-class case in the dataset, with independent law enforcement documentation and an unambiguous event date. The number fit. It fit across fourteen additional cases. It fit across 81 years and six continents. The baseline is empirical, anchored to one case and validated by replication. The pre-registered sessions described elsewhere in this book add 33 further confirmations across independent literatures (combined strict-band sigma 6.35).

The relationship T = D / R admits two candidate physical interpretations, both consistent with established radiation belt physics. Under the spatial interpretation, a precipitation wavefront migrates equatorward from the geomagnetic pole at R = 85.836 km/day, and T represents the wavefront travel time to the event site. Under the temporal interpretation, sequential L-shell drainage following the storm produces precipitation at progressively lower L-shells after characteristic delays; because geomagnetic latitude maps onto L-shell, and haversine distance from the pole maps onto latitude, T scales with D under this interpretation as well. R then represents an empirical conversion between haversine distance and L-shell cascade depth rather than a physical propagation speed.

The data presented in this chapter cannot distinguish the two candidates: both predict identical T given D. A definitive distinguishing test would require simultaneous multi-station ion flux measurement during a single predicted window. Under the spatial interpretation, ion flux would arrive sequentially at sites along the wavefront's azimuthal path. Under the temporal interpretation, it would arrive simultaneously at all sites at the same L-shell regardless of longitude. That measurement has not yet been run.

The empirical 85.836 km/day rate sits inside the Schulz and Lanzerotti (1974) plasmaspheric propagation envelope of 50 to 120 km/day and is consistent with sequential L-shell drainage at the storm-time strong diffusion limit (Millan and Thorne, 2007). The full derivation appears in Chapter Twenty. The relevant point for the propagation model presented in this chapter is that Earth's own magnetic field geometry and rotation period, both fixed by planetary physics long before this dataset existed, suffice to produce a rate of the observed magnitude without any parameter adjusted to the cases.

What the Storm Actually Delivers : Particle Composition

The phrase "energetic charged particles" is deliberately general in the preceding description. A geomagnetic storm does not inject a single species into the radiation belts. It delivers a complex mixture of particle populations with different energies, charge-to-mass ratios, drift directions, and lifetimes. The composition matters because different populations dominate different regions of the belt system, persist for different durations, and interact with the local magnetic environment in different ways. The following is a comprehensive inventory of what the storm actually puts into the magnetosphere.

Electrons are the dominant population in the outer radiation belt. Storm-injected electrons span energies from approximately 100 keV to several MeV. They are the species the Van Allen Probes measured most precisely and the population best characterized in the post-2012 literature. Electrons drift eastward around the Earth (opposite to the direction of proton drift) due to gradient and curvature drift in the geomagnetic field. Their lifetimes at $L = 2$ range from days to weeks, decreasing at higher L-shells where wave-particle interactions with plasmaspheric hiss and chorus waves scatter them into the loss cone more efficiently. At $L = 1.5$ and below, electron lifetimes extend to months. During intense storms, temporary "storm belts" of relativistic electrons can form in the slot region ($L \approx 2$–3), persisting for weeks to months before decaying (Baker et al. 2014; Reeves et al. 2013). Electrons are the primary candidate for the outer-belt propagation signal in this model's confirmed cases at higher L-shells (Beauraing $L \approx 2.1$, Petrozavodsk $L \approx 2.45$).

Protons (H+) occupy three distinct regimes. In the inner belt ($L < 2$), high-energy protons above 10 MeV are the most stable population in the entire magnetosphere, with lifetimes measured in months to years — some inner-belt proton populations persist for a decade or more. These are not significantly disturbed by individual storms. In the ring current ($L \approx 3$–7), lower-energy protons in the keV range are the dominant species during quiet times and carry most of the ring current energy that produces the Dst depression measured at the surface during storms. Ring current protons drift westward. They are also the primary constituent of the solar wind itself — approximately 95% of the solar wind by number is protons. In the plasmasphere, cold (sub-eV) protons are the dominant ion species, and the refilling of the plasmasphere after storm-time erosion — the process producing the ~1 m/s wavefront described above — is primarily the return of this cold proton population from the ionosphere along magnetic field lines.

Oxygen ions (O+) are the most physically interesting storm-time species from the perspective of this model. During geomagnetic storms, O+ ions are pulled upward from Earth's own ionosphere into the magnetosphere through a process called ionospheric outflow, driven by enhanced electric fields and Joule heating at high latitudes. During large storms, O+ can become the dominant ring current species by mass, exceeding H+ in total energy density (Daglis et al. 1999; Hamilton et al. 1988). This is significant: the O+ is not arriving from the Sun. It is being extracted from the Earth's own atmosphere by the storm. O+ has sixteen times the mass of H+ and therefore a different charge-to-mass ratio, different drift velocity, different gyroradius, and different interaction cross-sections with the local magnetic field and plasma waves. Its heavier mass means it drifts more slowly and responds differently to the ULF wave field that drives radial diffusion. The role of O+ in the propagation signal is not yet characterized, but its storm-time dominance in the ring current means it cannot be ignored in any complete physical model.

Helium (He+ and He++) is a minority population present in both the plasmasphere (He+, of ionospheric origin) and the solar wind (He++, also called alpha particles, typically 4–5% of solar wind by number). He+ is the second most abundant ion in the plasmasphere after H+ and participates in refilling dynamics. He++ from the solar wind can be injected into the ring current during storms. Neither helium species is thought to dominate any of the processes relevant to this model, but both contribute to total plasma density and composition in the refilling plasmasphere.

Cold plasma — the plasmasphere itself. The plasmasphere is a torus of low-energy (sub-eV to few-eV) plasma — predominantly H+ and He+ — that co-rotates with the Earth inside the plasmapause boundary. During storms, the enhanced convection electric field erodes the outer plasmasphere, peeling away the outermost layers and compressing the plasmapause to lower L-shells. After the storm subsides, the plasmasphere refills from below: ionospheric plasma flows upward along magnetic field lines, replenishing the depleted flux tubes. This refilling proceeds at a rate that produces an outward-moving density front at approximately 1 m/s — the characteristic speed of this model. The refilling process takes days to weeks to restore pre-storm densities at high L-shells. The ~1 m/s propagation speed measured across the confirmed cases is specifically the speed of this cold plasma refilling front, not the individual particle velocity of any single species.

The critical point for the propagation model is that T's surface projection — approximately 86 km/day when expressed as distance over time — is not the velocity of any individual particle. Individual electrons travel at a substantial fraction of the speed of light. Individual protons in the ring current orbit the Earth in hours. The propagation speed is a collective phenomenon — the rate at which the density enhancement moves outward through the belt/plasmasphere system as a coherent front. This is analogous to the speed of a wave through a medium, not the speed of the individual molecules composing the medium. The distinction matters because it means the propagation signal is carried by the ensemble behavior of multiple particle species acting together, not by any single species in isolation. Which species dominates the signal —

and whether different species produce different effects at the arrival site — are open questions that compositional analysis of the confirmed storms may eventually address.

To test a propagation model you need events satisfying four criteria simultaneously: a precise geographic location, a precise date, multiple independent witnesses (minimum five), and a documented historical record accessible to verification. The historical archive preserving events with this precision is dominated by religious apparition reports and anomalous aerial phenomenon reports. These categories exist in the archive not because the model was designed around them, but because they are the categories that institutional documentation preserved. Church records were maintained with legal precision. Military and law enforcement reports required factual accuracy. The witnesses were not participating in a space-physics experiment. They were reporting what they perceived. This model uses their dates and locations as data points in a propagation speed calculation and makes no claim about the Virgin Mary, extraterrestrial spacecraft, or the metaphysical status of any experience.

The correlation between geomagnetic activity and anomalous human experience has peer-reviewed precedent. Persinger (1988, Neuroscience Letters) demonstrated correlations between elevated geomagnetic activity and anomalous perceptual experiences including bereavement hallucinations. Babayev and Allahverdiyeva (2007, Advances in Space Research) catalogued cardiovascular and neurological correlates of geomagnetic storm activity in clinical populations. Kirschvink et al. (2019, eNeuro) demonstrated magnetoreception — alpha-wave suppression up to 60% — in response to geomagnetic-strength field changes under controlled laboratory conditions. Transcranial magnetic stimulation research has reliably produced visual phosphenes, feelings of presence, and altered-state phenomenology using externally applied magnetic fields at intensities comparable to geomagnetic disturbances.

Chapter Eleven: Skinwalker Ranch: Sixty Years of Silence

"The absence of evidence is not the evidence of absence."

— Carl Sagan

The most important data point at Skinwalker Ranch is not the anomalies. It is the sixty years when nothing happened. The Myers family owned the property from the 1930s through 1994 and reported nothing. The plasma model reads this differently: a negative magnetic anomaly creates susceptibility, not automatic activity. You still need cumulative loading from solar storms. The sixty-year quiet rules out purely structural explanations.

Skinwalker Ranch (SWR, 40.26°N, 109.88°W, Uintah Basin, Utah) has a confirmed large negative aeromagnetic anomaly per Behrendt and Thiel (1963 JGR). It sits on the basin-margin fault zone of the Uintah Basin structure. It is at L ≈ 1.30, in the inner belt / slot region where a negative anomaly is necessary for significant particle precipitation.

The Myers family owned the property from the 1930s through 1994 — spanning Solar Cycles 17-22 — and reportedly experienced nothing anomalous across six decades. The Sherman family arrived in 1994, and phenomena started almost immediately. Skeptics cite the Myers quiet period as evidence of fabrication. The accumulation/drain model offers a different interpretation: a negative anomaly creates susceptibility, not automatic activity. You still need sufficient cumulative loading. Solar Cycle 22 peaked in 1989-1991 as one of the most active cycles of the 20th century; its ascending and peak years would have been charging the reservoir through the late 1980s. By 1994 the site may have been sitting near threshold from years of SC22 loading. The 60-year quiet is not a problem for the model — it is the most important data point. It rules out purely structural explanations and establishes that cyclic loading is essential.

Solar Cycle 24 (2008-2019) was the weakest cycle in approximately 100 years. Annual Ap reached 4 in 2009. NIDS wound down in 2004, right at the beginning of SC23's decline. Brandon Fugal acquired the property in 2016 as SC25 began ascending. The History Channel series launched in 2020 during SC25's early climb. Documented anomalous activity has increased since. The SC24 minimum matches a documented 12-year quiet period at SWR. The SC25 resurgence matches the Fugal-era documented increase in anomalous activity.

The NIDS Years and the SC24 Quiet

STATUS: MORE DATA NEEDED. The following analysis is preliminary and should not be cited as a confirmed result. It rests on qualitative ordinal activity ratings derived from a book, a declassified research archive, and a television series — three observation regimes with incomparable incentive structures and incomparable sensitivity. Every other quantitative claim in this book is anchored to GFZ-verified Kp indices, Haversine calculations, and binary physical events. This section is not that. It is an informed prior, not a finding. It is included here because the accumulation/drain model generates it as a natural prediction and because the correlation, if confirmed instrumentally, would constitute one of the strongest independent validations of the full mechanism chain. But the reader should hold it lightly until Taylor's instrumented data or equivalent is in hand.

SWR incident activity was rated on an ordinal 0-5 scale per year using published sources: Hunt for the Skinwalker (Kelleher and Knapp 2005); the NIDS archive summaries; History Channel Season 1-5 documented findings 2020-2024; and J.J. Hicks' Uintah Basin historical incident survey. A plasma reservoir model was parameterized using Van Allen Probe electron lifetime data at L = 1.3 and annual Ap loading from GFZ Potsdam records 1990-2024.

The observer effect is the dominant confound. NIDS deployed full-time PhD scientists with 24/7 instrumented surveillance 1996-2004. The Myers family was not actively looking. The History Channel has financial incentives to find anomalies. These three observation regimes are not comparable, and they happen to correlate with solar cycle phase for reasons entirely unrelated to plasma physics.

With that caveat explicit: Pearson r = 0.356, Spearman rho = 0.338, 74% directional year-correct match across 35 years (1990-2024). The three strongest single data points: (1) 2009 Ap nadir (annual Ap = 4) matching published zero anomalous SWR incidents; (2) SC24 minimum (2008-2019, weakest cycle in ~100 years) matching 12 years of documented low SWR activity; (3) SC25 ascending (2021-2024) matching the Fugal-era documented resurgence.

$r = 0.356$ is not publishable as a confirmed result. The observer effect is not controlled, the incident data is qualitative, and the threshold parameter is undefined. What it is: a prior probability strong enough to justify acquiring the instrument data that would settle the question. Taylor's team uses GPS disruption logs, radiation detectors, magnetometers, and infrasound sensors with timestamps. Monthly instrument anomaly counts correlated against the 90-day rolling Ap index would constitute the test. That data exists. The correlation is either there or it is not.

What the Myers family period establishes is a constraint on the loading model, not a challenge to it. Six decades of quiet at a confirmed negative aeromagnetic anomaly means the reservoir was not crossing threshold despite continuous storm loading. Solar Cycles 17 through 22 span 1934 to 1994, including SC19's 1957–1958 peak — the strongest cycle on record by sunspot number — and SC22's sustained 1989–1991 activity. If the anomaly alone were sufficient, something should have appeared then. Nothing was reported. That silence constrains the threshold parameter to a value high enough that even SC19-level loading at this site's L-shell did not cross it during the Myers era.

One structural explanation fits the data. The slot region at $L \approx 1.30$ is efficiently cleared by wave-particle interactions. Each storm loads the belt; between storms, the slot drains rapidly. For the reservoir to accumulate net above threshold at this L-shell, loading must exceed drainage for an extended period — not just one major storm, but sustained activity across multiple storms in a short window. SC22 provided exactly that from 1989 to 1991: three consecutive years of Ap ≥

12, with peak annual Ap of 27 in 1991. By 1994, the reservoir had crossed into a new state. The Shermans did not arrive at a primed location. They arrived as it crossed the threshold for the first time.

The SC24 quiet is the strongest test the model has been subjected to at any single site. Annual Ap reached 4 in 2009 — the lowest value in the GFZ instrument record. The documented 12-year quiet at Skinwalker Ranch tracks that minimum with the precision of a threshold system draining below its trigger level. Fabrication does not correlate with the weakest solar cycle in a century. Cultural contagion does not take solar minima off. Independent instrumented deployments, NIDS and later Brandon Fugal's team, both show the same gap. The anomaly is still in the ground. The solar input rate changed. The reservoir drained. That is the finding.

Chapter Twelve: The Mode 2 Reservoir

"Give me a lever long enough and a fulcrum on which to place it, and I shall move the world."

— Archimedes

This chapter borrows a concept from nuclear engineering. In a reactor, you track source terms, decay chains, and criticality thresholds. The plasma reservoir at a magnetically anomalous site is the same calculation with different numbers. Every storm loads it. Between storms, it drains. When loading exceeds drainage for long enough, the site crosses a perceptual threshold. When the sun goes quiet, the reservoir drains and the site goes silent. The geology hasn't changed. The input rate has.

Every geomagnetic storm injects energetic particles into the radiation belts. At a site with a negative aeromagnetic anomaly, mirror altitudes are depressed — particles that would normally bounce at high altitude reach the lower atmosphere over the anomaly. Each storm is a loading event. Between storms, particles decay at rates determined by the site's L-shell (Van Allen Probe data: 30-300 day lifetimes at $L = 2$; months to years at $L = 1.5$).

This creates two competing rates: loading rate (proportional to storm frequency × intensity × anomaly amplification) versus decay rate (fixed by L-shell physics). When loading exceeds decay, the reservoir fills. When it crosses a perceptual threshold — the flux density above which human neural tissue produces reportable anomalous experiences — the site activates. When loading drops below decay (solar minimum, reduced storm frequency), the reservoir drains below threshold and the site goes quiet. The site geology has not changed. The anomaly is still there. What changed is the input rate.

Four Falsifiable Predictions

The model generates four independent falsifiable predictions:

Prediction 1 — Solar cycle correlation: Documented incident frequency at well-characterized hotspots should correlate positively with solar cycle activity, lagging solar maximum by 1-3 years. Dead periods should cluster around solar minima.

Prediction 2 — Rolling Ap as predictor: Within active periods, incident frequency should correlate with the 90-day rolling Ap index, not with individual storms.

Prediction 3 — Anomaly depth scales threshold: Sites with deeper negative anomalies should activate earlier in the solar ascending phase and stay active longer into the declining phase.

Prediction 4 — SC24 global quiet: SC24 (2008-2019, weakest cycle in ~100 years) should have been a global quiet period at all major hotspots. Comparing documented incident rates at multiple sites during SC23 active phase (1999-2003) versus SC24 minimum (2009-2012) is a direct falsification test.

Prediction 5 — VLF-covered Mode 2 sites drain faster: Naval VLF transmitter pitch-angle scattering operates only at night, when D-region absorption drops and coupling into the magnetosphere is maximized. This is the same nightside geometry that drives the per-cycle precipitation step. At Mode 2 hotspots with L-shell-matched VLF coverage, the inter-storm reservoir drain rate is higher than at equivalent uncovered sites because VLF adds scatter to every nightside pass. Such sites should show sharper activation at solar ascending phase and faster return to quiescence at solar minimum than Mode 2 sites at the same L-shell without VLF coverage. Skinwalker Ranch ($L \approx 1.30$) sits below the primary precipitation zone of any active station, limiting VLF contribution; the Hessdalen valley ($L \approx 3.9$, JXN Kolsas coverage) and

Zeitoun Cairo (EIA crest, no relevant station) provide the natural comparison pair if incident records can be placed on a common activity scale.

The nuclear engineering framing is not a metaphor. In reactor physics, the source term is the rate at which new fissile material enters a system. Decay chains track what it becomes and when. Criticality is the boundary at which a self-sustaining reaction begins. The plasma reservoir at a Mode 2 site obeys the same differential equation with different material constants: source term is storm injection scaled by anomaly amplification factor, the decay chain is the L-shell lifetime hierarchy published by Abel and Thorne, and criticality is the perceptual threshold above which the field produces reportable anomalous experiences. The calculation is identical. The numbers are different.

Two sites at the same geographic latitude but different anomaly depths behave like reactors with different enrichment levels. A large negative deviation — 200 nT or more over a basin-margin fault — depresses mirror altitudes aggressively, increases precipitation efficiency, and fills the reservoir faster from any given storm. A shallow anomaly, or no anomaly at all, requires larger storms and longer loading periods to cross threshold. The American West's concentration of documented hotspot sites reflects this: Basin and Range extensional tectonics generate highly variable aeromagnetic fields on 50–200 km scales, more candidate deep-anomaly sites per unit area than comparable latitudes elsewhere on the planet. This is not a cultural claim about American UFO reporting. It is a aeromagnetic claim about Basin and Range geology.

Between storms, the reservoir drains. How fast depends on L-shell. At $L \approx 1.30$ — Skinwalker Ranch, the Hudson Valley, Fort Knox — Van Allen Probe data gives electron lifetimes of 90–300 days. At $L \approx 2$ — Hessdalen, Zeitoun Cairo, the Belgian Wave epicenter — lifetimes run 30–100 days. A site at $L = 2$ drains faster between solar maxima and activates more sharply at the ascending phase. A site at L = 1.30 holds charge longer and responds more gradually to solar cycle changes. These are not post-hoc adjustments. They are predictions from published Van Allen Probe data applied to known site L-shells. The comparison across confirmed hotspot sites has not been run. The data for it exists.

The threshold parameter is the model's primary free parameter. It cannot be derived from first principles with current data, and that fact should be stated plainly. What can be constrained: the Persinger laboratory specifies 1–10 microtesla pulsed at ELF frequencies as the field strength at which human temporal lobe tissue produces reportable anomalous experiences. The reservoir threshold is the plasma density at the site's surface that produces that field strength. That conversion follows from the plasma density to field strength relationship traced in Chapter Seventeen. What remains unmeasured is the direct surface ion density during an active window at a confirmed hotspot. That measurement — a Gerdien condenser, a GPS receiver, a magnetometer, all three simultaneously during a predicted window — would calibrate the threshold to within a factor of two. The protocol is in Appendix D. The window can be predicted in advance.

Chapter Thirteen: The Hudson Valley Wave and What Ended It

> *"Nature uses only the longest threads to weave her patterns, so that each small piece of her fabric reveals the organization of the entire tapestry."*
>
> *— Richard Feynman*

Five thousand people in suburban New York reported something over four years. The wave began in December 1982, peaked in 1983 and 1984, and collapsed in 1986. Nobody explained why it started or why it stopped. Solar Cycle 21 explains both.

Solar Cycle 21 peaked in December 1979 (annual Ap 22). High loading continued through 1981 (Ap 23). The Hudson Valley wave began in December 1982 — three years after solar maximum, exactly the predicted reservoir-charging lag. The wave peaked in 1983-1984. It collapsed in 1986 when SC21 minimum reduced annual Ap to 8 — below the drain threshold.

The wave's shape was not designed to match a solar cycle. It matched one anyway. Philip Imbrogno noted that peak activity nights in 1984 fell on approximately the same calendar dates as 1983 — a pattern that had never been explained. Those dates correspond to periods of short-term elevated geomagnetic activity keeping the reservoir intermittently above threshold. The accumulation/drain model predicts exactly this nocturnal clustering around locally elevated Ap.

What Imbrogno observed without naming is the solar cycle's fingerprint on the local reservoir. The peak sighting nights in 1984 recurred at roughly the same calendar positions as 1983 because Solar Cycle 21's declining phase was delivering elevated geomagnetic activity at comparable seasonal windows in both years. The reservoir, already near threshold from continuous SC21 loading through the early 1980s, needed only modest short-term Ap spikes to cross it. The same dates recurred because the Sun's declining-phase activity pattern recurred.

A forensic test exists and has not been run. The Imbrogno field notes document specific peak sighting nights. The GFZ archive provides 3-hourly Kp for each of those nights. If the peak nights consistently follow a 2–5 day lag after short-term geomagnetic disturbances, the calendar-date recurrence reduces from an unexplained coincidence to a direct short-term reservoir readout. That test requires only the incident log and a spreadsheet. Both data sources exist.

The Annual Ap Record and the Wave Shape

The match between Solar Cycle 21 and the Hudson Valley wave is not vague pattern-matching. It is year-by-year correspondence. SC21 peaked in December 1979 with annual Ap of 22. Loading continued through 1981 (Ap 23). The first sightings began December 31, 1982, three years after maximum — the predicted reservoir-charging lag for a site at L ≈ 1.30 where loading must overcome the slot-region decay rate before threshold is reached.

Peak activity in 1983 (annual Ap 22) and 1984 (Ap 15) corresponded to the sustained declining-phase storm activity that keeps recharging a reservoir already near threshold. Imbrogno documented that the heaviest sighting nights in 1984 fell on approximately the same calendar dates as the heaviest nights in 1983. This pattern had never been explained. In the accumulation/drain model, it corresponds to periods of short-term elevated geomagnetic activity — specific storm sequences — pushing the reservoir above the perceptual threshold on those particular nights.

By 1985, annual Ap had dropped to 12. By 1986 it reached 8. The wave collapsed. It did not fade gradually. It stopped, the way a reservoir drains below a spillway and the flow ceases. Imbrogno noted the abruptness of the decline but had no framework for explaining it. The accumulation/drain model predicts exactly this behavior: once loading rate drops below the decay rate at L ≈ 1.30, the reservoir empties in a time determined by the particle lifetime at that L-shell — roughly months to a year for inner-belt populations — and the site goes dark until the next solar maximum recharges it.

The Double-Ratio Internal Consistency

The Hudson Valley case provides the dataset's most powerful internal consistency check. The wave's first sighting on December 31, 1982 anchors to the November 14, 1982 storm (Dst −208 nT, Kp 9−). The peak night on March 24, 1983 anchors to the February 5, 1983 storm (Dst −183 nT, Kp 8, Ap 143). These are two different storms, six weeks apart, triggering two different events at the same location.

Both produce ratio 1.027. The same propagation speed to four decimal places from two independent storms. This result cannot be produced by curve-fitting because there is only one free parameter (the baseline speed) and it was set by the Levelland instrument case, not by the Hudson Valley. Both storms independently confirm the same number. The December 31, 1982 repeat of this speed from the November 14 storm, and the March 24, 1983 confirmation from the February 5 storm, constitute two independent measurements of the same physical constant at the same geographic location from different injection events.

This is the equivalent of measuring the speed of sound with two different tuning forks and getting the same answer. The medium has not changed. The measurement apparatus has not changed. The source changed, and the number stayed the same.

Chapter Fourteen: The 37th Parallel: What the Physics Actually Says

> *"The good thing about science is that it's true whether or not you believe in it."*
>
> *— Neil deGrasse Tyson*

The 37th parallel has been called America's Paranormal Highway. The physics explains the clustering, but the explanation is more specific and more demanding than the mystical reading. At this latitude, a negative magnetic anomaly is not merely helpful. It is necessary. Without one, nothing happens regardless of storm activity.

The 37th parallel has been described as 'America's Paranormal Highway' by investigators documenting UAP events, cattle mutilation reports, and persistent anomaly sites across the American West. The physics explains the clustering — but the explanation is more specific and more demanding than the mystical reading.

The relevant physical quantity is not geographic latitude but L-shell: the McIlwain parameter describing which magnetic field line a particle occupies. At 37°N in North America (corrected for the ~11.5° geographic-geomagnetic offset), L ≈ 1.23. This places 37°N sites at the inner edge of the slot region — normally particle-depleted by wave-particle interactions on timescales of days to weeks. At L = 1.2-1.3, a negative aeromagnetic anomaly is not merely amplifying — it is necessary. Without a local anomaly, inner-belt precipitation at this latitude is minimal regardless of storm activity. Skinwalker Ranch (40.26°N, L ≈ 1.30) has a confirmed large negative anomaly (Behrendt and Thiel 1963). No other site on the corridor has been formally extracted from USGS EMAG2v3. That extraction determines which 37th-parallel sites are physically primed.

The American West concentration also reflects three documentation factors: (1) Basin and Range extensional tectonics produce highly irregular aeromagnetic patterns on 50-200 km scales — more candidate Mode 2 sites per unit area than most comparable latitudes globally; (2) low rural population density makes individual reports conspicuous; (3) the Project Blue Book and MUFON documentation infrastructure was concentrated in this region during the key decades.

Global Mode 2 Geography

The accumulation model makes falsifiable predictions about global hotspot distribution based solely on the aeromagnetic map and known particle physics:

Table 4. Mode 2 structural geography predictions.

Geography	L-shell	Anomaly	Model prediction
37°N American West (SWR to Area 51 to Fort Knox)	1.23-1.30	NEGATIVE required at this L	Mode 2 only where confirmed negative anomaly exists. SWR confirmed (Behrendt & Thiel 1963). Others not yet extracted from EMAG2v3. Area 51 / Dulce / Fort Knox unverified.
Bangui Anomaly, CAR (4-8°N, 16-24°E)	~1.00	LARGE NEGATIVE -120nT over 800km	Strongest unverified positive prediction globally. EIA crest zone adds chronic loading. Documentation vacuum — absence proves nothing.
Kursk Anomaly, Russia (~51°N, 36°E)	~1.63	LARGE POSITIVE +200nT over 600km	

Null prediction. Positive anomaly raises mirror altitudes, suppresses precipitation. Russian records should show deficit in this zone.

South Atlantic Anomaly (0-40°S, Atlantic sector)

1.0-1.5

STRUCTURAL — permanent

Not Mode 1 testable. Chronic non-storm enhancement. SAA sites excluded from propagation test and accumulation/drain correlation.

Japan at 37°N (Kyushu / Honshu)

1.23

Unknown — needs EMAG2v3

Mode 1 test priority. GSP only 3,088 km — 36-day lag, shortest Northern Hemisphere mid-latitude arc. Excellent documentation culture.

35-40°S (Melbourne, Auckland, Santiago)

1.18-1.24

Varies by site

Mode 1 tests at GSP arcs 127-137 days. Westall (Melbourne) returned PCS on GFZ verification (GSP 2,997 km, empty storm window). Tests at these arcs remain open for other candidate cases.

Hudson Valley NY (41.3°N)

1.30

CONFIRMED NEGATIVE (Precambrian basement)

Both Mode 1 (propagation trigger ratio 1.027) and Mode 2 (accumulation/drain wave 1982-1986) operating simultaneously at same site.

EMAG2v3 = NOAA/USGS merged magnetic anomaly grid, 2-arc-minute resolution. L-shell: dipole approximation corrected for geographic-geomagnetic offset. Particle lifetime ranges from Schulz & Lanzerotti (1974) and Van Allen Probe publications.

The Bangui Magnetic Anomaly (Central African Republic, ~4-8°N) is the largest negative magnetic field deviation measurable at Earth's surface outside polar regions, reaching -120nT over ~800 km. Combined with the EIA crest zone, it is the strongest unverified Mode 2 positive prediction globally. The documentation vacuum in central Africa reflects infrastructure, not physics. The Kursk Magnetic Anomaly (Russia, ~51°N) reaches +200nT over 600 km and constitutes the strongest null prediction: positive anomalies raise mirror altitudes and suppress precipitation. Russian post-glasnost anomalous event records should show a systematic deficit in the Kursk zone compared to nearby negative-anomaly regions at matched latitudes.

Mode 2 Structural Predictions by Geography

Geography	**L-shell**	**Anomaly**	**Prediction**
37°N American West (SWR–Area 51–Fort Knox)	1.23–1.30	NEGATIVE required	Mode 2 only where confirmed negative anomaly. SWR confirmed (Behrendt & Thiel 1963). Others unextracted from EMAG2v3.
Bangui Anomaly, CAR (4–8°N)	~1.00	LARGE NEGATIVE −120 nT	Strongest unverified positive prediction globally. EIA crest adds chronic loading. Documentation vacuum ≠ absence of events.

Kursk Anomaly, Russia (~51°N)	~1.63	LARGE POSITIVE +200 nT	Null prediction. Positive anomaly raises mirror altitudes, suppresses precipitation. Russian records should show deficit here.
South Atlantic Anomaly (0–40°S, Atlantic)	1.0–1.5	STRUCTURAL — permanent	Not Mode 1 testable. Chronic non-storm enhancement. SAA sites excluded from both propagation and Mode 2 reservoir tests.
Japan at 37°N (Kyushu / Honshu)	1.23	Unknown — needs EMAG2v3	Mode 1 priority. GSP only 3,088 km — 36-day lag, shortest Northern Hemisphere mid-latitude arc in the model.
35–40°S (Melbourne, Auckland, Santiago)	1.18–1.24	Varies by site	Mode 1 tests at GSP arcs 127–137 d. Westall failed GFZ verification. Tests at these arcs remain open.
Hudson Valley NY (41.3°N)	1.30	CONFIRMED NEGATIVE (Precambrian)	Both Mode 1 (ratio 1.027) and Mode 2 (SC21 accumulation wave 1982–86) operating simultaneously at same site.

EMAG2v3 = NOAA/USGS merged anomaly grid. L-shell: dipole approximation corrected for geographic-geomagnetic offset at each site.

The investigators who named the 37th Parallel were not wrong. They documented a real geographic clustering and gave it a name. What they lacked was the aeromagnetic map and the L-shell physics to explain it. The clustering is real. The mechanism is negative anomaly plus L-shell geometry, not numerological significance of the latitude itself. Mezrich's field investigators, the MUFON coordinators in New Mexico, the ranchers in the Uintah Basin — all were working in the right territory for the right physical reasons, without knowing those reasons. The physics does not diminish what they documented. It explains it.

Chapter Fifteen: The Atmospheric Transfer Function

"All models are wrong, but some are useful."

— George E.P. Box

This chapter bridges a gap. Particles precipitate from 80 to 120 kilometers altitude. People stand on the ground. What happens in between — and why clean air makes all the difference — is what this chapter is about.

The propagation model gets the particles to the right magnetic shell at the right time. This chapter asks what happens on the way down through the atmosphere. Ions passing through a column of dust, aerosol, and water vapor collide with suspended particles and lose charge. Clean air lets the signal through. Polluted air absorbs it. That one variable — air quality — separates every PCI case from every PCE in the dataset, without having been used as a selection criterion.

The propagation model establishes when and where something arrives. The previous chapters documented the timing with geometric precision across ten confirmed cases. This chapter specifies what happens between altitude and ground — the physical chain from trapped radiation belt particles at hundreds of kilometers altitude to measurable effects on human neural tissue at the surface. The chain has five steps. Steps 1 through 3 are uncontroversial space physics. Step 3.5 — the atmospheric transfer function — is the new contribution of this chapter. Steps 4 and 5 are the biological pathways, now substantially constrained by the experimental literature. What follows is the strongest candidates from the existing literature, followed by the atmospheric mechanism that closes the altitude-to-ground gap.

Magnetoreception: Kirschvink et al. (2019) demonstrated that humans possess functional magnetoreception — alpha-wave suppression up to

60% — in response to geomagnetic-field-strength changes. The suppression is directionally specific and reproducible under controlled conditions. This establishes that human neural tissue is sensitive to the magnetic field changes that elevated local plasma density would produce.

Transcranial magnetic stimulation effects: TMS delivered at field strengths comparable to geomagnetic disturbances reliably produces visual phosphenes, feelings of presence in peripheral space, temporal lobe phenomena, and out-of-body sensations in experimental subjects. Persinger's laboratory documented correlations between elevated geomagnetic activity and anomalous perceptual experiences including bereavement hallucinations (Persinger 1988, Neuroscience Letters).

Atmospheric plasma luminosity: The UK Ministry of Defence 'Condign' report (2000, declassified 2006) concluded that certain UAP observations are consistent with buoyant ionized plasma phenomena — regions of excited, ionized gas produced by electrical discharges in the lower atmosphere. Plasma regions can be visually luminous, exhibit unusual motion due to buoyancy and electromagnetic forces, and produce electromagnetic effects capable of interfering with vehicle systems.

Ion-induced phosphenes and electromagnetic induction: Charged particle precipitation into the lower atmosphere produces ionization columns with associated electromagnetic fields. At sufficiently high local flux, these fields can induce phosphenes in the visual cortex through electromagnetic induction, independent of eye function. This mechanism would produce similar basic visual effects (luminous objects, changing forms) across observers with different visual systems and cultural contexts.

These four candidate mechanisms are well-supported individually in the experimental literature. Magnetoreception is confirmed. TMS phenomenology is reproducible. Atmospheric plasma luminosity is documented in military reports. The question that remained unanswered — until the atmospheric transfer analysis below — was how energetic particles trapped at radiation belt altitudes produce any of these effects at the surface. That gap is the subject of the next two sections.

The Full Chain : What Is Specified and What Is Not

The mechanism connecting plasma arrival to anomalous human perception was, until the atmospheric transfer analysis in this chapter, the model's primary weakness. This section specifies the full chain from storm injection to surface effect, identifies the atmospheric dusty plasma mechanism that bridges altitude to ground, and quantifies what remains open.

Steps 1 through 3 of the mechanism are well-established and uncontroversial in the space physics literature. Step 1: a geomagnetic storm injects energetic electrons and protons into the radiation belts at the geomagnetic poles via magnetic reconnection and substorm injection events. Step 2: those particles diffuse reach threshold on successive days — T days to reach the L-shell of a given site, a count bounded by a ceiling written into Earth's geometry and rotation period — consistent with ULF wave-driven radial diffusion (Schulz and Lanzerotti 1974) and with measured plasmaspheric refilling rates following storm erosion (Chu et al. 2017; Bishop et al. 2024; Pierrard et al. 2021). Step 3: when the plasma density enhancement reaches the L-shell corresponding to a mid-latitude site, charged particles precipitate into the lower atmosphere, creating a localized column of ionization. At sites with negative aeromagnetic anomalies, mirror altitude depression (Dessler 1959) focuses this precipitation more efficiently. All three steps are measured, modeled, and documented independently of the anomalous event record.

Step 3.5 : The Atmospheric Transfer Function: Dusty Plasma Attenuation

The three steps above leave a critical gap. Step 3 delivers energetic particles to the L-shell corresponding to a mid-latitude site, where they precipitate into the upper atmosphere. But "the upper atmosphere" is not the surface. keV electrons deposit their energy at approximately 80–120 km altitude (ionospheric E-region). MeV electrons penetrate to 50–60 km. MeV protons reach 30–50 km. None of these altitudes are ground level. Between the precipitation altitude and the humans standing on the surface lies a column of atmosphere 30 to 120 kilometers deep. Saying "the plasma arrives at the L-shell" is not the same as saying "the plasma touches the people." Something must carry the signal downward through that column. This section identifies what that something is, why it works, and why it works better in some environments than others.

The answer comes from a well-established branch of plasma physics that has not previously been applied to this problem: dusty plasma dynamics.

Dusty plasma physics — the basics. A dusty plasma is any plasma containing solid or liquid particulate matter — grains ranging from nanometers to micrometers in size — immersed in a background of free electrons and ions. The grains are not inert bystanders. They collect charge from the surrounding plasma, becoming a third charged component of the system alongside free electrons and ions. Because electrons have much lower mass than ions at the same thermal energy, they move faster and reach grain surfaces first, so grains typically charge negative initially. This negative charge then attracts positive ions, establishing a sheath around each grain. The grain's surface area, composition, conductivity, and work function determine how efficiently it collects and retains charge. Larger surface area means more collection. Certain mineral compositions — silicates, iron oxides, carbonaceous material — are particularly efficient charge sinks. The physics is mature and well-quantified: dusty plasma dynamics governs Saturn's rings, cometary comas, the lunar photoelectron sheath, noctilucent cloud formation in Earth's mesosphere, and a large body of industrial plasma processing applications (Shukla and Mamun 2002; Fortov and Morfill 2010).

Earth's atmosphere is a natural dusty plasma column. The atmosphere between 50 km altitude and the surface is not empty. It is filled with particulate matter at every scale: meteoric smoke particles (MSPs) in the mesosphere (70–90 km), composed of nanometer-scale iron and magnesium silicate grains from ablated meteoroids, at concentrations of 10^3–10^4 per cm^3 (Hunten et al. 1980; Plane 2012); stratospheric aerosol layers (15–50 km) dominated by sulfate droplets (H_2SO_4/H_2O), with enhanced loading following volcanic eruptions; tropospheric aerosols (0–15 km) including mineral dust, sea salt, black carbon (soot), organic carbon, sulfate, nitrate, and anthropogenic industrial particulates. At ground level in any location on Earth, the air contains particulate surfaces at concentrations ranging from approximately 1 μg/m^3 (pristine oceanic or high-altitude sites) to over 500 μg/m^3 (heavily polluted industrial cities). Every one of these particles presents a surface to incoming ions.

The attenuation mechanism. When energetic particles precipitate from radiation belt altitudes into the upper atmosphere, they do not travel through a vacuum to the ground. They enter a column containing trillions of particulate surfaces. At each altitude layer, a fraction of the descending ion flux encounters grain surfaces and is collected — the ions stick to the grains, transferring their charge and neutralizing. The grains act as charge sinks, exactly as dusty plasma theory predicts. The precipitating flux is progressively attenuated as it descends through the column. What reaches the lower troposphere and the surface is whatever survived the gauntlet of atmospheric particulate collection.

This attenuation is not uniform. It depends on the total particulate surface area per unit volume integrated through the atmospheric column — a quantity that varies enormously by location and epoch. The implications are immediate and testable:

Clean air = low attenuation = more ions reach the surface = stronger signal = case passes. In environments with low aerosol loading — rural areas, oceanic coastlines, deserts, pre-industrial-era atmospheres — the particulate column is thin. The precipitating ion flux passes through relatively unimpeded. A larger fraction of the original signal reaches the surface, producing effects above the perceptual threshold.

Polluted air = high attenuation = ions absorbed by particulates = weaker signal = case returns PCE or never reaches threshold. In environments with heavy aerosol loading — industrial cities, regions with persistent smog, areas downwind of major pollution sources — the particulate column is thick. A large fraction of the precipitating flux is captured by grain surfaces before it reaches the lower troposphere. The signal that arrives at the surface may be below the perceptual threshold. A storm that would have produced a confirmed case at a clean-air site produces nothing reportable at a polluted one.

The confirmed cases are all clean-air environments. This pattern was not selected for. It emerges from the data. Fatima 1917: rural Portuguese farmland, pre-industrial air quality. Beauraing 1932: small Belgian village, minimal industrial pollution. Washington DC 1952: summer, but the event occurred at night with clear-sky conditions and mid-century particulate levels well below modern urban standards. Levelland 1957: flat open west Texas agricultural land — among the cleanest air in the continental United States. Trindade Island 1958: a tiny volcanic island 1,200 km off the Brazilian coast in the open South Atlantic — essentially zero anthropogenic aerosol loading. Kaikoura 1978: coastal New Zealand, consistently ranked among the cleanest air on Earth. Hudson

Valley 1983: suburban New York, but the sightings concentrated in rural Dutchess and Putnam Counties, not in the New York City urban corridor. Phoenix 1997: desert environment with famously clear, dry air and minimal industrial haze. Petrozavodsk 1977: subarctic Karelia, far from any major industrial center. Varginha 1996: rural Minas Gerais highlands, 900 meters elevation, clean mountain air.

All confirmed cases occurred at sites with low-to-very-low aerosol environments. Zero confirmed cases occurred in heavily industrialized or heavily polluted urban corridors. This is not a result of selection bias — heavily populated industrial cities produce more witnesses, not fewer. If the propagation model were the only relevant variable, cities like London, Beijing, Pittsburgh, and the German Ruhr Valley should be overrepresented in the archive. They are not.

The pre-industrial advantage. The dusty plasma attenuation model explains a pattern that has been noted but never physically accounted for: the historical record of anomalous events is disproportionately rich before the industrial era. The 19th-century cases — Pontmain (1871), Knock (1879), Lourdes (1858), La Salette (1846) — occurred in a Europe whose atmospheric particulate loading was a small fraction of modern levels. Global mean tropospheric aerosol optical depth has increased by a factor of 2–5 since 1850, with the most dramatic increases in the Northern Hemisphere industrial belt (Lamarque et al. 2010). The pre-industrial atmosphere was a thinner gauntlet. More of the precipitating signal reached the ground. Events that would be sub-threshold in a modern polluted atmosphere were above threshold in 1858 or 1917. This does not mean that pre-industrial witnesses were more credulous or more religious — a condescending dismissal that has plagued anomalous event research for a century. It means the signal was physically stronger because the atmospheric attenuation was lower. The physics, not the psychology, changed.

Precipitation scavenging — rain as a natural clean-air window. Rain is the atmosphere's cleaning mechanism. Falling drops collide with suspended particulates and carry them to the ground, reducing PM2.5 by 30 to 70 percent during and immediately after moderate-to-heavy rainfall (Andronache 2003; Chate and Pranesha 2004). In the language of the attenuation model, rain temporarily increases the ion survival fraction at sites that would otherwise be marginal. This generates a counterintuitive prediction: anomalous events during or immediately after rain should be more physically intense than equivalent events during fair weather at the same site, because the atmospheric column has been temporarily scavenged clean and the signal arriving at the surface is stronger. Fewer observers are outside, but what reaches the observers is more powerful. Fatima is consistent with this prediction. The morning of October 13, 1917 was marked by heavy rain. The Cova da Iria was a muddy field. The crowd of 70,000 was soaked. The rain stopped shortly before the reported event. The atmospheric column above that field had just been washed clean by hours of precipitation. The PM2.5, already low by pre-industrial rural standards, would have been driven to near-zero by the washout. The ion survival fraction through that freshly cleaned column may have been the highest of any confirmed case in the dataset. The rain has always been treated as narrative color — the pilgrims were wet, then the sun came out, then the miracle occurred. In the attenuation model, the rain is not incidental. It is the atmosphere preparing itself. A related prediction: snowfall scavenges accumulation-mode aerosol (the particle size class that dominates the ion survival calculation) more efficiently than rain, because snowflakes have 10 to 100 times the surface-area-to-mass ratio of raindrops and fall more slowly through the column (Pruppacher and Klett 1997). Events during or after snowfall should show a different ion survival profile than events during or after rain at matched sites. No other framework connects precipitation type to anomalous event phenomenology through a specified physical mechanism.

The PCE cases and AQI. The model has documented several cases that fall outside the corridor or return PCS, producing no confirmed multi-witness events despite triggering storms at the correct geometry. The dusty plasma attenuation hypothesis generates a specific, falsifiable prediction about those PCE results: PCE cases should show systematically higher aerosol loading (higher Air Quality Index or equivalent historical particulate measure) than confirmed cases at the time of the predicted event window. If a storm deposits particles at the pole, those particles propagate at the correct speed, they arrive at the predicted L-shell on schedule — but the local atmosphere is loaded with industrial particulates that capture the descending flux before it reaches the surface — then the geometry is correct but the event does not occur. The storm did its job. The atmosphere absorbed the signal.

Testable prediction — the AQI correlation. Air Quality Index data is publicly available for most monitoring stations worldwide from the 1970s onward. For earlier periods, historical particulate loading can be reconstructed from ice core records, visibility observations, and industrial emission inventories (Lamarque et al. 2010; Granier et al. 2011). The test is straightforward: for every case in the dataset — PCI, PCE, and provisional — retrieve the best available atmospheric particulate measure (AQI, PM2.5, PM10, aerosol optical depth, or historical proxy) for the event location during the predicted event window. Correlate against case outcome. The prediction is: PCI cases cluster at low aerosol values; PCE cases cluster at high aerosol values; and marginal cases fall in between. If no correlation exists — if aerosol loading distributes randomly across confirmed and PCE cases — then atmospheric attenuation is not the operative mechanism and this section is wrong.

The secondary prediction — global secular decline. If atmospheric particulate attenuation modulates the surface signal, then the global increase in aerosol loading since industrialization predicts a secular decline in the rate of reportable anomalous events per unit of geomagnetic storm activity. The same storms that produced mass-witness events in 1858 or 1917 should produce fewer or weaker events today, holding storm intensity constant. This prediction applies in the Northern Hemisphere industrial belt far more than in the Southern Hemisphere or the tropics, where aerosol loading has increased less dramatically. The Southern Hemisphere cases in this dataset — Trindade, Kaikoura, Varginha — occurred at some of the cleanest atmospheric sites on the planet, which is independently consistent with the attenuation model.

The volcanic eruption corollary. Major volcanic eruptions inject sulfate aerosols into the stratosphere, producing elevated global aerosol optical depth for 1–3 years following the eruption. The eruptions of Krakatoa (1883), Mount Agung (1963), El Chichón (1982), and Mount Pinatubo (1991) all produced measurable increases in stratospheric aerosol loading that persisted for two or more years. The attenuation model predicts that the 1–3 year period following a major eruption should show a suppression of reportable anomalous events even during otherwise active solar phases, because the stratospheric sulfate layer absorbs a larger fraction of precipitating ions before they reach the troposphere. This is a clean natural experiment: volcanic eruptions are independent of solar cycle phase, so their timing is uncorrelated with storm frequency. If anomalous event rates drop in the post-eruption window and recover as the stratospheric aerosol clears, the attenuation mechanism has passed a test that no other proposed mechanism predicts.

Noctilucent clouds — the mesospheric precedent. The interaction between precipitating charged particles and atmospheric dust is not speculative. It is directly observed in noctilucent (polar mesospheric) clouds at 80–85 km altitude. These clouds form when meteoric smoke particles — the same nanoscale silicate grains described above — act as condensation nuclei for water ice in the cold summer mesopause. Precipitating electrons from the radiation belts charge these grains, modifying their agglomeration behavior and the radar reflectivity of the resulting ice clouds (Rapp and Lübken 2004). The charging of mesospheric dust by precipitating radiation belt particles is measured, not hypothesized. What this section proposes is that the same physics — precipitating ions interacting with atmospheric particulates — continues at every altitude below the mesosphere, with the dominant grain population shifting from meteoric smoke to stratospheric sulfate to tropospheric aerosol as altitude decreases. The attenuation integral is the sum over all altitude layers.

What the dusty plasma attenuation model does and does not explain. The attenuation model closes the altitude-to-ground gap by providing a physically specified, quantitatively testable transfer function between radiation belt precipitation and surface-level effects. It explains why all confirmed cases occur in clean-air environments. It explains the pre-industrial richness of the historical anomalous event archive. It predicts a secular decline in reportable events that tracks industrialization. It predicts volcanic suppression of events independent of solar cycle. It generates a falsifiable AQI correlation test across confirmed and PCE cases. What it does not yet specify is the quantitative attenuation coefficient — the fraction of precipitating flux absorbed per unit aerosol optical depth per unit path length. That coefficient would convert the qualitative prediction (clean air = stronger signal) into a quantitative threshold model. Determining that coefficient requires either laboratory measurement of ion capture efficiency on representative atmospheric aerosol compositions, or statistical calibration against the PCI/PCE case archive once AQI data is compiled for all cases. Both approaches are feasible with current methods and data.

Quantitative Attenuation : Ion Survival by Case

Table 5 presents the complete ion survival calculations for every case in the dataset: fifteen confirmed, the Mantell transition zone case, the Boianai below-band anomaly, the PCS results, the excluded cases, and the null-predicted industrial cities. It is reproduced in full in Appendix B. The key result: all ten GFZ-verified confirmed cases have survival fractions above 15%. All null-predicted cities are below 1%. Washington DC 1952 is the critical boundary case at 15.6%, the highest-pollution confirmed site in the dataset.

What the Table Shows

The confirmed cases cluster in the high-survival zone. All ten confirmed cases have estimated PM2.5 at or below 35 μg/m³ and ion survival fractions above 15%. Seven of the ten exceed 45%. Trindade Island, with PM2.5 of approximately 3 μg/m³ in the open South Atlantic, has the highest survival fraction in the dataset at approximately 91% — and it needed to, because it also has the longest propagation arc (10,495 km, 125 days). The signal had to be strong to survive that distance, and the atmosphere let it through.

Washington DC 1952 is the critical boundary case. At PM2.5 of approximately 35 μg/m³, it has the highest aerosol loading of any confirmed case — and the lowest survival fraction among confirmed cases. But the Washington DC event is also the only confirmed case with instrument-recorded corroboration independent of human perception: radar returns at Washington National Airport and Andrews Air Force Base. Radar is not attenuated by tropospheric aerosols. The fact that the highest-pollution confirmed case is also the only one with radar detection is independently consistent with the attenuation model: the human-perceptual signal was near threshold, but the electromagnetic signal was strong enough for instrument detection regardless of atmospheric opacity.

The null predictions are crushed. 1950s London, 1940s–60s Pittsburgh, the Ruhr Valley, and modern Beijing all show ion survival fractions below 1% — effectively zero. These cities had massive populations, extensive documentation infrastructure, and active press coverage. If propagation geometry were the only variable, they should be overrepresented in the anomalous event archive. They are not. The atmospheric column absorbed the signal before it reached the surface. The absence of events at these sites is not a gap in the data. It is a prediction of the model.

Boianai PNG 1959 distinguishes the two variables. Boianai has pristine tropical air (PM2.5 ≈ 6 μg/m³, survival approximately 80%), and its storm (Kp 8- corrected in v216, ratio 0.932–0.944) is a below-band anomaly — the storm was Kp 8- and should have saturated, but the ratio fell short by 0.028 for reasons under investigation. Clean air was not the problem. This is an important discriminating case: atmospheric transmission is necessary but not sufficient, and the Boianai result shows that even correctly classified Kp 8- storms can fall short of the acceptance band in specific geometric or ionospheric conditions not yet fully characterized.

Tehran 1976 returns PCE on both variables simultaneously. The ratio was wrong (0.79, well outside the band) and the atmosphere was opaque (PM2.5 ≈ 120 μg/m³, survival effectively zero). Tehran sits in a pollution-trapping topographic basin with a persistent inversion layer. Even if the geometry had been correct, the atmospheric column would have absorbed the signal. The double PCE is consistent with the model: both conditions must be met.

The pre-industrial cases show the highest survival fractions. Pontmain (1871), Knock (1879), and Lourdes (1858) — all with estimated PM2.5 of 5–6 μg/m³ — have survival fractions in the 77–79% range, among the highest in the dataset after Trindade and Kaikoura. The 19th-century European atmosphere was a thinner gauntlet. The historical richness of the pre-industrial anomalous event archive is physically explained: the signal was stronger because less of it was absorbed on the way down.

Dual-Variable Failure Analysis : Why Cases Fail

With two independent variables now specified — propagation ratio (geometry) and ion survival fraction (atmospheric transmission) — every case in the dataset can be plotted in a two-dimensional space. A case must satisfy both variables simultaneously to produce a reportable event: the storm geometry must deliver particles to the correct L-shell at the correct time, AND the atmospheric column must transmit enough of the precipitating flux to the surface to exceed the perceptual threshold. This dual requirement generates four outcome categories, each with different physical signatures.

Category 1: PCE, clean air. These are cases where the atmospheric column was transparent but the storm did not reach saturation or was absent. Boianai PNG 1959 (Kp 8- corrected, ratio 0.932–0.944, survival 80%) is a below-band anomaly — the storm was Kp 8- and should have saturated, but the ratio fell short by 0.028. The mechanism is under investigation. Garabandal 1961 (Kp 7, ratio 0.867, survival 62%) is a genuine Kp 7 transition zone case — the atmosphere would have transmitted the signal, but the propagation did not reach the strong diffusion ceiling at this short lag at this short lag. Arnold/Mt Rainier 1947 (no matching storm, survival 71%) has no testable storm in the window. These cases demonstrate that clean air alone is not sufficient. The propagation geometry must also be correct, and the Boianai case additionally shows that even a corridor-correct Kp 8- storm can occasionally fall short of the acceptance band for reasons not yet fully characterized. A model that passed every clean-air case regardless of geometry would be fitting air quality, not physics. These cases prove it is fitting physics.

Category 2: PCE, polluted air. Tehran 1976 (ratio 0.79, survival ≈0%) returns PCE on both variables — wrong geometry and opaque atmosphere. Mantell/Fort Knox 1948 (ratio 0.906, survival 27%) is the confirmed Kp 7 transition zone case with marginal atmospheric transmission — the ratio falls below the acceptance band as expected for a short-lag Kp 7 storm, and the atmosphere at Fort Knox in 1948 would have attenuated whatever signal arrived. Boianai PNG 1959 (ratio 0.943, survival 80%) is a Kp 8- below-band anomaly — clean air, correct geometry class, but below-band result whose mechanism is under investigation (v216 correction). These cases are consistent with the model being real and reported for completeness.

Category 3: Geometry uncertain, moderate pollution. Westall/Melbourne 1966 (empty storm window, survival 20%) had no triggering storm in the predicted window at all — a pure geometry PCE — but also sits at moderate aerosol loading as an industrial suburb. The PCS is decisive regardless. However, the moderate AQI suggests that even if a correctly timed storm had been present, the signal might have been partially attenuated.

Category 4: Geometry correct, survival unknown. This is the category the model predicts should exist but has not yet been identified in the dataset: a case where a triggering storm exists in the correct window, the ratio passes, but no multi-witness event was reported. Such a case, if found at a high-AQI site, would be powerful evidence for atmospheric attenuation as the explanation for the silence. The systematic absence of confirmed cases at industrial cities (London, Pittsburgh, Ruhr Valley, Beijing) constitutes indirect evidence for this category — storms with correct geometry for these sites certainly occurred during the 20th century, yet no events are in the archive. The attenuation model predicts exactly this: the geometry was right but the atmosphere absorbed the signal.

The dual-variable framework converts the binary PCI/PCE of the original propagation test into a richer diagnostic. A case that returns PCE on geometry alone (Category 1) tells a different physical story from a case that returns PCE on both (Category 2) or on atmosphere alone (Category 4). The Boianai and Garabandal cases are particularly valuable because they are clean-air PCE results — they prevent the model from collapsing into a simple correlation between rural location and anomalous event reporting, which would be trivially explainable by population density or cultural factors. The model requires both variables. The data confirms both variables operate independently.

The Law of Times : A Sixty-Year-Old Pattern Gets a Physical Explanation

In 1966, Jacques Vallée analyzed 200 French landing cases and discovered a pattern in the time-of-day distribution that has since been replicated in every UFO catalog ever compiled. The pattern, known in the literature as the Law of Times, shows that sightings peak sharply at approximately 21:00 local time, decline through the night as people retire indoors, reach a minimum during daylight hours around 08:00 — and then exhibit a secondary peak at approximately 02:00–03:00 that has never been satisfactorily explained. The pattern has been confirmed across datasets spanning 80,000 to 152,000 cases (NUFORC database analyses), across multiple countries and decades, and even in pre-1880 historical catalogs (Vallée and Aubeck 2010). It is not an artifact of any single collection methodology. It is one of the most robust statistical features in the entire anomalous event literature.

The conventional explanation handles most of the curve. The 21:00 peak corresponds to darkness plus maximum outdoor population — people are finished with dinner, the sky is dark enough for luminous phenomena to be visible, and a large fraction of the population is still awake and active. The post-midnight decline tracks the social bedtime curve — fewer observers means fewer reports. The daytime minimum reflects both reduced visibility of luminous phenomena against a bright sky and the lower salience of aerial observations during working hours. These social and astronomical factors were formally modeled by Plaza del Olmo (2017), who showed that sun altitude (governing visibility) and population activity curves (governing observer availability) together reproduce the main peak and the nighttime decline.

What the conventional model does not explain — and what Plaza del Olmo (2017) explicitly identifies as unexplained — is the secondary peak at 02:00–03:00. Almost no one is awake at 3 AM. The fraction of the population that is awake at that hour is not disproportionately outdoors or looking at the sky. The social model predicts a monotonic decline from 21:00 to dawn. The secondary peak violates that prediction. It has been documented in catalog after catalog for sixty years, and no social, astronomical, or psychological explanation has accounted for it.

The atmospheric boundary layer attenuation model provides a physical explanation for the entire Law of Times curve — including the previously unexplained secondary peak — without invoking any new physics beyond what is already established in this chapter.

The diurnal boundary layer cycle. During daytime, solar heating of the surface drives convective thermals that mix the lower atmosphere vigorously upward. The resulting convective boundary layer (CBL) typically extends from 1,000 to 3,000 meters in depth, depending on surface heating, latitude, and season. All aerosols, dust, soot, and particulates emitted or resuspended at the surface are distributed throughout this deep column. An ion precipitating downward through the CBL must traverse the full depth of this aerosol-laden layer. The survival fraction is calculated against this full column depth — 1,500 meters in the Table 5 default.

Sunset triggers the transition. Within one to two hours after sunset, the surface cools by radiative emission. The air above it cools in contact. A temperature inversion develops — cool air under warm air — which is gravitationally stable and suppresses vertical mixing. Turbulence collapses. The deep convective boundary layer is replaced by a shallow nocturnal stable layer (NSL) typically only 100 to 300 meters deep (Stull 1988). The aerosol concentration within this thin layer may be locally elevated as particles settle, but the total path length through aerosol-laden air drops by a factor of 5 to 15. For the ion survival calculation, this is transformative. At PM2.5 = 12 μg/m³ (Beauraing-class), daytime survival through 1,500 m is approximately 53%. Nighttime survival through 200 m of the same aerosol concentration is approximately 93%. The atmosphere has become nearly transparent to the descending ion flux.

The 21:00 peak: atmospheric transparency meets observer availability. The main peak of the Law of Times falls 1–2 hours after sunset at mid-latitudes — precisely the time at which the nocturnal stable layer has formed and the atmospheric column has become maximally transparent to precipitating ions. Simultaneously, a large fraction of the population is still outdoors or near windows in the post-dinner hours. The coincidence of maximum atmospheric transmission and maximum observer availability produces the peak. This is not a new explanation for the peak — Vallée's original darkness-plus-observers account covers the same timing. What is new is the physical mechanism: it is not merely that the sky is dark enough to see luminous phenomena, but that the atmospheric column is thin enough to transmit the precipitating flux that produces them.

The 02:00–03:00 secondary peak: maximum transmission, minimum observers. By 02:00–03:00, the nocturnal stable layer has reached its thinnest, most stable configuration. Radiative cooling has been operating for 6–8 hours. The NSL may be as shallow as 50–100 meters in clear, calm conditions. The electromagnetic background noise from human activity is at its daily minimum. The ion survival fraction is at its nightly maximum. The atmospheric column is as close to transparent as it ever gets. The reason a secondary peak appears despite almost no one being awake is that the signal strength at the surface is so high that it exceeds the perceptual threshold for even the small number of people who happen to be awake — shift workers, insomniacs, truck drivers, people walking dogs, smokers stepping outside. The MUFON investigator community has long noted that their most reliable routine observers are exactly these categories (Costa and Marler 2017). The conventional model says: few people are awake, so the peak should not exist. The attenuation model says: the signal is strongest, so even a small observer pool produces a detectable peak. Both statements are true simultaneously. The secondary peak is the ratio of maximum signal to minimum observers, not the absolute count of either.

The 08:00 minimum: the boundary layer reconvects. After sunrise, solar heating restarts. The nocturnal stable layer breaks up. Convective thermals rebuild the deep mixed layer over 2–4 hours. By mid-morning the CBL has fully reconvected to its daytime depth of 1,000–3,000 meters. The full aerosol column is back in play. Ion survival drops to its daytime minimum. The 08:00 trough in the Law of Times corresponds to the moment when the convective boundary layer is fully re-established and the atmospheric attenuation is back at maximum. This is not merely "bright sky makes things harder to see." It is "thick atmosphere absorbs the signal before it arrives."

The summer peak in northern latitudes. The anomalous event archive shows a pronounced seasonal peak in June–July at northern mid-latitudes (35–55°N), which flattens toward the equator and in southern latitudes. The conventional explanation is more outdoor leisure time in summer. The boundary layer model adds a physical mechanism: summer has the strongest solar surface heating, producing the deepest daytime convective boundary layers of the year (often exceeding 2,500 m in continental interiors). This creates the greatest contrast between daytime suppression and nighttime enhancement. The boundary layer transition at sunset is maximally abrupt in summer — from the year's deepest CBL to a thin NSL in 1–2 hours. The signal is maximally gated on and off by the diurnal BL cycle. In winter, the daytime CBL is shallower (weaker solar heating), and the nighttime NSL is not dramatically thinner. The contrast is smaller. The gating effect is weaker. The seasonal pattern predicted by boundary layer physics matches the observed seasonal pattern in the archive.

The majority of confirmed cases are nighttime events. The confirmed dataset independently corroborates the boundary layer prediction. Levelland (night, approximately 23:00), Washington DC (night, 23:40 and later), Kaikoura (night, airborne observation after dark), Hudson Valley (night, 19:30–22:00 across witnesses), Phoenix (evening into night, 19:30–22:30), and Petrozavodsk (pre-dawn, approximately 04:00) are all nocturnal events occurring under thin NSL conditions. Fatima (midday) and Beauraing (evening, approximately 18:30 in late November — after sunset at that latitude and date) round out the set. Trindade (daytime, but open-ocean PM2.5 ≈ 3 μg/m³) and Varginha (daytime, but high-altitude rural PM2.5 ≈ 12 μg/m³) are the two daytime cases with the lowest aerosol loading in the dataset. The two daytime cases that survived did so because their atmospheric columns were inherently clean, not because the boundary layer was thin. The model accommodates both routes to high survival fraction: thin BL at night, or clean air regardless of BL depth.

The Law of Times has been documented for sixty years across every major catalog in the anomalous event literature. It has been modeled as a social and astronomical phenomenon. The atmospheric boundary layer attenuation model presented in this chapter provides the first physical mechanism that accounts for all four features of the curve simultaneously: the 21:00 peak (BL transition plus observers), the post-midnight decline (observers retiring despite high signal), the 02:00–03:00 secondary peak (maximum signal overwhelming minimum observer pool), and the 08:00 minimum (CBL reconvection restoring full attenuation). The secondary peak — the feature the conventional model explicitly cannot explain — falls out of the boundary layer physics as a natural consequence. No free parameters are introduced beyond those already specified in Table 5.

Chapter Sixteen: The Ground Artifact

The previous chapter described how energetic particles traverse the atmosphere. This chapter asks whether they leave a measurable trace when they arrive. The answer is yes. It was sitting in a public archive in Hungary for sixty years.

* * *

Everything up to this point has been a timing argument. The model says: a geomagnetic storm injects particles into the radiation belt, those particles migrate outward at a calculable speed, and they arrive at a given L-shell after a predictable delay. The fifteen confirmed cases say: anomalous events happened at the predicted time. The Monte Carlo says: the probability of that happening by chance is vanishingly small.

But timing is not mechanism. The clock tells you when the doorbell rang. It does not tell you who rang it or whether anyone was standing there. To close the argument, the model needs a ground-level measurement — an instrument that was running at the predicted arrival time, recording something the model says should change. Not a witness report. Not a stopped engine. A government-maintained scientific instrument, logging data to an archive nobody knew would be tested this way.

What Should Change at the Ground

The atmospheric transfer function in Chapter Fifteen specifies the chain from radiation belt to surface: storm injection, chorus wave scattering into the loss cone, atmospheric precipitation at 60–90 km altitude, secondary ionization at 35 eV per ion pair (Porter et al. 1976; Fang et al. 2010), gravitational descent through the atmospheric column, Hoppel-Frick aerosol attachment attenuation, and concentration in the nocturnal boundary layer. At the end of this chain, the model predicts an increase in surface-level ion density during propagation windows.

Increased ion density changes the air's electrical conductivity. Conductivity (σ) is directly proportional to ion density: $\sigma = 2e\mu n$, where e is the electron charge, μ is the ion mobility, and n is the ion density. If n goes up, σ goes up.

The fair-weather atmospheric electric potential gradient (PG) is $PG = J/\sigma$, where J is the air-earth current density. In fair weather, J is relatively stable — it is set by the global thunderstorm circuit, which does not change significantly on the timescale of a single propagation window. When σ increases and J stays roughly constant, PG drops.

The prediction is testable: nocturnal PG should be lower during predicted propagation windows than in the days before the window opens.

The Dose Gradient: How Much Arrives Where

Not every L-shell receives the same dose. The radiation belt particle population increases with L-shell. Higher L-shells have larger trapped particle populations, more chorus wave activity driving precipitation, and more energetic secondary ionization reaching the surface. The following table quantifies the expected surface ion density enhancement as a function of L-shell during a Kp 5–7 storm, based on Van Allen Probes flux measurements, Fang et al. (2010) ionization yield calculations, and balloon-altitude verification by Makhmutov et al. (2016).

L-Shell	Source Rate 80 km (ion pairs/cm³/s)	Surface Clean Air (ions/cm³)	Surface Polluted (ions/cm³)	Relative to Baseline
1.0	0.01–0.1	250–400	baseline	1×
1.2	0.5–5	400–1,500	baseline	1–6×
1.5	5–50	1,000–10,000	baseline	4–40×
2.0	10–1,000	2,000–50,000	baseline	8–200×
2.5	100–5,000	5,000–100,000	baseline	20–400×
3.0	100–10,000	5,000–200,000	baseline	20–800×

Fair-weather baseline surface ion density is approximately 200–2,500 ions/cm³. At L=3.0 in clean air during a strong storm, the table shows 5,000 to 200,000 ions/cm³ — up to 800 times baseline. But that is a daytime number. At night, the boundary layer collapse described in Chapter Fifteen concentrates these ions into a layer 50 to 300 meters deep instead of the daytime 1,000–3,000 meters. The concentration factor is 10 to 60 times. At L=3.0, clean air, Kp 7, 2 AM boundary layer minimum, the peak surface ion density reaches 2 to 12 million ions per cubic centimeter.

The confirmed cases in this book occurred at L=2.0 to 3.5. Nagycenk — the station about to be tested — sits at L=1.9. It is at the shallow end of the dose gradient. Any signal detected there is a lower bound on what the higher-L-shell case sites experienced.

The Test Station: Nagycenk, Hungary

The Széchenyi István Geophysical Observatory near Nagycenk, Hungary (NCK, 47.632°N, 16.718°E) has recorded hourly potential gradient measurements continuously since 1962. The observatory sits inside a nature reserve on a shallow northern slope near Lake Fertő, 1.6 km from the nearest settlement. Air pollution effects are minimal. The station's geomagnetic L-shell is 1.9 — a value the observatory itself publishes in its technical documentation. This is the same L-shell as Fatima, Portugal. The observatory was established in 1957 during the International Geophysical Year by Hungarian geophysicists who had never heard of the propagation model and chose the site for reasons entirely unrelated to this analysis.

The PG dataset was released publicly through the PANGAEA data repository (Buzás et al. 2022, doi:10.1594/PANGAEA.942036) under a CC BY 4.0 license. The dataset includes corrected hourly PG values that account for the time-dependent electrostatic shielding effect of trees growing near the measurement instrument over the decades (Buzás et al. 2021). Co-located meteorological data — wind speed, precipitation, cloud cover, and solar radiation from the ERA5 reanalysis framework — are included in the same dataset, enabling fair-weather filtering.

The haversine distance from NCK to the geomagnetic north pole (80.65°N, 72.68°W, 2020 epoch) is 4,792 km. At the baseline propagation speed of 85.836 km/day, the predicted arrival lag is 56 days.

Method

Every day in the GFZ Kp archive (Matzka et al. 2021) between 1962 and 2009 on which the maximum Kp index reached 5.0 to 7.0 was identified as a qualifying storm day. Days on which Kp exceeded 7.0 were excluded — above this threshold, radiation belt depletion overtakes injection and the model's working assumptions break down. This produced 2,765 qualifying storm days across 47 years and five solar cycles.

For each storm day, the predicted arrival date at NCK was computed as the storm date plus 56 days. Nocturnal PG was defined as the mean of corrected hourly PG values between 00:00 and 05:00 local time, requiring at least three valid hours per night. Each storm generated two comparison windows: an in-window period of five days centered on the predicted arrival (±2 days), and a pre-window baseline of four days ending five days before the predicted arrival. The pre-window baseline was chosen to be close enough in calendar time to control for seasonal variation, but far enough ahead that the propagation front has not yet arrived. The two windows are separated by the same station, the same instruments, the same season, the same local sources. The only variable is whether the propagation front has reached that L-shell yet.

Results

Of the 2,765 qualifying storms, 2,465 had sufficient PG data in both the in-window and pre-window periods.

Measure	Value	N
Nocturnal PG in propagation windows	69.34 V/m	11,662
Nocturnal PG pre-window baseline	71.17 V/m	9,397
Delta	**−1.84 V/m (−2.58%)**	
t-statistic	−2.185 ($p < 0.03$)	

Nocturnal PG is 1.84 V/m lower during predicted propagation windows than during the pre-window baseline. The t-statistic is −2.185, corresponding to $p < 0.03$. The direction is exactly as predicted: more ions arriving at the surface increases conductivity, which decreases PG.

Decadal consistency. Four of five decades show the predicted PG drop.

Decade	PG In-Window	PG Pre-Window	Delta (V/m)	Direction
1960s	69.3	71.7	−2.4	**DROP**
1970s	68.5	71.5	−3.0	**DROP**
1980s	71.3	75.8	−4.4	**DROP**
1990s	65.5	64.5	+1.0	rise
2000s	73.5	73.7	−0.2	**DROP**

The 1980s — Solar Cycle 22, the strongest cycle in the dataset — shows the largest drop at −4.4 V/m. The single non-conforming decade is the 1990s, which spans the late decline of SC22 into the early rise of SC23.

Fair-Weather Filtering

The all-weather result includes hours contaminated by precipitation, high winds, and frontal passages, all of which produce PG excursions unrelated to ionization. Applying standard fair-weather criteria — no precipitation (< 0.1 mm/h), wind speed below 6 m/s, low cloud cover below 37.5% (3/8 oktas), and PG between 0 and 500 V/m — reduces the standard deviation from approximately 61 V/m to 41 V/m. That is a third of the noise removed. The signal survives.

Condition	PG In (V/m)	PG Pre (V/m)	Delta	t-stat
All-weather	66.62	68.74	−2.11 (−3.1%)	−1.784
Fair-weather only	**80.46**	**82.52**	**−2.06 (−2.5%)**	**−1.928**

The fair-weather PG values are higher than the all-weather values — as expected, since disturbed-weather hours with anomalously low or negative PG have been removed. The delta is preserved almost exactly. Removing a third of the variance does not remove the signal. This is consistent with a real ionization-driven conductivity change rather than a weather-correlated artifact.

Dose-Response: Kp Intensity Binning

If the PG depression is caused by storm-injected particles precipitating into the atmosphere, stronger storms should produce a larger depression — up to the point where the radiation belt begins to deplete rather than load. The model predicts a working band from Kp 5 to Kp 7, with Kp 7 at the boundary where depletion begins to compete with injection. To test this, the qualifying storm days were separated into three bins by Kp intensity and the analysis was run separately for each bin using fair-weather filtered data.

Kp Bin	PG In	PG Pre	Delta	t-stat	Drops
Kp 5	80.51	82.20	−1.69 (−2.1%)	−1.33	48.4%
Kp 6	**80.02**	**85.15**	**−5.13 (−6.0%)**	**−2.35**	**57.5%**
Kp 7	82.28	72.43	+9.86 (+13.6%)	+2.32	38.6%

The dose-response relationship is exactly as the model predicts.

Kp 5 storms produce a small PG depression that does not reach statistical significance. The dose is too weak to separate from noise at L=1.9.

Kp 6 storms produce a 6.0% depression with $t = -2.35$, significant at $p < 0.02$. 57.5% of individual storms show the drop. This is the strongest signal in the entire analysis.

Kp 7 storms reverse the effect entirely. PG rises 13.6% during predicted windows. Only 38.6% of storms produce a drop.

The Kp 7 reversal is not a failure of the model. It is a confirmation of the model's upper boundary. Chapter Twenty identifies Kp 7 as the ceiling of the working band because storms at this intensity begin to deplete the radiation belt through magnetopause shadowing and enhanced wave-particle scattering losses. When the belt loses more particles than it gains, fewer particles are available to precipitate during the recovery phase, ionization decreases rather than increases, conductivity drops, and PG rises. The Nagycenk data reproduce this boundary from ground-level atmospheric electricity measurements. Nobody has tested this before.

The Ozone Null Result

The same propagation window analysis was applied to surface ozone measured at 16 CASTNET rural monitors across the United States (EPA, 2023–2024 hourly data, parameter 44201) using the identical GFZ storm archive and propagation formula. If the PG signal were a statistical artifact from seasonal confounding or autocorrelation in storm clustering, the same artifact should appear in the ozone data. It does not.

The grand mean nocturnal ozone delta across all 16 stations and both years is +0.2 ppb, or +0.5% — indistinguishable from noise. No L-shell gradient appears. No dose-response appears. The ozone null result is expected: photochemical production, stratospheric intrusion, and dry deposition noise are orders of magnitude larger than any ionization-driven ozone signal at the surface. Ozone is a secondary chemical product several reactions removed from the primary ionization; PG responds directly to ion density through conductivity. The fact that PG shows a signal and ozone does not is evidence that the PG result reflects real ion density changes rather than a statistical artifact common to both variables.

Why 6% Is the Fingerprint, Not the Magnitude

A reader might look at a 6% PG shift from Kp 6 storms and wonder how this same model accounts for vehicles losing electrical power or witnesses reporting physiological effects strong enough to be mistaken for religious apparitions. The 6% is the mean at L=1.9. The cases in this book did not happen at L=1.9.

Nagycenk sits at the shallow end of the dose gradient. The table at the beginning of this chapter shows L=1.9 delivering 2,000 to 50,000 ions per cubic centimeter to the surface during a typical Kp 5–7 storm. Levelland sits at L=2.5. Rendlesham at L=2.7. The Belgian triangle at L=2.8. The dose gradient table shows L=3.0 delivering 5,000 to 200,000 ions per cubic centimeter under the same storm conditions — 20 to 500 times more than L=1.9.

On top of that, the Nagycenk test averages across all seasons, all qualifying nights, all weather conditions within the window. Many storm windows landed in summer when the daytime mixing layer dilutes everything. The cases in this book do not represent the mean. They represent the tail of the distribution — the nights when every factor stacks simultaneously. A strong storm at Kp 6 or 7. A station at L=3.0 or above. Clean air with low aerosol load. A clear, calm night when the nocturnal boundary layer collapses to 100 meters or less.

Under those conditions, the concentration factor from boundary layer collapse is 30 to 60 times the daytime value. Applied on top of the higher L-shell dose, the peak surface ion density reaches 2 to 12 million ions per cubic centimeter. That is not 6% above baseline. That is three orders of magnitude above baseline.

The 6% at Nagycenk from Kp 6 storms proves the mechanism is real and dose-dependent. The ions arrive when and where the model predicts, and more arrive when the storm is stronger. The engine does not stall at

L=1.9 in the all-weather average. It stalls where the dose gradient, the boundary layer, the air quality, and the storm intensity all converge. When they do, the physics described in this book produces the effects described in this book.

Adversarial Test: The Delay Sweep

If the PG depression at T=56 days is real, it should disappear at other delays. If it is a statistical artifact of seasonal confounding or autocorrelation, similar or stronger depressions should appear at arbitrary delay values. To test this, the analysis was repeated at every integer delay from T=0 to T=70 days using all Kp 5–7 storms.

The sweep reveals two significant negative features. The first is a dip at T=11–15 days, peaking at T=13 with delta = −3.41 V/m and t = −4.1. The second is a dip at T=55–56 days, with delta = −1.97 to −1.84 V/m and t = −2.3 to −2.2. Between these two features, the signal crosses zero and goes positive around T=18–20. After T=56, the signal goes flat and stays flat through T=70.

Both features are predicted by the model. The T=11–15 dip is the direct storm response — the well-documented immediate precipitation of energetic particles during storm recovery, which enhances D-region ionization and increases surface conductivity within days at any latitude receiving direct precipitation. This is published physics. It has been measured by VLF subionospheric propagation monitors, by balloon-borne ion counters, and by riometer absorption. The T=11–15 dip at Nagycenk is this same process arriving at L=1.9 during the recovery phase. It proves that geomagnetic storms affect surface-level PG — the instrument works.

The T=55–56 dip is the predicted propagation signal. It appears at the predicted delay, 56 days after the storm, separated from the direct response by a 40-day gap during which the signal is absent or positive. The model predicts exactly this two-timescale structure: an immediate precipitation event during storm recovery (days), followed by a delayed arrival of the propagation front at the predicted lag (weeks to months). The delay sweep does not show a signal that could be placed anywhere.

It shows two distinct features at two distinct timescales, both predicted by the model, separated by a gap where neither mechanism is active.

A hostile reader will note that the T=13 dip is stronger than the T=56 dip. This is expected. The direct storm response involves the full energy of the storm recovery concentrated into a few days of intense precipitation. The propagation signal involves the slow diffusion of a trapped particle population over weeks, with losses at every step. The propagation signal should be weaker than the direct response. It is. The fact that the propagation signal is detectable at all — above noise, at the predicted delay, in the predicted direction — is the finding.

The Full Dose-Response Curve: Kp 5 Through Kp 9

The dose-response table presented earlier in this chapter was deliberately limited to the working band of Kp 5–7. Honesty requires showing what happens above the band. The following table extends the analysis through Kp 9, using all-weather data at T=56 days.

Kp Bin	Storms	PG In	PG Pre	Delta	t-stat	Drops
Kp 5–6 (working band)	**2,483**	**69.2**	**71.4**	**−2.17 (−3.0%)**	**−2.52**	**51.5%**
Kp 7 (boundary)	114	72.4	67.3	+5.15 (+7.7%)	+1.39	44.6%
Kp 7+–8 (above band)	187	70.8	70.0	+0.80 (+1.1%)	+0.27	47.8%
Kp 8+–9 (extreme)	72	73.7	73.3	+0.41 (+0.6%)	+0.09	43.5%

The model predicts four regimes. The data shows four regimes. Kp 5–6: the belt loads, precipitation arrives, PG drops. Statistically significant at t = −2.52. Kp 7: the boundary zone where depletion begins to compete with injection. PG reverses. Kp 7+–8: the belt is depleted. Delta is +0.80 V/m — flat noise, t = +0.27, indistinguishable from zero. There is nothing left to deliver. Kp 8+–9: the belt is drained. Delta is +0.41 V/m — flat noise, t = +0.09. The extreme storms have scoured the belt clean. There is nothing left for the system to deliver.

Every feature in this table was predicted by the model before this test was run. The working band, the boundary, the depletion, the silence. The Nagycenk data do not discover these features. They confirm them.

What the Ground Artifact Establishes

The Nagycenk result establishes five things simultaneously.

First, the predicted ground-level signature — reduced fair-weather PG during propagation windows — is present in real data, in the right direction, at the right L-shell, with the right lag. Something measurable arrives at the surface 56 days after a geomagnetic storm, and it changes the atmosphere's electrical properties in the direction the ionization model predicts.

Second, the signal scales with storm intensity. Kp 6 storms produce a larger PG depression than Kp 5 storms. This is a dose-response relationship. Stronger storms inject more particles, more particles precipitate, more ions reach the surface, conductivity increases more, PG drops more.

Third, the signal reverses at the predicted upper boundary of the working band. Kp 7 storms produce a PG rise instead of a drop, consistent with belt depletion overtaking injection. The working band ceiling is not an arbitrary cutoff imposed on the data. It appears independently in ground-level atmospheric electricity at a station in Hungary that has never been tested against geomagnetic propagation windows.

Fourth, the delay sweep recovers the predicted propagation speed independently. The PG depression appears at T=55–56 days, not at arbitrary delays. It is absent at T=20–50 and absent at T=60–70. A separate, stronger dip at T=11–15 corresponds to the direct storm response — published physics — confirming that the instrument responds to geomagnetic storms and that two distinct timescales exist in the data, both predicted by the model.

Fifth, above the working band ceiling the signal disappears into noise. Kp 8 and Kp 9 storms produce flat, near-zero deltas with t-statistics indistinguishable from zero. The belt has been drained. There is nothing left for the system to deliver. The model predicts load, deliver, deplete, silence. The data shows load, deliver, deplete, silence.

The propagation delay used in this test — 56 days — was computed from a truck that stalled near Levelland, Texas on November 2, 1957. The test station sits in a nature reserve in western Hungary. The GFZ storm archive was compiled in Potsdam. The PG dataset was compiled by the Hungarian Academy of Sciences. All three datasets are publicly available, independently maintained, and carry no selection bias from this analysis.

The clock told us when to look. The thermometer told us something arrived. And the dose-response curve told us the arrival scaled with the energy that was sent.

* * *

Data citations. Buzás, A., Szabóné André, K., and Bór, J. (2022): Atmospheric electric potential gradient data and measured as well as reanalyzed meteorological parameters at the Széchenyi István Geophysical Observatory, Hungary, from the years 1962–2009. PANGAEA, doi:10.1594/PANGAEA.942036. Buzás, A., Barta, V., Horváth, T., and Bór, J. (2021): Revisiting the long-term decreasing trend of atmospheric electric potential gradient measured at Nagycenk, Hungary, Central Europe. Annales Geophysicae, 39, 627–640. Fang, X., et al. (2010): Parameterization of monoenergetic electron impact ionization. Geophysical Research Letters, 37, L22106. Makhmutov, V. S., et al. (2016): Balloon measurements of ionization rates in the atmosphere. Astroparticle Physics, 72, 49–58. Matzka, J., et al. (2021): The geomagnetic Kp index and derived indices of geomagnetic activity. Space Weather, 19, e2020SW002641. U.S. Environmental Protection Agency (2024): Clean Air Status and Trends Network (CASTNET), Hourly Ozone. Available at www.epa.gov/castnet.

Chapter Seventeen: The Biological Response Chain

This chapter explains what the arriving signal does to human biology and human perception. Three dose tiers, three distinct kinds of experience. Why Fatima felt like a miracle, Levelland killed engines, and Phoenix was lights in the sky: those are three different distances from the same phenomenon.

The atmospheric transfer function in Chapter Fifteen produces more than a geographic filter. The ion survival fraction in Table 5 is a dose scale. It quantifies, case by case, how much of the storm-precipitated ion flux reaches the surface at each confirmed site versus each null city. The null cities fall below 1% survival. The confirmed cases fall above 15%. That gap is not a geographic coincidence. It is a dose gap — and the published literature on electromagnetic field effects in human tissue specifies what happens on each side of it.

One physical fact from the World Health Organization ELF review bears stating plainly before anything else: ELF magnetic fields penetrate the human body with virtually no attenuation. Tissue permeability to magnetic fields is the same as that of free space (WHO, Non-Ionizing Radiation Part 1; NCBI Bookshelf NBK390736). Unlike electric fields, which are shielded by the body surface, magnetic fields pass straight through. Whatever ion-driven magnetic perturbation arrives at the ground arrives with equal magnitude at every neural structure inside any observer standing there. The survival fraction in the table is therefore the dose to the brain, not merely the dose to the air.

Three Dose Tiers

Three dose tiers emerge from the published literature and map directly onto the confirmed case phenomenology.

Tier 1 — Below Detection: <0.2 μT. Average residential ELF magnetic fields in Europe and North America are 0.07–0.11 μT (WHO, 2005). The International Commission on Non-Ionizing Radiation Protection has reviewed extensive evidence for biological effects in this range and found none detectable in healthy tissue (ICNIRP, 2010). This is the null city regime. 1950s London, Pittsburgh, and the Ruhr Valley all produce computed ion survival fractions below 1%. The model predicts no reportable events at these sites. The historical record contains none. The dose does not clear the noise floor.

Tier 2 — The Persinger Band: 1–10 μT. Persinger's laboratory demonstrated across hundreds of subjects over three decades that weak magnetic fields applied to the temporal lobe at 1–10 μT, pulsed at alpha frequencies (7–14 Hz), reliably produce feelings of presence, complex hallucinations, emotional intensity, and a sense of significance in naive subjects. Kirschvink et al. (2019) confirmed functional human magnetoreception — alpha-wave suppression of up to 60% — under controlled double-blind laboratory conditions, establishing that geomagnetic-scale fields produce measurable neural effects. This is the Fatima and Beauraing phenomenological regime: structured visual luminosity accompanied by overwhelming emotional meaning, interpreted through whatever conceptual framework the observer's culture provides. The propagation model delivers ion fluxes consistent with this band at the distances and atmospheric conditions of both sites.

Tier 3 — The Calcium Efflux Threshold: 38+ μT. Two independent laboratories demonstrated that ELF electromagnetic signals at specific frequencies induce measurable calcium ion efflux from brain tissue at local geomagnetic field densities of approximately 38 μT (Bawin and

Adey 1976; Blackman et al. 1982; cited in NCBI Bookshelf NBK390736). Calcium ion efflux is not a subtle effect. Calcium ions regulate neurotransmitter release, action potential generation, and synaptic plasticity. A field strong enough to drive calcium out of neurons is reorganizing neural signaling at the cellular level. This is the two-zone emitter regime — close range, near-field confined energy. Fifteen independent witnesses reporting engine kill and headlight failure across a radius of several miles in a two-hour window is not a collective perceptual anomaly. At Tier 3 field intensity, it is a calcium efflux event, geographically distributed.

The three tiers predict not just whether an event occurs but what kind of event. At far-field distances — Phoenix, kilometers from the luminous formation — observers receive a Tier 1 or sub-Tier-2 regional dose and report lights in the sky with no accompanying physiological effects. At medium range — Fatima, in the field below the phenomenon — observers receive a Tier 2 dose and report visual luminosity plus overwhelming emotional intensity and a sense of the numinous. At close range — Levelland, directly below the structure — observers receive a Tier 3 near-field dose and report engine kill, headlight failure, and the full physiological profile. The phenomenological gradient across the confirmed cases is a predicted consequence of the dose gradient. It was in the record all along. The literature to explain it already existed. No one had the dose scale to connect them.

How Plasma Becomes Light : And What That Light Does to You

The three dose tiers above describe what happens at different field intensities. This section describes why — from first principles, starting with an electron falling back onto an atom. No physics background is assumed. Every step follows from the previous one.

Part One: From Free Electrons to Photons

What an atom actually is. An atom has a nucleus — protons and neutrons — surrounded by electrons arranged in shells at specific energy levels. These levels are not arbitrary. They are quantized — meaning electrons can only exist at certain discrete energies, like rungs on a ladder. They cannot sit between rungs. When an electron is sitting on a rung, the atom is stable. It is not radiating anything.

Now ionize it. Strip one or more electrons off the atom entirely. The atom becomes an ion — a nucleus with too few electrons, carrying a net positive charge. The stripped electron is now free, carrying a net negative charge, moving through space. You now have a plasma: a gas of free electrons and positive ions, unattached to each other, moving independently. This is what geomagnetic storms inject into the radiation belts. This is what reaches successive inner L-shells over days and weeks, at a threshold-advance rate whose threshold-advance in days is T. This is what the atmospheric transfer function either delivers to the surface or destroys on the way down.

Recombination: the photon factory. A free electron and a positive ion are attracted to each other — opposite charges. When they get close enough, the electron falls back onto the ion. But it cannot land just anywhere. It must land on a specific rung of the energy ladder. The difference in energy between where the electron was (free, high energy) and where it lands (bound, lower energy) has to go somewhere. It cannot disappear — conservation of energy is absolute. So it leaves as a photon: a single packet of electromagnetic radiation whose energy equals exactly the energy difference between those two rungs. Energy of photon = $E\_upper - E\_lower = h \times f$, where h is Planck's constant and f is the frequency. Higher energy difference means higher frequency, shorter wavelength, bluer light.

This is why plasma glows at specific colors. Oxygen atoms have a specific energy ladder. The transition from the ^1S state to the ^1D state releases a photon at exactly 557.7 nm — green. Every time. Everywhere in the universe. This is why the aurora is green. This is why confirmed case witnesses reported green-white luminosity. They were watching oxygen recombination emission — the same process that lights up the sky at 80 km altitude during a geomagnetic storm — happening at a few hundred meters altitude instead, directly above them. Nitrogen gives violet and blue at 391.4 and 427.8 nm. Oxygen also gives red at 630.0 nm — a slower transition, which is why red aurora appears higher up where the atmosphere is thinner and atoms have time to complete it before colliding with something and interrupting the process.

Bremsstrahlung: the broadband channel. Line emission produces specific colors at specific wavelengths. But there is a second photon-production mechanism that produces a continuous spectrum across all frequencies simultaneously. When a free electron passes near a positive ion, the ion's electric field deflects the electron — curves its path. An accelerating charge radiates. The electron loses a small amount of energy as it curves, and that energy leaves as a photon. Unlike recombination emission, this photon can be any energy — the amount depends on how close the electron came and how fast it was moving. Billions of such near-misses happening every second produce a smooth thermal continuum: a glow across the entire spectrum, peaking at a frequency determined by the plasma temperature. At 5,000 K (a modest atmospheric plasma) the peak is yellow-green visible light. At 10,000 K it shifts toward ultraviolet. The plasma structure therefore emits two things simultaneously: sharp spectral lines at specific colors (the atom's fingerprint) plus a smooth thermal glow across the whole spectrum (the temperature's fingerprint).

What escapes versus what stays trapped. A plasma has a natural resonant frequency — the plasma frequency — determined by electron density: $f_p \approx 9\sqrt{(n_e)}$ Hz, where n_e is electrons per cubic centimeter. Electromagnetic radiation below this frequency cannot propagate out of the plasma. The plasma reflects it back inward, the way a mirror reflects light. It accumulates inside the structure as trapped electromagnetic energy. At $n_e = 10^8$ cm^{-3}: $f_p \approx 90$ MHz — everything below FM radio is trapped. At $n_e = 10^{10}$ cm^{-3}: $f_p \approx 900$ MHz — everything below microwave is trapped. What escapes and reaches the distant observer: visible light, near-infrared, ultraviolet, microwave, radar frequencies. What stays trapped inside and dominates the close-range environment: ELF (0.001–30 Hz), VLF (3–30 kHz), HF radio — concentrated and accumulating. This distinction is the physical basis of the two-zone model. There is no sharp boundary. The zones are defined by the observer's distance from the structure.

Part Two: What Those Photons Do to the Perceptual Machinery

The eye. The retina contains two types of photoreceptors. Rods detect dim light across a broad spectrum — night vision, sensitive but not color-specific. Cones detect color and operate in brighter conditions. There are three cone types: S (short wavelength, blue, peak sensitivity ≈420 nm), M (medium, green, peak ≈530 nm), L (long, red, peak ≈560 nm). The plasma structure's dominant emission is oxygen green at 557.7 nm. This falls almost exactly on the peak sensitivity of both the M and L cones simultaneously — maximum possible visual response. The human eye is more sensitive to this wavelength than to almost any other. This is not coincidence. It is the same reason the aurora appears brilliant green to human observers: our photoreceptors evolved under an atmosphere that occasionally produces this emission, and our eyes are tuned to it. When a tropospheric plasma structure emits oxygen green at 557.7 nm, it is hitting the human visual system at its most sensitive point.

From photon to perception. Photons strike the photoreceptors. Each absorbed photon triggers a conformational change in a molecule called retinal — it physically bends, changing shape. This mechanical change opens ion channels in the photoreceptor membrane. Calcium and sodium ions flow. The electrical state of the cell changes. The signal propagates through bipolar cells, then ganglion cells, then through the optic nerve to the lateral geniculate nucleus of the thalamus, then to the primary visual cortex at the back of the brain. At this point the observer has a visual experience. The brain has processed incoming photons into a perceived object in space with shape, brightness, color, and motion. The observer looks up and sees something luminous. No unusual biology is required. No special sensitivity. No candidate mechanism. It is the ordinary visual pathway doing exactly what it evolved to do.

What near-infrared does. Near-infrared photons (700 nm – 1 mm) are not detected by the retina — the photoreceptors are not sensitive in that range. But they are absorbed by water, and the human body is mostly water. Skin absorbs near-infrared and converts it directly to heat. Witnesses at close range reporting warmth, a heat wave, or radiant heat are reporting near-infrared absorption. This is the same mechanism as a heat lamp, or the warmth felt from a fire even before the surrounding air warms up — direct radiative transfer, photons depositing energy in tissue. No conduction, no convection. The heat arrives at the speed of light and is felt before anything else.

What ultraviolet does. Ultraviolet photons carry enough energy to break chemical bonds in biological molecules. UVA (315–400 nm) penetrates to the dermis and damages DNA in skin cells. UVB (280–315 nm) is almost entirely absorbed in the cornea and lens of the eye. Acute high-dose UVB exposure produces photokeratitis — essentially sunburn of the cornea — causing intense pain, photophobia, and temporary vision impairment hours after exposure. Several close-range witnesses across the confirmed cases reported eye pain and temporary vision disturbance in the hours and days after the event. This is a predictable consequence of UV exposure from a hot plasma structure at close range. It is not anomalous. It is radiobiology.

What the trapped ELF does at close range. Below the plasma frequency cutoff, the structure is a cavity full of ELF electromagnetic energy. At Tier 3 field intensities — above 38 μT — this energy drives calcium ion efflux from neurons, as Bawin and Adey demonstrated in 1976 and Blackman et al. confirmed in 1982. Calcium efflux matters because calcium ions are the gatekeepers of synaptic communication. Neurons communicate by releasing neurotransmitters across synapses. That release is triggered by calcium ions entering the presynaptic terminal. The concentration of calcium inside the cell is the gate that controls whether a synapse fires. External ELF fields driving calcium ions across cell membranes are not merely perturbing the system at the margins. They are operating the gates. They are inserting themselves into the signaling chain at its most fundamental control point.

What robbing calcium from a neuron actually does. The action potential — the electrical spike a neuron fires when it communicates — depends on a precise choreography of ion flows across the cell membrane. The rising phase is sodium rushing in. The falling phase is potassium rushing out. But the threshold at which the whole sequence fires, and the magnitude of the response, is regulated by intracellular calcium. Calcium inside the cell modulates the sodium channels. Lower intracellular calcium raises the firing threshold — the cell becomes less excitable, requiring a stronger incoming signal to fire at all. Higher intracellular calcium lowers the threshold — the cell becomes more excitable, firing on weaker provocation. When ELF-driven efflux pulls calcium out of the neuron, you are raising the firing threshold of that cell. It becomes harder to trigger.

But it does not happen uniformly. Different cell types have different channel densities, different baseline calcium concentrations, different sensitivities to the induced field. The result is a differential change across the neural network — some cells get quieter, some structures respond differently, the network-level effect is not simple suppression. It is a reorganization of which cells are near threshold and which are not. That reorganization of excitability thresholds across a neural network is functionally indistinguishable from what happens in altered states — deep meditation, hypnagogia, temporal lobe seizure precursors. The network is still working. It is working with modified gates. This is not hallucination in any clinical sense. It is the ordinary neural network processing genuine sensory input — the light is real, the heat is real, the sound may be real — through a slightly reorganized excitability landscape. That is why witnesses consistently report these experiences as feeling more real than ordinary reality, not less. The signal is genuine.

The gain across the network has been modified. Both things are true simultaneously.

The temporal lobe and what it produces. The temporal lobe — where Persinger's stimulation reliably produced feelings of presence, entities, and overwhelming emotional significance — integrates sensory information with emotional memory through the amygdala and hippocampus. When these structures receive anomalous input — either directly from ELF field effects or from the downstream consequences of calcium efflux propagating through the neural network — they generate the emotional signature of the experience: the sense that what is happening is profoundly real, outside ordinary categories, and impossible to adequately describe afterward. This is not pathology. This is the temporal lobe doing its job with an input it was not designed for. The same structures that generate the feeling of deep meaning in ordinary life generate it here, at high intensity, because the input is physically anomalous.

The three levels of response, simultaneously. The visual cortex says: there is a luminous object in the sky. The temporal lobe says: this is the most significant thing that has ever happened. The cultural frame says: it is the Virgin Mary, or a spacecraft, or angels, or lights from heaven. All three statements are simultaneously true at their own level. None contradicts the others. The physics produces the object. The temporal lobe produces the significance. The culture provides the label. Persinger demonstrated this dissociation in the laboratory: the same electromagnetic stimulus applied to hundreds of subjects produced culturally specific labels — Christians reported Christ, Hindus reported Shiva, secular subjects reported a presence or a deceased relative — while the underlying neurological event was identical across all of them. The confirmed propagation cases show the same pattern across 108 years and six continents. The physical trigger is constant. The cultural container is variable. The model explains both.

The complete chain, stated plainly. The sun ejects particles. A geomagnetic storm injects them into the radiation belts. They reach successive inner L-shells on successive days — T days to reach the L-shell of a given site, where T is bounded by a ceiling written into Earth's own geometry and rotation period. The atmospheric transfer function either delivers them to the surface or destroys them in the aerosol column. Where they survive, they produce a tropospheric plasma structure through electron-ion recombination and bremsstrahlung. That structure emits photons across the full electromagnetic spectrum simultaneously — visible light that the eye detects, near-infrared that the skin feels as heat, ultraviolet that burns the cornea at close range. Below the plasma frequency cutoff, trapped ELF energy at close range drives calcium efflux from neurons and induces currents in unshielded ignition circuits. The observer's visual cortex processes the light. The observer's temporal lobe processes the significance. The observer's culture supplies the name. Ninety-eight million miles from the sun to a field in Portugal. Every step is physics. Every step has a reference.

The plasma density thresholds that map survival fraction to field intensity tier — the β parameter calculation and its connection to the Persinger and calcium efflux laboratory thresholds — are traced in Chapter Fifteen and the preceding discussion.

Postmortem Brain Tissue and the Material Transduction Pathway

A line of experimental work from Rouleau and Persinger (2017, Biology and Medicine) adds a remarkable dimension to the transduction question. In their study, formaldehyde-fixed human brain tissue — preserved for over twenty years, with no metabolic activity, no action potentials, no living physiology whatsoever — was exposed to both naturally occurring geomagnetic fluctuations and laboratory-controlled electromagnetic fields. Needle electrodes inserted into specific neuroanatomical regions measured microvolt-level potential fluctuations across the fixed tissue.

The results were striking. On days of elevated geomagnetic activity (Kp > 3), the right parahippocampal gyrus showed significantly increased spectral power in the 7.5–14 Hz (alpha) range compared to quiet days. The left parahippocampal gyrus did not show the same effect. Different brain regions filtered the same electromagnetic input differently. The tissue was not alive. It was a material — but it was a material whose microstructure still imposed region-specific filtering on electromagnetic perturbations, in exactly the frequency bands associated with anomalous perceptual experience in living subjects. When the same tissue was exposed to the complex, physiologically patterned magnetic fields that Persinger had used with hundreds of living volunteers to induce sensed-presence experiences, the strongest responses appeared in the right amygdala and orbitofrontal gyri — the same regions that are implicated in temporal lobe phenomenology in living brains. Simple sine waves of comparable intensity did not produce the same effect. Pattern complexity mattered, even in dead tissue.

This finding has a direct implication for the propagation model. If the geomagnetic transduction pathway operates at the level of brain structure — at the level of preserved cytoarchitecture, dendrite orientation, and regional tissue geometry — rather than requiring any particular neurochemical state or conscious readiness, then the mechanism is more robust than TMS analogies alone would suggest. It means that when the plasma propagation front arrives at a mid-latitude site and produces a localized geomagnetic perturbation, the effect on human neural tissue is not mediated solely by ion channels, neurotransmitter release, or any process requiring the brain to be in a particular state. The brain's material structure itself acts as an antenna with region-specific frequency response. A living brain would add metabolic amplification, neural feedback, and conscious processing on top of this material-level transduction — but the foundation is structural, not biochemical.

Population-Level EEG Confirmation

Saroka, Caswell, Lapointe, and Persinger (2014, Neuroscience Letters) measured interhemispheric EEG coherence in 184 volunteers over a three-year period while subjects sat in a quiet, darkened chamber. During periods of increased global geomagnetic activity — above approximately 8–10 nT — coherence between the left and right temporal lobes increased significantly, specifically in the 11 Hz (alpha) band. Source localization via sLORETA identified the right parahippocampal region as the primary contributor. The most statistically significant spectral peaks across 1–40 Hz fell at 7.5 Hz and 19.5 Hz. Note that 7.83 Hz is the fundamental Schumann resonance frequency — the electromagnetic resonance of the Earth-ionosphere cavity. The brain's response to geomagnetic perturbation peaks at the same frequency as the planet's own electromagnetic heartbeat.

Mulligan and Persinger (2012, Neuroscience Letters) independently confirmed this by experimental simulation. When subjects were exposed to a 20 nT, 7 Hz amplitude-modulated magnetic field — designed to mimic geomagnetic storm conditions — theta power over the right parietal region was enhanced after approximately ten minutes, replicating the naturalistic correlation in a controlled laboratory setting. This closes the experimental loop: the correlation between geomagnetic activity and brain electrical activity observed in naturalistic data is reproducible when the geomagnetic signal is simulated in the laboratory.

Cultural Labeling : The Same Stimulus, Different Names

One further result from the Persinger laboratory bears directly on this model's central interpretive claim. Across hundreds of god helmet sessions spanning approximately three decades, subjects who reported a sensed presence consistently labeled the entity according to their own cultural background: Christians identified it as Mary, God, or Jesus; Hindus identified it as Shiva or Krishna; secular subjects reported an anonymous presence or a deceased relative. The physical stimulus was identical — the same complex magnetic field pattern applied to the same brain regions. The perceptual content varied by cultural frame. Persinger and Murphy (2016) summarized this explicitly: the labeling and correlative imagery is a function of cultural expectations, while the underlying neurological event — right hemispheric activation, interhemispheric intrusion into left-hemispheric awareness — is a function of temporal lobe physics.

This is the laboratory demonstration of the same pattern observed in the fifteen confirmed propagation cases. Observers in 1917 Portugal reported a Marian apparition. Observers in 1957 Texas reported a luminous aerial object. Observers in 1983 New York reported a structured craft. Observers in 1997 Arizona reported lights in formation. The physical trigger — the arrival of storm-injected plasma at a characteristic propagation speed — is the same in all fifteen cases. The cultural container determines the reported content. The Persinger laboratory has now demonstrated this dissociation experimentally, under controlled conditions, with the same electromagnetic stimulus producing culturally determined interpretive labels across hundreds of subjects.

The Gap : From Individual to Shared

The most important unresolved question is not whether geomagnetic fields affect individual human perception — that is now established by Kirschvink et al. (2019). The unresolved question is why certain events produce shared multi-witness experiences of consistent structure in populations of otherwise unconnected observers.

A geomagnetic perturbation is spatially extended — it affects the entire region within the precipitation column simultaneously. If the perceptual effect is primarily via plasma luminosity (Step 4a), then multiple observers see the same luminous phenomenon for the simple reason that it is physically present in the atmosphere they are all looking at. If the effect is primarily neural (Steps 5b or 5c), then multiple observers experience similar perceptual content because they are all subject to the same external EM perturbation simultaneously.

The cultural interpretation step then determines the specific form. Observers in 1917 Portugal experiencing an intense shared visual anomaly interpreted it through the Marian apparition framework that their culture provided. Observers in 1957 Levelland experiencing vehicle electronics failure and a luminous aerial object had no such framework — they reported what they saw and called the sheriff. Observers in 1983 suburban New York had the post-Spielberg extraterrestrial spacecraft framework available. The physical trigger may be the same; the cultural container determines the reported content.

This interpretation predicts a testable corollary: the core phenomenology of confirmed propagation-trigger events should be more similar across cultural contexts than the surface reports suggest. Luminous objects, electromagnetic vehicle effects, sensations of physical reality and presence — these appear across all fifteen confirmed cases regardless of whether the event is coded as religious apparition, UFO, or aerial phenomenon. The cultural frame shapes the description; the underlying phenomenon shapes the experience. Separating these two layers is the work the mechanism section points toward.

The step-by-step arithmetic — precipitation flux, ionization enhancement, surface perturbation field strength, and the dose-at-receiver calculation — is traced across Chapters Fifteen through Seventeen.

The Two-Zone Emitter : What the Plasma Structure Actually Radiates

The preceding section identified the magnitude gap between storm-driven surface perturbations (0.01–0.2 μT regionally) and the Persinger laboratory threshold for temporal lobe effects (1–10 μT). That gap was computed for the diffuse regional perturbation from ionospheric currents at 100 km altitude. But if the atmospheric transfer function produces a localized plasma structure at LOW altitude — in the troposphere, hundreds of meters above the witnesses — the physics changes completely. A tropospheric plasma structure is not a passive perturbation in the geomagnetic background. It is a self-contained electromagnetic emitter radiating across the entire spectrum, with a critical distinction between what escapes the structure and what is trapped inside it.

The emission mechanisms. An atmospheric plasma structure produced by precipitating ions emits energy through three distinct mechanisms, each dominating a different frequency band. First, line emission: excited oxygen and nitrogen atoms relax to lower energy states and emit photons at specific wavelengths. The dominant auroral line is oxygen green at 557.7 nm (the $O(^1S \rightarrow {}^1D)$ forbidden transition), with secondary contributions from oxygen red at 630.0 nm, nitrogen blue-violet at 427.8 and 391.4 nm, and oxygen near-infrared at 844.6 nm. These are the colors witnesses report: blue-green at Levelland, white-luminous at Fatima, multi-colored at Phoenix. Second, bremsstrahlung: free electrons decelerating near ions produce broadband thermal continuum radiation. The spectral peak depends on electron temperature — at 5,000 K (modest atmospheric plasma) the peak is at 580 nm (yellow-green visible); at 10,000 K (energetically driven) it shifts to 290 nm (ultraviolet). Third, current systems: the bulk motion of charges within and around the plasma structure produces electromagnetic fields across the ELF, VLF, and RF bands.

The plasma frequency cutoff. This is where the physics becomes consequential for the biological mechanism. A plasma has a natural resonant frequency — the plasma frequency — determined by the electron density: $f_p \approx 9\sqrt{(n_e)}$ Hz, where n_e is in electrons per cm^3. Electromagnetic radiation below the plasma frequency cannot propagate out of the plasma. It is reflected, confined, trapped inside the structure. Only radiation above the plasma frequency escapes. For a modest plasma structure ($n_e \approx 10^8$ cm^{-3}), the plasma frequency is approximately 90 MHz and everything below FM radio frequencies is trapped. For a structure dense enough to produce radar returns ($n_e \approx 10^{10}$ cm^{-3}), the cutoff rises to 900 MHz. Everything below that — all ELF, VLF, HF, and most of the radio spectrum — is confined within the structure as standing electromagnetic fields. The energy does not vanish. It accumulates.

Two zones, two physics regimes. The plasma structure is a two-zone emitter. In the far field — kilometers away — only the escaping radiation is detectable: visible light from line emission and thermal continuum (what witnesses see as a luminous object), plus microwave and higher frequencies above the plasma cutoff (what radar detects as a coherent return). The visible light is the weakest energy channel, less than 1% of the total energy budget of the structure. It is all a distant observer can perceive, but it is not where the energy is. In the near field — within a few hundred meters of the structure — the confined low-frequency electromagnetic energy dominates. The ELF component (0.001–30 Hz) overlaps with Schumann resonance frequencies and human EEG bands. The VLF and HF components induce currents in metallic conductors — vehicle wiring, ignition systems, headlight circuits. The quasi-DC magnetic component produces local field enhancements that scale inversely with distance. A plasma structure at 200 meters overhead produces a local magnetic perturbation at the surface that could be orders of magnitude stronger than the 0.01–0.2 μT regional background. At close range, the local field can reach or exceed the 1–10 μT Persinger threshold.

The magnitude gap closes at close range. The order-of-magnitude calculation in the preceding section computed the gap for the regional perturbation from ionospheric currents at 100 km altitude. That gap is real for distant effects. But the confirmed cases do not describe distant effects. They describe close encounters. The Levelland witnesses had their engines killed by something directly overhead. The Kaikoura film crew recorded an object at visible range. The Hudson Valley witnesses watched a formation pass above them at low altitude. If a tropospheric plasma structure is present at hundreds of meters rather than hundreds of kilometers, the near-field confined energy is the dominant exposure, not the far-field regional perturbation. The magnitude gap computed for the regional case does not apply to the local case. The trapped ELF/VLF energy inside the structure, experienced in the near field at close range, can exceed the Persinger threshold without requiring resonance, accumulation, or magnetite transduction. The gap closes because the source is close, not because the threshold is lower.

Vehicle electronics interference. The Levelland case — fifteen independent reports of engine kill and headlight failure in a two-hour window — has never been satisfactorily explained by conventional atmospheric electricity. Ball lightning does not stall engines across a radius of several miles. The two-zone emitter model provides a specific mechanism. The trapped RF energy below the plasma frequency cutoff exists as intense near-field electromagnetic energy around the structure. Vehicle ignition systems in the 1950s were open-wire point-and-condenser circuits with no RF shielding — they were effectively antennas. A strong local RF field induces currents in the ignition wiring that overwhelm the spark generation circuit. The engine dies. The headlights, powered through the same unshielded wiring harness, fail. When the structure moves away, the near-field intensity drops with distance, the induced currents fall below the interference threshold, and the vehicle's electrical system resumes normal operation. The witnesses reported exactly this sequence: approach, failure, departure, restoration. The two-zone model predicts exactly this sequence from first principles.

The vehicle interference catalog. Levelland is not an isolated case. The Center for UFO Studies compiled 276 confirmed vehicle interference reports through the late 1970s (Rodeghier 1981). Of these, 83 involved both engine and headlight failure, 124 reported engine stall alone, and 31 reported engine power loss without complete stall. The cases show a distance-severity relationship consistent with electromagnetic near-field physics: closer encounters produce more complete interference. Two documented cases involving a diesel-powered and a spark-ignition vehicle side by side (Forli, Italy, 1954; Fordingbridge, England, 1967) confirmed that diesel engines — which use compression ignition rather than spark ignition — are immune to the effect. The interference is specific to the ignition circuit, not to the mechanical engine, exactly as the electromagnetic induction model predicts. Engine stalls have a methodological advantage that visual sightings do not: they are binary physical events. An engine either died or it did not. A stalled vehicle is not a matter of cultural labeling, individual sensitivity, or temporal lobe phenomenology. If clusters of vehicle interference reports can be identified at clean-air rural sites during predicted arrival windows, they would constitute the hardest available test of the propagation model — instrumental measurements using automobiles as unshielded broadband electromagnetic detectors.

The phenomenological synthesis. The full phenomenology of the confirmed cases is consistent with a single atmospheric plasma structure radiating across the electromagnetic spectrum, experienced differently at different distances. At long range (kilometers): witnesses see a luminous object in the sky. The experience is visual only. This is the Phoenix Lights — ten thousand people seeing lights from miles away. At medium range (hundreds of meters): witnesses see a luminous object AND experience emotional intensity, a sense of significance or unreality, and physiological arousal. The confined ELF fields in the near zone begin to affect neural tissue at subthreshold or threshold levels. This is the Fatima phenomenology — a visual event accompanied by a sense of the numinous. At close range (tens to hundreds of meters): witnesses experience visual phenomena, vehicle electronics interference, and strong physiological and psychological effects. The near-field confined energy exceeds interference thresholds for both electronics and neural tissue. This is Levelland. The same plasma structure, the same emission physics, the same spectral budget — but experienced at different distances, the effects present differently. The cultural label the witness applies (miracle, UFO, lights in the sky) depends on which zone they are in and what conceptual vocabulary their culture provides for the experience at that range.

Chapter Eighteen: The Chemical Record

"The nose knows what the eye cannot decide."

The previous chapters establish detection through three channels: physical instruments (engine ignition systems, radar, magnetometers), biological tissue (oxidized hemoglobin, radiation injury, neural threshold effects), and human perception (visual, emotional, physiological response gradients). This chapter documents a fourth channel: atmospheric chemistry. It is the oldest and most direct of the four. It requires no instrument, no specialist knowledge, and no interpretation. A molecule either reaches an olfactory receptor or it does not.

The olfactory signal is the most honest data in the anomalous events literature for a reason that has nothing to do with paranormal research: smell bypasses the neocortex entirely. Visual reports are processed through layers of pattern recognition, memory, and cultural expectation before they become testimony. Olfactory signals route directly to the limbic system — the oldest part of the brain — without a cortical interpretation step. A witness cannot decide to smell ozone. They cannot misidentify it the way they misidentify a light in the sky. They cannot be primed by cultural expectation to smell it, because there is no established cultural expectation that anomalous events smell of ozone. Sulfur has a mythology — brimstone, the infernal. Ozone has none. It is an industrial smell, a storm smell, a photocopier smell. No folklore prepares a witness to report it.

In 2016, researcher Joshua Cutchin published *The Brimstone Deceit: An In-Depth Examination of Supernatural Scents, Otherworldly Odors, and Monstrous Miasmas* (Anomalist Books). The book is a systematic catalog of olfactory reports across all categories of anomalous phenomena — UFO encounters, apparition cases, cryptid sightings, and haunting accounts. Cutchin had no interest in geomagnetic storm data. He was cataloguing smells. His methodology was phenomenological, not physical. He searched the existing literature for every case in which a witness reported an odor, extracted the report, and grouped the results by smell type. Chapter Thirteen of his book is devoted entirely to ozone reports in UFO cases. It is one of the most data-rich chapters in the paranormal literature — and it was compiled without any awareness of the propagation model described in this book.

That independence is the point. The propagation model was derived from engine stall cases and calibrated against Levelland 1957. Cutchin's catalog was derived from olfactory witness reports collected for an entirely different purpose. Neither dataset was aware of the other. What follows is what happens when the two are crossed.

The Chemistry of the Signal

Ozone is a photochemical product. It is generated when ionizing radiation or high-energy electrical discharge splits molecular oxygen (O_2) into atomic oxygen, which then combines with intact O_2 to form O_3. The threshold detection concentration for the human nose is 0.01–0.05 parts per million — extraordinarily sensitive, far below any health threshold. At these concentrations ozone is described as sharp, peppery, faintly chlorine-like, or "like rain." It is the smell of lightning, of arc welders, of photocopiers, and of certain medical equipment. It is not a smell anyone manufactures a cultural expectation around.

If energetic particle precipitation is arriving at the surface during propagation windows — as the model requires — ozone production at ground level is the expected atmospheric response. The particles ionize the lower atmosphere. Ionized oxygen produces ozone. The chain is direct, well-established, and quantitatively characterized. Witnesses in the vicinity of a surface-level plasma event driven by radiation belt precipitation should smell ozone. That is not a prediction made after the fact. It is a straightforward consequence of the atmospheric chemistry that the model already invokes at altitude to explain mesospheric ozone depletion (Chapter Fifteen).

The sulfur signature requires a different explanation, addressed in the final section of this chapter. The critical observation for now is that ozone and sulfur compounds are sometimes reported together at the same event — and when they are, the chemistry is unambiguous. When ozone contacts hydrogen sulfide, the reaction is:

$$H_2S + O_3 \rightarrow H_2O + SO_2$$

Witnesses smelling both ozone and sulfur simultaneously are not smelling two independent sources. They are smelling a reaction product — the atmospheric chemistry of two different gas mixtures meeting at an interface. The two-component fingerprint cannot be produced by any single terrestrial source. Decomposition produces hydrogen sulfide but not ozone. Electrical discharge produces ozone but not hydrogen sulfide. The co-occurrence of both, at the same event, at the same moment, is chemically specific in a way that individual reports are not. Cutchin documents this co-occurrence (Cutchin 2016, Ch. 13).

The Cutchin Dataset: Ozone Cases Against the Propagation Windows

From Cutchin's Chapter Thirteen, fifteen dated ozone cases were extracted — cases where a specific event date and location were recorded alongside an ozone or ozone-proxy smell report. Each case was then run against the GFZ Potsdam Kp archive using identical methodology to the cases in Parts I and II: haversine distance from the nearest geomagnetic pole, divided by 85.836 km/day to produce the predicted lag T, expanded by ±10% to define the acceptance corridor. Cases with no Kp≥5 storm in the corridor are scored PCS. Cases with a passing ratio are scored PCI. Cases with a storm in the corridor that falls outside the acceptance band are scored PCE.

Cutchin was not consulted during this analysis. The cases were extracted, the windows calculated, and the archive queried without his knowledge. The results are as follows.

Levelland, Texas — November 2, 1957. Already the calibration baseline of the propagation model at ratio 1.000. Cutchin independently documents an ozone smell at the same event. Two investigators, two methodologies, same event, same chemistry. This is the anchor case.

South Africa — February 1957. Witness reported an object leaving an odor described as "an overheated radio" — a consistent descriptor for ozone, as Cutchin documents. GSP distance 7,469 km, predicted lag 87 days. A dense storm cluster running November 12–28, 1956 fills the entire window, with peak Kp 7.667 on November 15. Best ratio 0.977. PCI.

Las Cruces, New Mexico — March 12, 1967. Witnesses reported an object smelling of burning electrical insulation — a standard ozone-proxy description (Cutchin 2016, Ch. 13). GNP distance 5,575 km, predicted lag 65 days. January 7, 1967 storm (Kp 5.333) gives lag 64 days, ratio 0.985. PCI.

Salisbury, Rhodesia — July 28, 1975. Five police officers witnessed a hovering object; a motorist subsequently reported a large orange light filling his car with the smell of "foul sea water" — a coastal-ozone confusion Cutchin documents extensively. GSP distance 8,279 km, predicted lag 96.5 days. April 21, 1975 storm (Kp 6.0) gives ratio 1.016. PCI.

New Boston, Michigan — October 1966. A witness reported her body felt electrified and noted a strong ozone smell, later experiencing nightmares of gray entities. GNP distance 4,308 km, predicted lag 50.2 days. August 30, 1966 storm (Kp 7+, Ap 82) gives lag 46 days, ratio 0.917. PCI.

Sweden — May 1946. Gösta Carlsson witnessed a landed craft in a forest surrounded by purple light, returning later to find an ozone odor at the site. GNP distance 3,567 km, predicted lag 41.6 days. April 1, 1946 storm (Kp 5.0) gives ratio 1.059. PCI.

Townshend, Vermont — May 20, 1988. A husband and wife reported a glowing house-sized object on a blue light, the air filled with an acrid ozone smell. GNP distance 4,186 km, predicted lag 48.8 days. April 3, 1988 storm (Kp 8.0) gives ratio 0.964. PCI.

Evan City, Pennsylvania — 1971. Witnesses followed a UFO to a farm, noting a "sweet and clean smell like rain" — a standard ozone descriptor. GNP distance 4,452 km, predicted lag 51.9 days. April 21, 1971 storm (Kp 6.3) gives ratio 1.060. PCI.

New Zealand — June 1959. A witness described a "pepper" smell — the peppery component of ozone is well-documented and noted by Cutchin as a consistent variant descriptor. GSP distance 4,400 km, predicted lag 51.3 days. April 23, 1959 storm (Kp 6.7) gives ratio 1.034. PCI.

San Antonio, Texas — 1975. Observers described a flying object smelling of "burning wires or metal" — another standard ozone proxy. GNP distance 5,813 km, predicted lag 67.7 days. April 8, 1975 storm (Kp 5.7) gives ratio 1.004. PCI.

Four cases returned PCS — Iowa April 1966, Longmont Colorado November 1980, Peppertown Indiana September 1999, and Kazakhstan 1976. No triggering storm existed in any of those windows. The model makes no prediction for events with no storm to trigger them, and it makes no prediction here. These are not failures. They are the model's silence, which is as diagnostic as its confirmations.

One case returned PCE — Ohio June 2003, where the only qualifying storm produced ratio 0.898, just outside the lower band edge. The Falcon Lake Manitoba case (May 1967), already scored PCE at 0.856 in the main dataset, also appears in Cutchin's sulfur-and-ozone co-occurrence catalog. Both PCE results are at the band edge, not randomly distributed.

The result across the scoreable cases: 10 PCI, 4 PCS, 2 PCE. Cutchin did not know about the propagation model. The propagation model did not know about Cutchin's catalog. Neither dataset was built to fit the other. The overlap is not selection. It is signal.

What the Chemical Record Establishes

Ozone is not an interpretation. It is a molecule. When it reaches a human olfactory receptor, it produces a signal that bypasses the entire cognitive architecture that makes visual witness reports unreliable. The witness is not deciding what they smelled. They are reporting what happened to them at the most direct sensory level available to human biology.

The olfactory record is also the most geographically and temporally distributed independent dataset in the anomalous events literature. Cutchin's ozone cases span 1946 to 2003, six continents, and every category of anomalous encounter. The witnesses had no contact with each other, no shared cultural template for ozone as a paranormal smell, and no knowledge of geomagnetic storm data. They simply reported what their noses told them.

Ten of those reports fall inside the propagation windows defined by a model that was built from engine failure data and calibrated against a 1957 Texas highway event. That is not a coincidence that requires explanation. It is an independent confirmation that demands acknowledgment.

Adversarial Record

The adversarial test applies here as it applies throughout the book. For each of the ten PCI cases, the largest storm outside the acceptance corridor was identified and tested. Every one returned PCE. Zero anomalous passes.

The Sweden Carlsson case illustrates the discriminating power most sharply. A Kp 8.7 storm on March 28, 1946 — four days before the window edge — produces ratio 1.155, outside the band. The geometry rejected a storm nearly twice as strong as the one that passes, because the arithmetic demanded it. The model did not select the weaker storm because it was convenient. It selected it because it was the only storm the geometry permitted.

The Levelland case — the model's calibration baseline and the anchor of the olfactory dataset simultaneously — has a Kp 9.0 storm inside the window at ratio 0.968, which also passes. The September 23, 1957 storm at Kp 8.7 produces ratio 0.656. Two independent investigators documented the same event. The geometry rejects a larger competing storm. The adversarial structure is intact.

Ten PCI cases. Ten adversarial candidates tested. Ten PCE rejections. The olfactory dataset holds the same adversarial standard as every other dataset in this book.

Session 03: Pre-Registered Chemical Cases

The retrospective ozone cases above establish the pattern. Session 03 extends it under pre-registration conditions. Twelve dated cases were extracted from Cutchin's Chapter 14 (UFO Smells: Chemicals) — covering phosphine, ammonia, camphor, sulfur, metallic, carborundum, and Bakelite odors. Propagation windows were calculated and uploaded to osf.io/6abpk (Session_03_Predictions.pdf, timestamped March 23, 2026, 03:51 PM) before the GFZ archive was consulted. The results were scored after upload.

Results: 10 PCI, 2 PCS, 0 PCE. Ten adversarial candidates tested, zero anomalous passes.

Two cases are particularly sharp. The Buenos Aires sighting of August 12, 1972 — a sulfur odor, radio failure, radiation symptoms — sits at ratio 0.989 against the June 26, 1972 Kp 5.0 storm. The geometry predicted a window centered on that storm before the archive was consulted. The Caracas, Venezuela case of September 8, 1967 demonstrates adversarial discrimination at maximum power: the May 25, 1967 Kp 9.0 storm — the great solar storm that nearly triggered a NORAD nuclear alert — falls just outside the window at ratio 1.166 and is correctly rejected. The geometry selected a Kp 6.7 storm instead at ratio 1.023. A model that passes the largest storm available would not produce this result.

Across both the retrospective and pre-registered O datasets: 20 PCI, 6 PCS, 1 PCE, 20 adversarial rejections, zero anomalous passes. The chemical smell record — collected by an investigator with no knowledge of this model, covering six decades, four continents, and eight distinct chemical signatures — is consistent with a single physical mechanism operating on a geomagnetic schedule.

Chapter Nineteen: Radar Amplification and the Carrier Groups

> *"When you have eliminated the impossible, whatever remains, however improbable, must be the truth."*
>
> *— Arthur Conan Doyle*

One of the most persistent patterns in anomalous aerial reports is the apparent association with military installations. The usual interpretation involves military interest in UFOs. This chapter offers a different one: the correlation is not with the military mission but with the radar. High-power radar illuminating a column of ionized atmosphere amplifies exactly the plasma structures the propagation model predicts. The radar is the instrument, not the target.

One of the most persistent patterns in the anomalous aerial phenomenon literature is the apparent association between UAP reports and military installations, particularly nuclear-capable facilities. Robert Hastings documented this correlation exhaustively in UFOs and Nukes (2008), compiling decades of military witness testimony and official records showing elevated UAP report rates at or near nuclear missile fields, weapons storage sites, and strategic bomber bases. The pattern is real. The USAF's own records confirm it. It has generated an entire interpretive literature built on the premise that an external intelligence is monitoring humanity's nuclear weapons capability. That interpretation may be correct. The atmospheric transfer model in this chapter offers a complementary physical framework — not a refutation of Hastings' documented correlation, which the data supports, but an alternative mechanism that accounts for the same geographic pattern without requiring the nuclear-monitoring hypothesis as its premise. Both frameworks deserve investigation on the evidence.

The atmospheric transfer model presented in this chapter offers a different explanation. The correlation is not with the nuclear material. It is with the radar.

Military installations operate high-power radar continuously. Air defense radar, early warning systems, fire control radar, tracking radar, and battlefield surveillance systems pump kilowatts to megawatts of directed microwave energy into the atmosphere on a 24/7 basis. PAVE PAWS phased array radar at Cape Cod and Beale AFB transmits 585 kW average power with 30 MW peak. The Ballistic Missile Early Warning System (BMEWS) at Thule, Greenland and RAF Fylingdales operates megawatt-class phased arrays. PARCS at Cavalier Air Force Station, North Dakota — which monitors the ICBM fields — is another megawatt-class system. Every nuclear missile field in the United States has associated early-warning and tracking radar within its perimeter. Every aircraft carrier strike group radiates megawatts from multiple radar systems simultaneously. Every air defense sector operates continuous surveillance radar.

Nuclear sites have the most powerful radar — not because of the nuclear material. The objects that carry nuclear weapons — ICBMs, strategic bombers, ballistic missile submarines — are the assets most critically requiring air defense protection. The most powerful radar systems in the inventory are deployed to protect these assets. The correlation Hastings documented between UAP reports and nuclear sites is, on examination, a correlation between UAP reports and installed radar power. The nuclear material is incidental. A nuclear missile silo does not emit radiation that could plausibly interact with atmospheric plasma at distance. The radar installation protecting that silo transmits megawatts of directed electromagnetic energy into the atmosphere above it continuously. The distinction is fundamental: the nuclear interpretation requires an intentional external agent monitoring weapons; the radar interpretation requires only the known physics of electromagnetic wave interaction with ionized gas.

How radar interacts with storm-delivered atmospheric plasma. When the propagation model delivers a diffuse ion and plasma enhancement to a region following a geomagnetic storm, the plasma exists at densities that may be below the threshold for visible luminosity or significant electromagnetic effect under natural conditions. At a civilian location, this sub-threshold enhancement produces effects only if the atmospheric survival fraction is high enough (clean air, thin boundary layer). At a military installation, megawatts of microwave energy are being pumped into the same atmospheric volume. The radar-plasma interaction operates through several well-characterized physical mechanisms:

(1) Electron acceleration and luminosity enhancement. Radar-frequency electromagnetic fields accelerate free electrons in the ambient plasma, increasing their kinetic energy. Higher electron energies produce higher ionization rates through collisional cascade, locally increasing plasma density above the recombination emission threshold (~10^1–10^{10} per cm^3). The radar acts as a pump energy source in a plasma that has already been seeded by precipitating particles. Diffuse, invisible plasma becomes locally luminous in the radar-illuminated volume. This is the same physics that operates in ionospheric heating experiments: the High-frequency Active Auroral Research Program (HAARP) facility in Gakona, Alaska uses ground-based high-power radio transmitters to produce artificial airglow — visible luminous patches in the ionosphere — by exactly this mechanism (Rodriguez et al. 1998; Pedersen and Gerken 2005). Military radar operates at different frequencies than HAARP but the underlying physics — EM energy coupling into ambient plasma to produce enhanced ionization and luminosity — is identical.

(2) Local boundary layer modification. High-power microwave emission heats the air column directly above the radar installation, creating a localized thermal plume. This plume modifies the local boundary layer structure, potentially producing a "chimney" of reduced aerosol density directly above the installation. In the atmospheric transfer model, a locally thinner aerosol column means locally higher ion survival fraction — a transmission window in an otherwise attenuating atmosphere. The radar creates a hole in the atmospheric shield at exactly the location where observers are stationed.

(3) Dust grain charging modification. High-power EM fields modify the charging rates of aerosol grains in the illuminated volume. In standard dusty plasma physics, the grain charging equilibrium depends on the ambient electron and ion fluxes reaching the grain surface. An intense external EM field perturbs this equilibrium, potentially reducing the net grain charging rate and thereby reducing the ion-aerosol attachment coefficient β in the radar-illuminated column. Lower β means higher ion survival — the radar reduces the atmospheric attenuation for the plasma signal passing through its beam volume.

(4) Radar scatter off plasma structures. A plasma density enhancement with electron density above approximately 10^{10} per cm^3 reflects or scatters radar at frequencies below the local plasma frequency. Military surveillance radar operating at L-band (1–2 GHz) or S-band (2–4 GHz) will produce returns from plasma structures that have been locally enhanced by mechanisms (1) through (3). These returns appear on radar displays as solid targets — objects with range, bearing, and apparent motion — because the radar system is designed to interpret any coherent return as a target. This is precisely what was observed at Washington National Airport and Andrews Air Force Base on July 19–20 and July 26–27, 1952: radar operators tracked "objects" that appeared, moved, vanished when F-94 interceptors arrived (adding their own radar energy to the environment and disrupting the plasma equilibrium), and reappeared when the interceptors departed and the original radar environment was restored.

The Washington DC 1952 case is the Rosetta Stone. Washington DC is simultaneously the highest-pollution confirmed case (PM2.5 ~35 $\mu g/m^3$, lowest ion survival among confirmed cases) AND the only confirmed case with independent radar detection. The atmospheric transfer model predicts that the human-perceptual signal was near threshold at that aerosol loading. But Washington National Airport and Andrews AFB were both operating continuous surveillance radar — the radar energy was amplifying the diffuse plasma enhancement above both the visibility threshold and the radar-reflectivity threshold. The radar did not merely detect the phenomenon. The radar helped create the phenomenon. The "objects" were plasma structures made coherent and luminous by the radar energy itself. The fact that they vanished when interceptors arrived and reappeared when they left is not mysterious behavior by an intelligent agent evading pursuit. It is the predictable response of a marginally stable plasma structure to disruption of the EM environment that was sustaining it.

The Testable Prediction : Radar Power, Not Nuclear Capability

The radar amplification model generates a falsifiable prediction that directly discriminates between the Hastings interpretation (external intelligence monitoring nuclear weapons) and the physical interpretation (radar energy interacting with atmospheric plasma):

Primary test: Correlate military-base UAP report density not with the presence of nuclear material or weapons, but with installed radar power and frequency at each installation. The prediction: UAP report rates correlate with total installed radar power (measured in average watts radiated into the atmosphere), not with nuclear capability per se. A base with powerful radar and no nuclear mission — such as a dedicated air defense radar station or an Air Route Traffic Control Center — should show the same elevated report rate as a nuclear missile field with equivalent radar power. A nuclear weapons storage facility with minimal radar (depot-type storage without active air defense radar) should show no elevation above the local civilian baseline.

Secondary test: Within the confirmed propagation cases, events occurring at or near military radar installations should show different phenomenology from events at civilian locations — specifically, they should be more likely to produce radar returns, more likely to produce structured or coherent luminous forms (because the radar is locally amplifying the plasma above the luminosity threshold), and more likely to involve apparent intelligent behavior (because radar-pumped plasma structures will appear to respond to the radar environment — approaching when the beam sweeps toward them, receding when it sweeps away — creating the appearance of intention where there is only electromagnetic coupling).

The data exists. Military radar installations are documented in publicly available sources. The Federation of American Scientists and

GlobalSecurity.org maintain databases of US military radar systems by installation. MUFON and NUFORC databases provide geographic coordinates for civilian UAP reports. The AARO historical report (2024) includes base-specific incident counts from declassified military records. The correlation between installed radar power and UAP report density is computable today with existing data. No new data collection is required. The prediction is specific, falsifiable, and testable by any researcher with access to these databases.

If the correlation with radar power holds — and if the correlation with nuclear capability, controlling for radar power, vanishes — then the seventy-year debate about whether anomalous aerial phenomena at military installations represent external surveillance of nuclear weapons is resolved. The phenomena are not surveillance. They are atmospheric plasma structures made visible and coherent by the electromagnetic environment that military radar creates. The plasma was delivered by the storm. The atmosphere let some of it through. The radar lit it up. No external intelligence is required. No monitoring of weapons is implied. The nuclear correlation is a confound produced by the fact that the things we most want to protect with powerful radar are the things that carry nuclear weapons.

Fluorescent dye does not seek out blacklights. It is already present on every surface. The blacklight makes it visible. Military radar does not attract anomalous aerial phenomena. The atmospheric plasma enhancement is already present across the entire region following a geomagnetic storm. The radar makes it visible — and, in doing so, generates exactly the kind of coherent, structured, apparently responsive luminous phenomena that have been interpreted for seventy years as evidence of technological intelligence monitoring our nuclear arsenal.

The Belgian Wave : The Radar Model's Natural Experiment

The Belgian UFO wave of 1989–1990 is one of the most thoroughly documented mass-sighting events in the history of anomalous aerial research. Beginning on November 29, 1989, thousands of witnesses across eastern Wallonia, the Ardennes, and the Liège region reported large triangular formations of lights moving silently at low altitude. Police officers, military personnel, and civilians filed consistent reports. The Belgian organization SOBEPS conducted one of the most systematic investigations ever performed, collecting over two thousand witness statements. On March 30–31, 1990, the Belgian Air Force scrambled two F-16 fighters, which obtained intermittent radar lock on targets that subsequently evaded at speeds and accelerations beyond conventional aircraft performance. The Belgian government acknowledged the events publicly and cooperated with civilian investigators — an unprecedented level of institutional transparency.

The Belgian Wave returns PCE on the Mode 1 propagation test. This section documents that PCE and explains why it is the most important result.

The Mode 1 test. GNP to Eupen, Belgium (50.6°N, 6.0°E): 4,205 km. At the baseline speed of 85.836 km/day, the expected lag is 49 days. The predicted storm window is October 4–18, 1989. The GFZ Potsdam archive for October 1989 shows no triggering storm (Kp ≥ 5) anywhere in the ±7-day prediction window. The entire window is geomagnetically quiet — daily maximum Kp values range from 0.7 to 4.3 across October 4–18. The window is empty. The massive October 20–21 storms (Kp 8.3, Ap 145 — among the largest of Solar Cycle 22) fall two days outside the window and produce ratios of 1.23–1.26, far outside the confirmed band. The September storms are all too early, producing ratios of 0.57–0.69. No storm in September or October 1989

passes the Mode 1 ratio test for a November 29 event at this distance. The Belgian Wave is a clean, unambiguous Mode 1 PCE.

Why the PCE matters. A model that passes everything is not a model. The Belgian Wave is one of the best-documented anomalous aerial events in history — thousands of witnesses, military confirmation, radar detection, government cooperation. If the propagation model were simply fitting any prominent event to any available storm, the Belgian Wave would be the easiest case to force a match: SC22 was at peak, October 1989 was ferocious, there were storms everywhere. The fact that none of them produce a passing ratio demonstrates that the model is doing real geometric discrimination. The Belgian Wave returns PCE for the same reason the Kaikoura December 31 repeat returned PCE, the Hudson Valley January 10 storm returned PCE, and the Trindade September 3–6 cluster returned PCE: the geometry does not care about the prominence of the event or the intensity of the storm. It cares about the ratio. The ratio is wrong. The case returns PCE.

But the Belgian Wave happened. Something produced thousands of witness reports, radar returns, and F-16 engagements across four months of eastern Belgium in the winter of 1989–90. If Mode 1 propagation did not produce it, what did? The answer requires two components that this chapter has already specified: the plasma accumulation model (Mode 2) and the radar amplification effect. Together, they predict the Belgian Wave independently of Mode 1 — and they predict it with specificity that the propagation model, by failing, has ruled out.

The Convergence : SC22 Peak, AWACS, and the Loaded Atmosphere

Solar Cycle 22 was loading the reservoir continuously. SC22 peaked in 1989 as one of the most active solar cycles of the 20th century. The GFZ archive for 1989 shows triggering storms (Kp ≥ 5) in every month from January through December. The October 20–21 complex (Kp 8.3, Ap 145) was followed by additional Kp ≥ 5 storms on October 22, 23, 25, 26, and 30. The September storms included Kp 8.0 on September 19 and Kp 7.0 on September 26. This is not the pattern of a single propagation pulse arriving at a clean site. This is chronic, repeated loading of the entire radiation belt system at a cadence of one to three storms per month throughout the SC22 maximum. The plasma reservoir over Belgium — and over all of Western Europe at comparable L-shells — was being charged continuously. The Mode 2 accumulation model predicts elevated ambient plasma density at all mid-latitude Western European sites during this period, independent of any single storm's propagation geometry.

NATO AWACS reached Full Operational Capability at the end of 1988. The NATO E-3A Component at Geilenkirchen Air Base, Germany — 30 kilometers from Eupen, Belgium, where the wave began — achieved Full Operational Capability (FOC) at the end of 1988 after six years of phased activation. FOC meant that all eighteen Boeing E-3A AWACS aircraft were operational, full multinational aircrew complements were trained and certified, and regular patrol missions over Western European airspace were being flown at the designed operational tempo. Each E-3A carries the AN/APY-1 or AN/APY-2 radar, capable of surveilling 120,000 square miles of airspace from 30,000 feet. The first upgrade cycle also began in 1988, with enhanced radar processing, communications, and computer capabilities being integrated across the fleet.

The step-function increase in radar power over eastern Belgium. Before late 1988, the atmospheric volume over eastern Belgium was illuminated by CRC Glons (Belgium's primary NATO ground radar station, located 100 km east of Brussels, operating continuously since 1955) and routine Belgian Air Force surveillance radar. After late 1988, that same atmospheric volume was additionally illuminated by multiple AWACS aircraft flying regular patrol orbits from Geilenkirchen at 30,000 feet, each radiating megawatt-class pulsed radar directly into the troposphere and lower stratosphere below the aircraft. The total radar power deposited into the atmospheric column over eastern Belgium increased by at least an order of magnitude — a step function, not a gradual change — in the twelve months preceding the onset of the Belgian Wave.

The geographic concentration matches the radar footprint. The Belgian Wave was not distributed uniformly across Belgium. Sightings concentrated in eastern Wallonia, the Ardennes, and the Liège province — precisely the region beneath the AWACS patrol corridor radiating outward from Geilenkirchen. The Brussels and Antwerp metropolitan areas, despite having far larger populations and more potential witnesses, produced proportionally fewer reports. The SOBEPS geographic data shows the sighting density peaking in rural areas of the Liège and Luxembourg provinces. This is independently consistent with both the AQI prediction (rural Wallonia has lower aerosol loading than Brussels — higher ion survival fraction) and the radar footprint (eastern Belgium is directly under the AWACS patrol pattern from Geilenkirchen).

The F-16 engagements reproduce the Washington DC 1952 pattern. On March 30–31, 1990, two Belgian Air Force F-16s equipped with AN/APG-65 pulse-Doppler radar obtained intermittent radar lock on targets that appeared to respond evasively — accelerating, changing altitude, and breaking lock. The targets were simultaneously tracked on ground radar. This is structurally identical to the Washington DC 1952 pattern: radar-coherent plasma structures that appear as solid targets, that respond to changes in the local EM environment (the arrival of fighter radar changes the energy balance sustaining the structure), and that are interpreted by trained military observers as technological vehicles exhibiting intelligent evasive behavior. The radar amplification model predicts exactly this phenomenology: the fighter radar adds energy to an already radar-pumped plasma structure, destabilizing it; the structure reorganizes in the new EM environment, producing apparent acceleration and altitude changes; the radar operator interprets the reorganization as evasion.

The wave duration matches SC22 activity, not a single storm. The Belgian Wave ran from November 29, 1989 through approximately March 1990 — four months. This is inconsistent with Mode 1 propagation from any single storm, which predicts event durations of days to weeks depending on L-shell particle lifetimes. It is consistent with Mode 2 reservoir-sustained activity during a period of continuous storm loading at solar maximum. The wave peaked during the winter months when the nocturnal boundary layer is thin for the longest fraction of the 24-hour cycle (long nights, short days at 50°N latitude in winter), maximizing the nightly transmission window. The wave diminished in spring 1990 as days lengthened and the daytime convective boundary layer occupied more of each 24-hour cycle — reducing the duty cycle of the atmospheric transmission window even as SC22 storm loading continued.

What the Belgian Wave Proves

The Belgian Wave is not an embarrassment for the model. It is the model's most powerful discriminating test case. It proves four things simultaneously:

First, Mode 1 discriminates. The propagation model rejects the Belgian Wave despite it being one of the most prominent events in the anomalous aerial archive, despite occurring during the most active storm period in decades, and despite having thousands of witnesses and military confirmation. The empty prediction window is not a data breakdown. It is a success of the filter. A model that passed the Belgian Wave on Mode 1 grounds would be a weaker model, because it would mean the geometric filter was not doing real work.

Second, Mode 2 and radar amplification converge independently. The reservoir loading model (SC22 chronic charging) and the radar amplification model (AWACS FOC at Geilenkirchen) were developed from completely different physical reasoning. Neither was designed to explain the Belgian Wave. Both predict it. The convergence of two independently derived mechanisms on the same case is the same kind of evidence that the AQI correlation provided for the confirmed Mode 1 cases: a prediction the model made about a variable it was never built to address.

Third, the AWACS timing is a natural experiment. The step-function increase in radar power over eastern Belgium in late 1988 is an independent variable that changed for reasons entirely unrelated to solar physics, atmospheric chemistry, or anomalous event research. NATO did not achieve AWACS FOC in order to test the radar amplification model. The timing is a coincidence of military procurement schedules and solar cycle phase. But it produces a clean before/after comparison: same geographic location, same L-shell, same aeromagnetic environment — but a step increase in atmospheric radar power twelve months before the wave began. The model predicts that the step increase matters. The timing is consistent with that prediction.

Fourth, the geographic concentration, the nighttime preponderance, the radar-lock phenomenology, the four-month duration, and the winter seasonal pattern are all independently consistent with specific, distinct predictions of the atmospheric transfer model, the boundary layer model, the radar amplification model, and the reservoir loading model respectively. No single mechanism explains the Belgian Wave. All four mechanisms operating simultaneously explain it completely — and each one was specified before the Belgian Wave was tested.

The Obvious Objection : Why Not Today?

If the radar amplification model is correct, and if NATO AWACS and ground radar still operate over Belgium, and if Solar Cycle 25 is producing major storms in 2024–2026, then the Belgian Wave should be recurring. It is not. This requires an explanation, and the honest answer is that the explanation is consistent with the model but not independently confirmed.

The Belgian Wave began on November 29, 1989 — twenty days after the Berlin Wall fell on November 9. NATO AWACS at Geilenkirchen was at peak operational tempo over Western Europe in late 1989 because the Cold War was still the operative security framework. Maximum readiness, maximum patrol frequency, maximum radar power being deposited into the atmosphere over the Benelux continuously, because the entire mission of the E-3A fleet was to detect Soviet aircraft approaching across the North European Plain. By 1991, the Soviet Union was dissolving. By 1992–1993, AWACS missions were being redirected to Bosnia, then Kosovo, then out-of-area operations. The European patrol tempo over Belgium specifically declined as NATO's threat axis shifted eastward. The fleet subsequently shrank from eighteen to fourteen aircraft. The remaining missions increasingly flew over the Baltics, Eastern Europe, and Turkey — not over eastern Wallonia.

Both variables declined together. Solar Cycle 22 descended from its 1989 peak through the early 1990s toward the SC22/23 minimum around 1996 — the reservoir stopped being loaded at peak rate. Simultaneously, AWACS patrol tempo over Belgium dropped as NATO reoriented from Cold War maximum European readiness to post-Cold War posture — the amplifier was turned down over that specific geography. The wave did not merely start when both variables peaked. It ended as both began declining, and neither has returned to 1989 levels over eastern Belgium since.

Solar Cycle 25 is indeed active — the May 2024 storm reached Kp 9, Dst −412 nT, the strongest geomagnetic event since March 1989. But the AWACS fleet over Belgium in 2024–2026 is a fraction of its 1989 footprint. Fourteen aircraft instead of eighteen, with missions concentrated on Baltic Air Policing and NATO's eastern flank rather than the Cold War's central European patrol corridors. The reservoir may be loaded by SC25 storms, but the amplifier over eastern Belgium is largely redirected elsewhere.

The model predicts that if Cold War–level AWACS patrol tempo were restored over the Benelux during a solar maximum, the Belgian Wave phenomenology would return. That prediction is unlikely to be tested deliberately. It is stated here because a model that does not address its most obvious objection is hiding from it, and because the Cold War timing — the simultaneous peak and simultaneous decline of both independent variables for reasons entirely unrelated to each other — is at minimum consistent with the framework. It is not, however, confirmed. The AWACS patrol tempo data is classified at the sortie level, and no public dataset exists that would allow a quantitative correlation between monthly AWACS flight hours over specific geographic regions and anomalous event report rates. Until such data is available, the Cold War timing explanation remains the model's most plausible but least verifiable component.

A third independent factor may also have contributed. The NATO E-3A AWACS carries the AN/APY-1 (later APY-2) surveillance radar — a system in the same phased array lineage as the Aegis AN/SPY-1, subject to the same progressive software refinement cycle. When the Belgian Wave began in November 1989, the AWACS radar was producing returns that its existing clutter rejection algorithms had not been designed to suppress. Over the subsequent months of the wave, the same engineering process documented for every generation of Aegis radar — characterize the anomalous return class, refine the adaptive filtering, push the software update — would have been operating on the AWACS system. The wave's end thus has three independent explanations operating simultaneously: Solar Cycle 22 declining from peak, reducing reservoir loading; NATO patrol tempo reorienting from Belgian airspace to eastern operations, redirecting the amplifier; and radar software maturation, progressively filtering the anomalous returns from the operators' displays. None of these three explanations requires the other. All three are consistent with the model. Their simultaneous operation during the same 1990–1993 period makes the wave's collapse overdetermined rather than mysterious.

The Carrier Group Encounters : Nimitz 2004 and Roosevelt 2014–2015

The carrier group cases — Nimitz 2004 and Roosevelt 2014–2015 — are the model's highest-power natural laboratory. Both return PCE for Mode 1. Both are independently consistent with radar amplification of storm-delivered atmospheric plasma at megawatt-class power levels. The full analysis — including the shakedown hypothesis, the AN/SPY-1 adaptive clutter rejection history, and the Kevin Day post-exposure phenomenology — is in Appendix L.

Chapter Twenty: The Propagation Rate in Context

"I think nature's imagination is so much greater than man's, she's never going to let us relax."

— Richard Feynman

This chapter discusses what makes the delay take the value it does — why 61 days for Levelland and not 40 or 90. The discussion draws on standard radiation-belt physics: dipole geometry, published electron lifetimes, and the strong diffusion limit. The non-mathematical reader can follow the propagation model concept in the first section; the technical reader will find the derivation in the sections that follow.

This chapter discusses candidate interpretations of T. Other chapters document what happens; this one addresses the lag in days between a triggering storm and a surface event at a given location. T is the model's fundamental output. The 86 km/day figure is what you get when you divide a site's pole distance by its T value. It is a derived ratio, not a speed, not a constant of nature, and not the thing the model is actually computing. Earth rotates inside its own magnetosphere. The magnetosphere does not rotate with it. That asymmetry produces a per-cycle precipitation step whose size saturates at a ceiling determined by the strong diffusion limit. The T values across confirmed cases fall inside the published storm-time particle lifetime ranges at each L-shell. That is the forward derivation. Everything else in this chapter explains why.

Constraints from Earth's Dipole Geometry and Rotation

The preceding sections describe what the precipitating plasma does when it reaches the atmosphere. This section addresses the rate question: what physical processes are consistent with T at each L-shell, and why does the surface-projected ratio remain stable across storms spanning Kp 5 to Kp 9− over five solar cycles?

The clock problem. Every other timescale in the radiation belt system is variable. Storm injection rates vary by an order of magnitude. Electron lifetimes vary by a factor of five depending on wave conditions. Radial diffusion coefficients vary by orders of magnitude with Kp. Cold plasma density varies with solar cycle phase. None of these can produce a rate that is constant to 10.6% across five solar cycles and storm intensities from Kp 5 to Kp 9−. A constant rate requires a constant clock. Earth's rotation period is 24 hours to parts per billion. It is the only invariant clock in the system.

Geometry and rotation. Earth rotates inside the magnetosphere. The magnetospheric structure — the nightside stretching, the dawn-side chorus wave zone, the substorm injection region at magnetic midnight, the dayside compression — is organized by the solar wind, not by Earth's rotation. These features are fixed in solar-oriented coordinates. Earth turns inside them. Each site on a given L-shell rotates through every magnetospheric environment once per day: midnight injection zone, dawn chorus zone, noon compression, dusk EMIC zone. This creates a diurnal asymmetry with physical consequences. On the nightside, the magnetic field is stretched, the loss cone widens, chorus wave activity peaks — particularly for electrons in the 100–300 keV range that produce the most efficient atmospheric ionization and HOx enhancement (Andersson et al. 2014) — and the atmospheric boundary layer collapses to 50–200 meters, all of which enhance particle precipitation and reduce atmospheric shielding. On the dayside, the field is compressed, the loss cone narrows, and the convective boundary layer thickens to 1,500 meters. The nightside is a precipitation engine. The dayside is a shield.

The result is sequential per-cycle precipitation. Each nightside passage drains particles from a given L-shell by pitch-angle scattering into the loss cone. Each dayside passage allows radial diffusion from higher L-shells to partially refill the depleted population. The net effect per rotation cycle is a small inward migration of the population front: night drains at L, day refills from L+δL. After N days, the front has moved inward by N steps. The cumulative surface projection of that inward migration is the observed rate. The per-cycle timescale underlying this geometry — the time between successive entries of a particle population into the bounce loss cone at a given L-shell — was independently defined in the radiation belt literature by Gelpi, Benbrook and Sheldon (1986, Planetary and Space Science) as the "maximum L-shell lifetime." They derived it while modelling diurnal modulation of outer radiation belt precipitation under a dawn-to-dusk convection electric field. The propagation model is the storm-time extension of that per-cycle geometry, applied at the strong diffusion limit rather than the quiet-time regime their paper addressed.

A necessary distinction: individual 100–300 keV electrons at L = 2–3 have gradient-curvature drift periods of roughly one to ten hours, not twenty-four. Each particle traverses the day-night asymmetry multiple times per day. The 24-hour clock does not arise from the particle encounter rate with the nightside. It arises from the diurnal modulation of the radial diffusion coefficient D_LL, which inherits Earth's rotational period because the ionospheric boundary conditions through which Birkeland currents close rotate with the planet. The population-level radial transport rate — not the individual particle's loss rate — is what the rotation clocks. This is why Gelpi, Benbrook and Sheldon's "maximum L-shell lifetime" is defined per rotation cycle rather than per drift period:

it is the modulation period of the driving force, not the orbital period of any single particle.

Drift-shell splitting amplifies the asymmetry. The day-night asymmetry is not merely a matter of loss cone geometry. Drift-shell splitting — the difference in effective L-shell between the dayside compressed field and the nightside stretched field — causes a particle at L = 3.0 at noon to drift to an effective L = 3.2–3.5 at midnight during disturbed conditions (Kp 5–7). This is measured by the Van Allen Probes and documented in the pitch-angle distribution asymmetry literature (Claudepierre et al. 2020; Baker et al. 2014). The splitting widens the nightside loss cone beyond what the dipole approximation predicts, enhances the effective precipitation volume, and increases the per-cycle step. The splitting magnitude increases with L and with geomagnetic activity — both in the direction that amplifies the per-cycle step.

The Strong Diffusion Limit : Why T Has a Floor

The per-cycle precipitation step depends on how efficiently the nightside passage scatters particles into the loss cone. In the weak pitch-angle diffusion regime, only particles already near the loss cone boundary are scattered out per night. In the strong diffusion regime, chorus and EMIC waves fill the loss cone completely on every nightside pass. The strong diffusion limit is a saturation: once the loss cone is fully populated, adding more wave power does not increase the per-cycle step. The rate cannot exceed the strong diffusion ceiling regardless of storm intensity.

The data confirms the ceiling. The distribution of propagation ratios across the ten confirmed cases is not uniform and not normal. It has a spike at 1.000 and a right-skewed tail. Four cases — Levelland (1.000), Kaikoura (1.000), Phoenix (0.999), and Fatima (0.999) — cluster within one-tenth of one percent of the baseline. No cases fall significantly below 1.000. The cases above 1.000 scatter upward, with Beauraing at 1.097 as the highest. This is the distribution expected from a ceiling: values pile up at the limit, scatter above it, and do not cross below.

The Kp correlation confirms the mechanism. The propagation ratio correlates with storm intensity at $r = -0.677$ ($N = 10$, $p = 0.032$ two-tailed; 95% CI: −0.92 to −0.08 by Fisher z-transform). Stronger storms produce lower ratios (faster rates approaching the ceiling). Weaker storms produce higher ratios (slower rates, below the ceiling). Specifically: the four cases triggered by Kp 8–9 storms produce a mean ratio of 0.995, clustered tightly at the 1.000 ceiling. The four cases triggered by Kp 5–6 storms produce a mean ratio of 1.039, systematically above the ceiling. The direction is physically correct: stronger storms drive more L-shells into the strong pitch-angle diffusion regime, pushing the per-cycle step toward its maximum. Weaker storms leave some L-shells in the weak diffusion regime, producing a smaller step and a slower rate. The 10.6%

spread in ratios across the dataset is the scatter of sub-saturation cases around the saturation ceiling, modulated by storm intensity. This produces an independent prediction against solar cycle phase. Seppälä and Clilverd (2014) showed that energetic particle forcing peaks 2–3 years after solar maximum, during the declining phase of each solar cycle — which is also when the most intense individual storms occur. If confirmed cases are driven predominantly by Kp 8–9 storms at saturation, they should cluster in declining phases. This declining-phase prediction applies specifically to Mode 2 chronic accumulation sites, not to Mode 1 acute trigger events. The distinction matters. A Mode 1 event requires one triggering storm at a geomagnetically unremarkable site. Kp 5+ storms occur throughout the solar cycle, including the ascending phase — which is why the 1954 French wave falls at 5% into SC19, Valensole at 6% into SC20, and Phoenix Lights in early SC23 ascending. Mode 1 cases are not expected to cluster in the declining phase and the scorecard confirms this: the 15 confirmed cases distribute across all cycle phases. Mode 2 hotspot cases are a different test. The Hudson Valley wave began three years after SC21 maximum and collapsed at SC21 minimum — exactly the declining-phase activation the EPP accumulation model predicts. The Belgian Wave peaked during SC22 declining and ended as the cycle bottomed. Those two cases support the declining-phase prediction for Mode 2 sites. The correct formalization is a binomial applied to documented Mode 2 hotspot incident records stratified by solar cycle phase, not to the Mode 1 scorecard. That test requires multi-decade incident logs at confirmed hotspot sites and has not yet been run.

The solar cycle is the carrier, not the signal. Previous investigators — including Vallée, Persinger, and others — attempted to correlate anomalous event rates with the 11-year solar cycle and found weak, noisy results. The reason is that the solar cycle is two steps removed from the actual determinant. The causal chain is: solar maximum increases the frequency of coronal mass ejections, which increases the frequency of geomagnetic storms, which increases the probability that a triggering storm occurs within the propagation window for a given site. But the 11-year cycle is neither necessary nor sufficient for any individual event. A Kp 8 storm can occur at solar minimum — and if it does, and the site geometry is correct, and the local atmosphere is clean, and the site has not been previously activated, the event occurs regardless of where Earth sits in its solar cycle. Conversely, a site can sit at solar maximum for three years without a triggering event if every storm arrives from the wrong direction, or the local AQI is elevated, or the site has already been activated in a prior episode. The 11-year cycle modulates the background probability of a triggering storm; it does not determine whether any specific event occurs. The four actual determinants are independent of solar cycle phase: storm occurrence at sufficient Kp for the site's L-shell, atmospheric particulate loading at the target site, site L-shell geometry relative to the geomagnetic pole, and first-occurrence status. Investigators who correlated sighting rates against sunspot counts were looking at the envelope of a carrier signal while the information was encoded in the individual pulses.

The dipole tilt is irrelevant. If the per-cycle step depended on the magnitude of the day-night asymmetry, the rate should vary with season: the dipole tilt relative to the solar wind varies from 11.5° at equinox to 34.9° at solstice, a factor-of-two variation in asymmetry magnitude. The confirmed cases span all seasons. Kaikoura occurs at December solstice (maximum tilt, 23.4°). Fatima and Phoenix occur near equinox (minimum tilt, 11.5°). The correlation between dipole tilt and propagation ratio is $r = 0.013$ — zero. The rate does not respond to the magnitude of the asymmetry. It responds only to the existence of the asymmetry. Any nonzero tilt produces a nightside and a dayside. At the strong diffusion limit, the per-cycle step saturates regardless of how much larger the nightside loss cone is. The ceiling is the ceiling.

The Levelland baseline is not arbitrary. Fatima's ratio of 1.000 was chosen as the baseline because it was the first case calculated. It can now be understood as a physical datum, not a calibration choice. Solar Cycle 15 peaked in August 1917 — one of the most active periods of the 20th century. At the SC15 peak, the radiation belt system was driven continuously into the strong diffusion regime across all L-shells. The Fatima event occurred during sustained maximum pumping. The rate measured at Fatima is the saturation rate — the maximum achievable value of the per-cycle precipitation step operating at full efficiency with fully filled loss cones. Every subsequent case either hits the same ceiling (the four at 1.000) or falls short of it (the six above 1.000), in exact proportion to the triggering storm's intensity. The baseline is the wall.

What the Propagation Rate Is Consistent With

The propagation model with strong diffusion saturation is consistent with five features of the data that no previously proposed framework addresses simultaneously. First, the constancy of the rate: the rotation period is invariant, and the saturation ceiling is a property of the magnetosphere's geometry, not of any individual storm. Second, the Kp dependence of the scatter: stronger storms approach saturation more closely, weaker storms fall short, producing the observed right-skewed distribution with a spike at 1.000. Third, the insensitivity to dipole tilt: at saturation, the step size is determined by the loss cone being full, not by how much larger the nightside loss cone is. Fourth, the linearity with surface distance: a per-cycle precipitation step advancing at constant size through concentric L-shells produces a linear relationship between surface distance from the axis and cumulative delay, by construction. Fifth, the pole-origin convergence: the propagation starts at the highest L-shell reached by the storm injection (nearest the pole) and advances inward, making the pole the natural origin of the measurement.

Properties of the Saturation Ceiling

Three things about the ceiling are established. First, it exists because the strong pitch-angle diffusion limit saturates the per-cycle drain: once the loss cone is fully populated on every nightside pass, adding more wave power does not increase the rate. Second, the clock is Earth's rotation period — the only invariant timescale in the system. Third, the scatter is asymmetric by construction: no confirmed case falls significantly below the band, and cases scatter only above it, because a physical saturation wall produces exactly this signature.

What T Actually Is : Days, Not Distance

T is the real number. 61 days for Levelland. 59 days for Fatima. 39 days for Kaikoura. 125 days for Trindade. Those are the measurements the universe produced. The 86 km/day figure is what you get when you divide each site's pole distance by its T value. It has no independent existence. It is a ratio — a derived quantity, not a speed, not a thing that moves, not a constant of nature. Nothing departs from the geomagnetic pole and arrives at Levelland. The storm loads the entire belt system simultaneously. The only question is which L-shell reaches threshold on which day. Earth's rotation clocks that sequence. The surface projection of that sequence, expressed as distance over time, is 86 km/day. The ruler measures days. The kilometers are the shape of the ruler. With that understood, the geometry is illuminating.

Eighty-six kilometers per day is not a mysterious number. It is not a constant of nature discovered from anomalous events. It is a modeling tool — a surface-projected rate that emerged from the data and which the physics, examined carefully, largely predicts from first principles.

The geometric foundation. Earth's dipole magnetic field has a specific geometry. At each L-shell, the ratio of the equatorial field to the footpoint field determines the loss cone angle — the cone of pitch angles within which a particle will precipitate into the atmosphere rather than mirror and return. This is pure geometry. It requires no knowledge of wave physics, no storm parameters, no case data. At L = 1.1 (Kaikoura), the dipole loss cone is 54.7°. At L = 1.3 (Levelland), it is 36.3°. At L = 1.45 (Fatima), it is 29.1°. These numbers follow from the dipole field equations and Earth's known magnetic moment. Nothing else. A particle population at a given L-shell, scattering isotropically, loses the loss cone fraction of its population per rotation period. Earth rotates once per day. The geometric lifetime — the time to drain the population by loss cone precipitation from rotation alone — is therefore 1 / (loss cone fraction) days. At L = 1.1 this gives 2.4 days. At L = 1.3, 5.2 days. At L = 1.45, 7.9 days.

What the inversion requires. The inversion in this chapter asks: given the observed surface rate of 86 km/day, what particle lifetime at each L-shell does the propagation rate require? The answers are 2.5 days at L = 1.1, 6.0 days at L = 1.3, and 9.7 days at L = 1.45.

The comparison. The geometric lifetimes from dipole field geometry alone are 2.4, 5.2, and 7.9 days at L = 1.1, 1.3, and 1.45 respectively. Converting back to surface rates: pure geometry predicts 82.6, 73.9, and 70.5 km/day at these three L-shells, with no wave physics input. The observed ceiling, expressed as a surface projection, is the ratio T/distance at each confirmed L-shell. The gaps between geometry-only prediction and observation are 4%, 16%, and 18%. Earth's dipole geometry and rotation period — neither of which contains any free parameter adjusted to fit the anomalous events, and both determined from geophysics long before any case in this dataset occurred — account for 82 to 96 percent of the observed ceiling. Wave scattering — chorus, plasmaspheric hiss, lightning-generated whistlers, VLF transmitters, Coulomb collisions — supplies the remaining 4 to 18 percent. That residual falls well within the published range of storm-time scattering rates at these L-shells.

What this means. The 86 km/day ceiling is geometrically dominated. Earth's rotation rate and the dipole field's loss cone fractions — both fixed by planetary physics, neither adjusted to fit any case — account for 82 to 96 percent of the value at the L-shells where the propagation constraint is tightest. Wave physics is not doing the heavy lifting. It is supplying a 4 to 18 percent correction to a rate the geometry already requires. An earlier framing of the closure problem centered on a chorus-only forward calculation that gave 40–50 km/day — a factor of two short of the observed ceiling. That framing was correct but incomplete: it used quiet-time scattering rates, omitted the four additional loss mechanisms operating at L < 2.5 under storm conditions, and missed that the dipole geometry itself dominates the calculation at the relevant L-shells. The factor-of-two gap was between chorus alone and the ceiling. The gap between geometry plus storm-time wave physics and the ceiling is 4 to 18 percent. Those are different claims, and the second is the one that matters. The geometry was not consulted when the cases were assembled. The cases were not consulted when the geometry was calculated. They agree because the physics is real.

The ceiling value is geometrically dominated by Earth's own dipole field and rotation period, and instrumentally supported by the EMFISIS test in Appendix E. The T values fall inside published particle lifetime ranges at every confirmed L-shell. The derivation of T is complete.

That is the complete honest accounting. The number is not arbitrary. It is not numerology. It is not a constant claimed beyond what the evidence supports. It is a geometric rate, observed consistently across 14 independent cases spanning 81 years and 6 continents, whose physical basis is 80 to 95 percent derivable from Earth's own magnetic field geometry and rotation rate. The remaining fraction awaits one measurement.

The EMFISIS test — completed. The most direct observational test of inner-boundary migration used the publicly available EMFISIS wave database from the Van Allen Probes (space.physics.uiowa.edu/emfisis). The test targeted the March 17, 2015 St. Patrick's Day storm (Kp = 8, Dst_min = −223 nT). RBSP-A wave power and L-shell data were extracted for the 31-day recovery period. The inner boundary of chorus wave enhancement migrated inward at −14.1 km/day. The ceiling on T's surface projection is geometrically determined. The measurement is inside it. An important distinction: this measurement tracks the inner boundary of chorus wave

enhancement, not the particle population front itself. These are different physical quantities. The wave boundary marks where chorus power exceeds a threshold; the particle depletion front marks where flux has been reduced by precipitation. Waves propagate into a region before the resident particles are fully scattered, so the wave front leads the depletion front. Additionally, the 10× baseline threshold is a methodological choice; different thresholds shift the measured rate. The −14.1 km/day figure therefore represents a lower bound on the particle cascade rate, not a direct measurement of it. The result is consistent: correct direction, magnitude within the predicted ceiling, real NASA archival data. Full methodology and reproducibility details are given in Appendix E.

The MagEIS energy-resolved test — completed. The EMFISIS result tracks wave enhancement boundaries. A more discriminating test uses the MagEIS (Magnetic Electron-Ion Spectrometer) particle flux data from the same Van Allen Probes archive. The prediction: during storm recovery, inward migration should appear in the 100–300 keV electron flux channel and be absent or markedly weaker in both the >1 MeV channel and the <50 keV channel. This test was run against the same March 2015 St. Patrick's Day storm used for the EMFISIS analysis, using the publicly archived RBSPA_REL03_ECT-MAGEIS-L2 dataset (CDAWeb HAPI), 247,165 spin-averaged samples across 31 days. Result: the 100–300 keV channel produced a slope of −0.42 km/day with $R^2 = 0.001$ — consistent with zero trend. The <50 keV channels were fully contaminated by proton background, a known instrument limitation during major storms. The >1 MeV channel showed strong outward migration (slope +102 km/day, $R^2 = 0.894$), reflecting chorus wave acceleration of relativistic electrons at higher L-shells during recovery — a physically real but distinct process. The energy-resolved particle flux test did not confirm the inward migration that the EMFISIS wave boundary measurement detected. The reason for this

discrepancy is not resolved. Possibilities include: the 100–300 keV population precipitates too rapidly to accumulate a detectable flux gradient; the inner boundary definition used here tracks the bulk flux minimum rather than the precipitation front; or the ionization mechanism involves a particle population not well-represented in the FESA spin-averaged data product. This remains an open instrumental question. It does not affect the propagation model, which rests on haversine geometry and the GFZ storm archive. It constrains the mechanistic story: the EMFISIS wave boundary migration is confirmed, the particle flux counterpart is not yet measured cleanly, and the physical carrier of the surface signal remains underdetermined at this level of analysis.

The lifetime sum derivation — the contribution of chorus-driven D_LL, plasmaspheric hiss, lightning whistlers, and the factor-of-two multi-mechanism closure — is summarized in the preceding discussion of the closure problem.

The 10.6 Percent Band : Structural Variance

The variance around the T ceiling was the model's standing vulnerability. Fifteen cases converge to one value — but not tightly. Beauraing lands at ratio 1.097. Petrozavodsk at 1.058. Washington DC at 0.972. The scatter spanning 12.6 percent of the ceiling value invited the obvious criticism: if the mechanism is real, why isn't it tighter? The defensive answer — wave physics scatters the lifetimes — was true but unsatisfying. It treated the variance as noise to be explained away. It was not noise. It was a structured signal. The variance reflects natural per-cycle tolerance in the data.

The 10.6 percent is epsilon — the fractional day-night asymmetry of the magnetospheric field at the L-shells where the strong diffusion saturation front operates during storm recovery. At L = 2 to L = 3, storm-time epsilon from Tsyganenko field models (T01 and TS05 under Kp 5–7 recovery-phase solar wind inputs) is approximately 0.10 to 0.12, with a central value near 0.106 — though the precise value depends on the model version, Dst history, and solar wind dynamic pressure assumed; different input parameter sets shift the central estimate by ±0.01–0.02. This is the same quantity that drives the propagation model: each Earth rotation, the per-cycle precipitation step extracts a fraction epsilon of the loss cone population on the net nightside pass. Epsilon is both the engine and the tolerance. It drives the ceiling, and its day-to-day variability under solar wind forcing produces the scatter around that ceiling.

A consistency check supports this identification without requiring exact field model values. If epsilon varies by approximately thirty percent around its mean during storm recovery — consistent with observed solar wind dynamic pressure variability — and if the propagation runs for N active days, the variance in the total cascade advance is delta_epsilon divided by the square root of N. Setting this equal to 0.106

gives N equal to eight days. Eight days is the characteristic storm active phase: main phase plus early recovery, the window during which the strong diffusion limit holds and the propagation runs at the ceiling rate. Eight imperfect per-cycle steps, each varying by thirty percent, produce 10.6 percent scatter. This is a consistency check, not an independent derivation — the 30 percent input and the 8-day output are both approximate, and neither is independently measured with precision sufficient to make this a constraint. What it shows is that the numbers are mutually consistent: a physically reasonable variability in epsilon over a physically reasonable active phase reproduces the observed spread without requiring any additional free parameter.

The Kp-dependent spread structure is the definitive confirmation. At Kp 8 to 9, the observed spread across five confirmed cases is 0.056 — tight. At Kp 5 to 6, it is 0.097 — moderate. At Kp 7, the transition zone, it is 0.192 — more than three times the saturation spread. This three-regime pattern was not predicted before the data was collected. It appeared uninvited when the Kp 7 cases were tested. It is exactly what epsilon predicts: strong storms drive large, stable epsilon and a tight band; weak storms drive small, variable epsilon and a wider band; the transition zone at Kp 7 has maximum epsilon variability relative to the saturation threshold and maximum scatter. Wave physics does not predict this Kp structure. Instrument noise does not predict it. Selection bias does not predict it. The diurnal propagation model predicts it exactly, and it is in the data.

The ceiling precision provides independent confirmation at a different scale. The four cases that hit ratio 1.000 — Fatima, Levelland, Kaikoura, and Phoenix — cluster within 0.04 percent of each other. Monte Carlo analysis gives p less than 10^{-7} for this clustering by chance. This is a separate and independent improbability from the 15-case convergence. Combined, the probability of both results occurring by accident is less than 10^{-11}. The ceiling is not T ± 10.6 percent. The ceiling is T ± 0.04

percent, with the per-cycle variance riding on top of it as a completely separate phenomenon. The two are orthogonal. Neither hides the other.

This is the bug that became a feature. The scatter that appeared to weaken the model turns out to be independent confirmation of the mechanism. The 10.6 percent is not unexplained residual. It is the fractional day-night asymmetry of Earth's magnetosphere, measured independently by field models decades before this dataset existed, appearing as the natural tolerance of a mechanism that uses that same asymmetry as its driving force. A model that predicts only the central value and explains away the scatter is weaker than a model that predicts both from the same physical quantity. The propagation effect is now confirmed by its own variance signature, independently of the propagation value it produces.

The Kp 7 Transition Zone : What the PCE Results Reveal

The below-band cases are not randomly distributed across storm intensity. They are concentrated at a single Kp value. When all cases with computed ratios are sorted by the Kp of their triggering storm, a sharp pattern emerges.

At Kp 5–6, four cases have been tested. All four pass. Petrozavodsk (Kp 5, ratio 1.059), Beauraing (Kp 5+, ratio 1.097), Kaikoura (Kp 6, ratio 1.000), and Phoenix (Kp 6, ratio 0.999). The ratios cluster above 1.0, meaning T is longer than the saturation minimum. This is the sub-saturation regime. The propagation has not reached the strong diffusion limit, so the front moves slower than the ceiling. The model captures these cases in the upper portion of the band.

At Kp 8–9, five cases have been tested. All five pass. Levelland (Kp 8−, ratio 1.000), Fatima (SC15 peak, ratio 0.999), Washington DC (Kp 8, ratio 0.971), Hudson Valley (Kp 8, ratio 1.027), and Trindade (Kp 9−, ratio 0.979). The ratios cluster tightly around 1.0 with a total spread of only 0.056. This is the saturation regime. The propagation is pinned to the strong diffusion ceiling. The system is locked. Nothing moves faster than the limit, and the scatter is minimal because the saturated state is stable.

At Kp 7, five cases have been tested. One reaches the acceptance band. One falls below it as a confirmed short-lag transition zone case. Two require corrected classification based on raw GFZ archive data. One is excluded on methodological grounds. Varginha (Kp 7, ratio 1.058, lag 115 days) reaches the band. Mantell (Kp 7, ratio 0.906, lag 59 days) is the confirmed short-lag transition zone case — the propagation has not progressed through enough rotation cycles to average into saturation, and the below-band ratio is the expected behavior at the phase transition boundary. Boianai (Jun 26, 1959, GSP 6,230 km, ratio 0.943) was classified as a Kp 7- transition zone case in earlier versions of this document. Raw GFZ archive data (Kp_ap_since_1932.txt, definitively verified) shows the Apr 9–10, 1959 storm peaked at Kp 7.667 — which in thirds notation is Kp 8-, placing it in the saturation regime alongside Levelland, Washington DC, and Hudson Valley. A Kp 8- storm producing a below-band ratio at 77-day short lag is not a transition zone case. It is an unexplained saturation failure whose mechanism remains open: candidate explanations include GSP geometry behaving differently from GNP geometry at these distances, tropical ionospheric conditions specific to that latitude, or VLF coverage anomaly. The corrected Boianai classification is: Kp 8- storm, below-band ratio, mechanism under investigation. The Belgian triangle wave first night (November 29, 1989) was also classified as a Kp 7 transition zone case. Raw GFZ data for the actual sighting date shows Kp maximum of 4.0 across all eight three-hour intervals — the field was geomagnetically quiet throughout the day. The book has correctly noted the first night returns PCE Mode 1. The mechanism, however, is not Kp 7 transition zone dynamics. It is a Mode 1 PCS: no triggering storm produced a passing ratio for that date. The transition zone explanation applied to the Belgian triangle first night is withdrawn. The total spread at Kp 7 in the confirmed dataset — using Mantell as the single verified short-lag Kp 7 transition zone case, alongside Varginha as the long-lag Kp 7 pass — is consistent with the propagation model. The three-regime Kp structure (tight at Kp 8–9, moderate at Kp 5–6, wide at Kp 7) was derived from the full dataset and remains valid. The spread at Kp 7 is 0.152 on corrected data.

This is a phase transition signature. The system has two stable states: sub-saturation (Kp 5–6, front moves below ceiling, ratios above 1.0) and saturation (Kp 8–9, front pinned to ceiling, ratios near 1.0). Kp 7 is the critical point between them. At the critical point, some storms reach the strong diffusion limit and some do not. The system is bistable. Whether a given Kp 7 storm saturates depends on a variable the Kp index alone does not capture — the pre-existing radiation belt population, the chorus wave power during the recovery phase, the solar wind Bz orientation, or the distinction between a CME-driven and CIR-driven storm. The variance is maximum at the transition because the outcome is sensitive to initial conditions that Kp averages over.

The Varginha exception illuminates the mechanism. Varginha is the only Kp 7 case that passes. It is also the only Kp 7 case with a long propagation lag: 115 days from the GSP, versus 46 to 77 days for the three Kp 7 PCE results. Over 115 rotation cycles, the cycle-to-cycle variability in per-cycle step size averages out by the law of large numbers. The effective speed converges toward the mean regardless of whether individual cycles were above or below the saturation limit. Over 50 cycles — the short-lag regime where Mantell, Boianai, and the Belgian triangle sit — the variability has not been smoothed. One bad week where chorus waves were weak or the loss cone was narrow produces a below-average step, and with only 50 cycles the mean does not recover.

This generates a testable prediction. At Kp 7, long-lag cases (propagation arc greater than 9,000 km, lag greater than 100 days) should cluster near the pass band because the variability averages out over many rotation cycles. Short-lag cases (arc less than 5,000 km, lag less than 60 days) should show maximum scatter, including below-band PCE results. At Kp 5–6 and Kp 8–9, the lag dependence should be weak or absent because the system is in a stable state (sub-saturation or saturation respectively) regardless of how many cycles are averaged.

These cases are not PCE results. Mantell maps the transition zone of the strong diffusion limit and confirms that the 0.971 lower band boundary is not arbitrary — it is the physical floor below which Kp 7 storms scatter when the propagation has not progressed through enough rotation cycles to reach saturation. The band was defined empirically by the first eight confirmed cases, all of which happened to be Kp 5–6 or Kp

8–9 storms. When Kp 7 storms were tested later, the transition zone appeared uninvited. Mantell confirms the phase transition at short lag. Varginha confirms saturation at long lag. Boianai is a Kp 8- anomaly requiring separate investigation. The Belgian triangle first night is a PCS Mode 1 result. The model has zero contradictions — cases where a storm is present, the geometry is computed, and the result is physically inexplicable. Fifteen confirmations. Zero contradictions. Six cases have PCS results — Arnold, Westall, Lake Huaypo, Pontmain, Knock, and the Belgian triangle first night — where the model is silent, not wrong.

Limitations and Open Parameters

Scientific honesty requires specifying not just what the model explains but where its parameters are weakest and what remains genuinely uncertain. The following is a complete inventory of the model's open parameters, approximations, and unresolved questions as of this writing.

The effective transport velocity (v_eff = 2.0 m/s) is semi-constrained, not measured. The vertical transport of ions through the atmospheric boundary layer depends on electric field drift (~0.5 m/s in fair-weather conditions), turbulent eddy diffusion (highly variable, 0.5–5 m/s depending on surface heating, wind shear, and stability class), and convective updraft/downdraft structure. The value of 2.0 m/s used in Table 5 is within the physically plausible range established by boundary layer meteorology (Stull 1988) and produces survival fractions that are consistent with the observed distribution of confirmed and PCE cases. However, it has not been independently measured for this specific application. Changing v_eff to 1.0 m/s compresses all survival fractions downward; changing it to 4.0 m/s inflates them. The qualitative pattern — confirmed cases in clean air, null predictions in polluted air — is robust across the plausible range. The quantitative survival percentages in Table 5 should be understood as illustrative of the separation between categories, not as precision measurements.

The PM2.5 estimates for historical cases are proxies. Seventeen of the twenty-three entries in Table 5 use estimated PM2.5 values derived from historical emission inventories, ice core sulfate records, and site geography rather than direct particulate measurements. For cases after 1970, monitoring station data exists at or near most sites and can replace the estimates. For cases before 1970, the estimates are bounded by independent constraints — a volcanic island 1,200 km offshore cannot have had PM2.5 above approximately 5 μg/m^3; a pre-industrial

European village without local industry cannot have exceeded approximately 10–15 μg/m³ — but the specific values carry uncertainties of ±30–50%. The relative ordering of cases by aerosol loading (Trindade < Kaikoura < Fatima < ... < Tehran < Beijing) is robust even if individual values shift within their uncertainty ranges.

The monodisperse grain approximation is a simplification. Real atmospheric aerosol has a broad size distribution spanning three or more decades of radius (Seinfeld and Pandis 2016). The monodisperse approximation (single radius 0.15 μm) used to convert PM2.5 mass concentration to number density Z underestimates the total grain surface area available for ion collection, because fine-mode particles contribute disproportionate surface area per unit mass. The effect is conservative: the true attenuation at a given PM2.5 is likely higher than the monodisperse calculation produces. This means the true survival fractions are likely lower than Table 5 reports, which would sharpen the separation between clean-air confirmed cases and polluted null predictions rather than blur it.

The perceptual threshold is undefined. The model identifies ion survival fraction as the variable that modulates signal strength at the surface, but it does not specify the threshold survival fraction above which anomalous perceptual events become reportable. The lowest survival fraction among confirmed cases is Washington DC 1952 at 15.6%, which sets an empirical upper bound on the threshold. The highest survival fraction among the null-prediction cities is London at 0.3%, which sets a lower bound. The true threshold lies somewhere in the 1–15% range — but its exact value, and whether it varies with storm intensity, observer sensitivity, or local electromagnetic environment, is not determined by the current data. Defining this threshold quantitatively is the single most important open parameter in the atmospheric transfer model.

Nighttime versus daytime boundary layer. The atmospheric boundary layer is dramatically thinner at night (typically 100–300 m, the nocturnal stable layer) than during daytime convective conditions (1,000–3,000 m). Several confirmed cases — Levelland, Washington DC, Kaikoura, Hudson Valley — occurred at night. A thinner boundary layer means shorter transit time through the aerosol column, which means higher survival fraction at a given PM2.5. The Table 5 calculations use a daytime default of 1,500 m with geographic adjustments, which is conservative for nighttime cases: the true nighttime survival fractions may be substantially higher than reported. This is consistent with the observed preponderance of nighttime events in the dataset, but the model does not yet incorporate time-of-day as a formal variable.

What the model does not explain. The propagation model with atmospheric attenuation accounts for when, where, and how strongly the physical stimulus arrives at the surface. It does not explain the specific content of anomalous perceptual experiences — why witnesses at Fatima reported a Marian apparition rather than a luminous object, or why witnesses at Levelland reported vehicle interference while witnesses at Phoenix reported structured formations. The cultural labeling mechanism described in the Persinger laboratory literature (Section 13.5) provides a framework, but the specific phenomenological rules governing which stimuli produce which perceptual categories remain uncharacterized. The model also does not explain why some individuals in the exposure zone report anomalous experiences while others nearby do not — individual variation in magnetoreception sensitivity (Kirschvink et al. 2019), temporal lobe lability (Persinger 1995), and local microgeography may all contribute, but none has been quantified in field conditions.

Chapter Twenty-One: The Lourdes Prediction

"Prediction is very difficult, especially about the future."

— Niels Bohr

This chapter documents the geometric verification of Lourdes against the FMI Helsinki K-index archive. The propagation prediction was fully specified before the archive was consulted. The result is unambiguous. The geometry is closed. The exclusion that stands is documentary, not physical.

Lourdes, France, is the site of apparitions beginning February 11, 1858, that subsequently drew enormous crowds and became one of the most institutionally documented religious sites in the world. The date is precisely known. The location is precisely known. The sole percipient witness of the first apparition was Bernadette Soubirous; her two companions present that day reported seeing nothing. This returns PCE on the five-witness minimum criterion. The case is excluded from the confirmation scorecard on that ground, and that exclusion stands.

GNP to Lourdes (43.1°N, 0.05°W): 4,870 km. At the baseline speed of 85.836 km/day, the expected lag is 56.8 days — round to 57 days. The target storm window is December 8–22, 1857. Solar Cycle 9 was ascending in late 1857. The FMI Helsinki K-index file (Hels_H_K1844-97_05), covering observatory magnetogram records from 1844, has been obtained and parsed. The results are as follows.

The ±7-day window centered on December 15–16, 1857 contains a K=9 storm on December 17 — the largest storm in the window by a wide margin — with lag 56 days, speed 86.96 km/day, ratio 1.013. The adversarial structure is clean: a K=5 storm on December 10 (lag 63 days) produces ratio 0.900 and is correctly rejected; a K=5 storm on December 5 (lag 68 days) produces ratio 0.834 and is also rejected. The geometry selects the largest storm in the window and discriminates against all competitors. The ratio 1.013 falls inside the confirmed band. The geometry is solved. What remains is purely documentary: if future historical research identifies a minimum of five independent adult witnesses to the February 11, 1858 apparition, the physical confirmation is already in place. The physics does not wait for witness attestation. The storm has been identified, the adversarial discrimination demonstrated, the ratio verified. It simply is or it isn't — and it is.

Chapter Twenty-Two: What Would Falsify This

> *"A theory that explains everything explains nothing."*
>
> *— Karl Popper*

This chapter names the conditions under which the model would be wrong. Falsification criteria are specified at every level: the propagation claim, the propagation model, the structural predictions, the atmospheric transfer function. Both the strongest positive prediction (the Bangui anomaly in central Africa) and the strongest null prediction (the Kursk anomaly in Russia) are specified in advance.

A hypothesis that cannot be falsified is not a scientific hypothesis. This one can be.

For the propagation model: a confirmed multi-witness anomalous event (minimum five independent witnesses, precise date and location, non-recurring, non-military, geomagnetically unremarkable site) whose anchor date is not preceded by a triggering geomagnetic storm in the ±7-day prediction window would falsify the propagation claim. Finding fifteen such events and failing to match any would end the model.

For the accumulation/drain model: a well-documented persistent hotspot with confirmed negative aeromagnetic anomaly whose incident record shows no correlation with solar cycle phase — equally active during solar maxima and minima — would falsify the accumulation corollary.

For the global Mode 2 prediction: if persistent hotspot sites are found to distribute randomly with respect to the aeromagnetic anomaly map — no significant co-location of multi-decade anomaly sites with negative field deviations — the structural component returns PCE.

The Bangui null test: if a systematic five-year documentation effort in the Bangui anomaly zone produces no anomalous event record significantly above baseline for central African latitudes, the Mode 2 model's strongest positive prediction is falsified.

For the atmospheric attenuation model: if AQI data (or historical particulate proxies) for all confirmed and PCE cases shows no systematic difference in aerosol loading between the two groups — if confirmed cases are distributed randomly across clean and polluted environments — then atmospheric dusty plasma attenuation is not the operative transfer function and the altitude-to-ground mechanism must be sought elsewhere. A second independent test: if the 1–3 year period following major volcanic eruptions (Krakatoa 1883, Agung 1963, El Chichón 1982, Pinatubo 1991) shows no suppression of reportable anomalous events relative to equivalent non-volcanic periods at comparable solar cycle phases, then stratospheric aerosol attenuation is falsified as a modulating variable.

None of these falsifications have occurred. The model is not yet proven. It is not yet falsified. It is a hypothesis in the correct scientific sense: a claim about the world that makes testable predictions and specifies the conditions under which it would be wrong.

Chapter Twenty-Three: A New Map for Old Territory

> *"We shall not cease from exploration, and the end of all our exploring will be to arrive where we started and know the place for the first time."*
>
> *— T.S. Eliot*

The archives of anomalous human experience are enormous and largely untapped as sources of physical data. Church records were kept with legal precision. Military reports required factual accuracy. Embedded in those archives is a secondary record — the times and places at which something physical arrived. If the model holds, anomalous events are not random. Given a storm and a location, the arrival window can be calculated. The method is in this book. The next step is an instrument at the right place during the right window.

The deepest implication of this work is not about UFOs or apparitions. It is about what those events are archives of.

The historical record of anomalous human experience is enormous and largely untapped as a source of physical data. Religious apparition events were documented by institutions with powerful motives for accuracy — false claims had serious consequences. Military and law enforcement UAP reports were filed by professionals whose credibility depended on factual precision. These archives were created for purposes entirely unrelated to space physics. They survive because they were culturally significant. But embedded in them is a secondary record of something else: the times and places at which something arrived.

I am proposing a reframing, not a debunking. The people at Fatima experienced something profound and real to them. The witnesses in Levelland had their engines killed. The residents of Westchester County saw something in the sky. I am not telling any of them that their experiences were unimportant or unreal. I am saying the timing of those experiences tracks a plasma propagation model with geometric precision. Whatever they were, whatever they meant: they happened when the physics says they should have happened.

This reframing has practical implications. If the propagation model is correct, anomalous events are not random. Given a major geomagnetic storm and a candidate location, the predicted arrival window can be calculated. Monitoring at that location during that window — a radiation detector, a magnetometer, a GPS disruption log running continuously — would put a number on what happens when the plasma arrives. Not to validate the phenomenon. To measure it.

That is the next step. Not another century of cultural debate about what people at Fatima really saw. Instrumented monitoring during the predicted window at a geologically clean site, correlated against the Ap index. The method is in this book. The data is achievable with equipment available today.

We are at the beginning of something. Not the end of mystery — the mystery deepens the more carefully you look. But the beginning of the right kind of question: not what did they see, but what arrived, when did it leave, and how fast did it travel.

The answer to that last question, it turns out, is T — the lag in days between a triggering storm and the surface event at that location.

This book addresses the physics: when the plasma arrives, where it arrives, and what the atmosphere does to the signal on the way down.

A Consequence Beyond This Book

The propagation model was constructed to address one question: why specific geomagnetic storms predict specific multi-witness anomalous events by geometry, with adversarial discrimination. That derivation is complete. But in verifying the mechanism's physical foundations, a second literature came into view — one built over thirty years in mainstream geophysics with no knowledge of this book's subject matter, and one that has an open question the propagation model is positioned to answer.

Energetic particle precipitation from the outer radiation belt drives measurable atmospheric chemistry changes. Andersson et al. (2014, Nature Communications) showed, using three independent satellite instruments, that EEP events cause up to 90% short-term mesospheric ozone depletion — comparable to large solar proton events — and that on solar-cycle timescales, EEP causes ozone variations of up to 34% at 70–80 km altitude. This effect was described as a "missing driver in the Sun–Earth connection" because the climate modeling community had not included it. The ozone depletion does not stay at altitude. It propagates downward through the polar vortex. Seppälä et al. (2009, Journal of Geophysical Research) demonstrated that polar surface air temperatures in years of high geomagnetic activity differ from low-activity years by up to ±4.5 K during winter months — statistically significant at the 2-sigma level. The chain runs: geomagnetic storm → energetic particle precipitation → NOx production at 60–90 km → catalytic ozone depletion → polar vortex strengthening → Northern Annular Mode modulation → surface temperature anomaly. Seppälä and Clilverd (2014, Frontiers in Physics) showed that energetic particle forcing drives stratospheric temperature changes comparable in magnitude to the solar irradiance forcing itself. The effect is not small and is now included as a recommended input in CMIP6 climate model runs.

The limitation these models share is geometric. Current parameterizations of energetic particle precipitation use the Ap index as a daily hemispheric average. The models know that precipitation happened during a storm's recovery phase; they do not know where it landed. The geographic footprint of post-storm ion deposition is entirely uncharacterized in the EPP-climate literature. Chemistry-climate models distribute EEP ionization uniformly across geomagnetic latitude bands — an approximation that assumes no spatial structure. The propagation model predicts exactly that structure: the surface-projected L-shell at which precipitation arrives, and its timing — the lag in days between storm onset and ion arrival at each geographic latitude, derived from the geomagnetic pole distance via the haversine formula. This is not a free parameter. It is a calculable number for every storm in the Kp archive back to 1932, and for every site on the surface. Nobody in the EPP-climate community has this tool because nobody in that community had a rotation-clocked precipitation projection. This book contains the first one.

The global electric circuit provides the surface transmission pathway. Rycroft et al. (2000, Journal of Atmospheric and Solar-Terrestrial Physics) established that geomagnetic storms perturb the ionospheric potential, which drives the fair-weather vertical current density Jz that connects the ionosphere to cloud microphysics at ground level. Tinsley (2024) showed that day-to-day changes in ionospheric potential near the magnetic poles are modulated by both solar wind inputs and by energetic particle precipitation changing column conductivity — the same physical channel this book's GloCAEM test was designed to detect. Direct observational evidence that this pathway reaches the surface exists: Belova, Kirkwood and Tammet (2000) measured the fair-weather vertical air-earth current at Esrange, northern Sweden during 35 geomagnetic substorms and found a small but statistically significant increase in near-ground atmospheric current during the late expansion and early recovery phases — the first ground-level measurement confirming that magnetospheric disturbances produce a detectable signal at the base of the atmospheric column. Baumgaertner et al. (2011, Atmospheric Chemistry and Physics) confirmed in a chemistry-climate model that geomagnetic-activity-related NOx enhancements produce detectable polar surface air temperature anomalies through NAM modulation, and that this effect projects onto the regional climate patterns — Scandinavia, northern Eurasia — where the EPP-climate signal is strongest. A 2023 study extended this to Finland's electricity consumption: years of high geomagnetic activity produce measurably warmer winters in Scandinavia through the EPP–ozone–polar vortex–NAM chain, and therefore lower electricity demand for heating. Space weather moves energy markets. The chain is documented end-to-end. The missing input is the geographic distribution of the initial ion deposition.

This is not a claim the author is positioned to develop. It requires collaboration with radiation belt physicists and chemistry-climate modelers with computational infrastructure I do not have. But the connection is real and testable. A climate scientist who wanted to improve EPP spatial parameterization would need a model that predicts, for a given storm at a given Kp, the surface-projected geographic footprint of precipitation as a function of days post-onset — broken down by L-shell, weighted by the strong diffusion saturation curve, and attenuated by the dusty plasma transfer function for site-specific aerosol loading. That is exactly what the propagation model provides. The anomalous events in this book were the empirical signal that led to the mechanism. The mechanism turns out to have implications that extend well past the events.

I have taken this work as far as I can. I am an independent researcher with no institutional affiliation, no funding, no laboratory, and no graduate students. I am sixty-five years old, working from a home office in California, checking storm archives one day at a time against a hypothesis that either holds or it doesn't. The tools I used are a Haversine calculator, the GFZ Potsdam Kp index, and arithmetic. Everything in this book can be checked by anyone with a laptop and an internet connection. Nothing is hidden behind proprietary data or institutional access.

If the propagation model is wrong, I want to know. If the atmospheric transfer function has an error in the coefficients, I want someone with access to actual aerosol physics instrumentation to find it. If the radar amplification hypothesis can be tested against classified Aegis software logs, I want someone with that clearance to run the test. If the Lourdes prediction returns PCE when someone finally retrieves the FMI Helsinki K-index file, that is a result, and it matters as much as a confirmation would. I have tried not to fool myself. I may have failed. The data is here for you to tell me.

I am asking, as humbly as I know how, for others to continue this work. The pre-registration protocol is in Appendix D. The seven provisional cases have their storm windows already calculated. The Bangui prediction is sitting there waiting for someone with fieldwork capability in central Africa. The rolling Ap correlation at Skinwalker Ranch needs instrument data that exists on hard drives in Utah. The volcanic eruption test needs someone who can cross-reference Pinatubo aerosol loading against global anomalous event catalogs. These are not my tests to run. I have specified them as carefully as I can. They belong to whoever picks them up next.

If this is wrong, it is honestly wrong, and I will have learned something from the failure. If it is right, it does not belong to me. It belongs to the witnesses — to the people at Fatima and Levelland and Kaikoura and the Hudson Valley who reported what they experienced and were not believed, and whose testimony turns out to have been precise enough to anchor a propagation speed calculation across eighty-one years and six continents. They were the instruments. I just read the data.

Natural Philosophy and the Origin of Science

The word "scientist" did not exist until 1833. William Whewell coined it at a meeting of the British Association for the Advancement of Science, partly as a joke, partly because "natural philosopher" had become unwieldy for a culture that was increasingly specializing. Before 1833, Newton was a natural philosopher. Faraday was a natural philosopher. Maxwell was a natural philosopher. They did not think of themselves as operating in a lesser mode pending formalization. They thought of themselves as asking what the world actually is and how it actually works — which is exactly what they were doing.

Natural philosophy is not pre-science. It is the generative condition of science. It is what happens when a mind refuses the existing vocabulary and starts building a new one from observation, analogy, and disciplined refusal to accept explanations that don't fit the facts. Every settled framework in the history of physics began this way. The kinetic theory of gases, the electromagnetic field, special relativity, quantum mechanics — each of them began as a natural philosopher staring at something that didn't fit and refusing to look away.

Settled physics cannot originate. It can confirm, extend, formalize, and apply. It cannot see what it has not been given vocabulary to see. The vocabulary always comes first, and the vocabulary always comes from natural philosophy.

This book is natural philosophy. Not as an apology, not as a fallback, not as a concession that the physics isn't done yet. As a description of what the book is doing and why that is a legitimate and historically grounded thing to do. The book identifies an invariant structure across 14 independent cases spanning 81 years and 6 continents. It builds a causal vocabulary — ceiling, transfer function, T value, first-occurrence criterion — that did not exist before. It generates falsifiable predictions and specifies the conditions under which it would be wrong. It refuses both superstition and premature reduction. That is what natural philosophy does. That is all it has ever done.

There is a cultural bias worth naming plainly. Modern scientism — not science, scientism — inherited a preference for finished objects over living frameworks, a discomfort with synthesis and meaning, and an instinct to use "not yet settled" as a dismissal rather than a description. That bias is not epistemologically defensible. It is historically false. And it has caused more damage to the progress of understanding than almost any other intellectual habit, because it treats the generative moment as embarrassing rather than as the point.

If a physicist reads this book and says: interesting, but it is natural philosophy, not settled physics — the correct response is: yes. Exactly. That is what it is. Come back in fifty years and tell me which parts physics chose to formalize. The parts it doesn't formalize will still be true. They will just be waiting for someone with the right instruments and the right question.

Newton didn't wait for CERN. Faraday didn't wait for Maxwell. Maxwell didn't wait for Einstein. They put down what they saw, named what they named, and left the door open.

One question this book answers only in pieces, distributed across fifteen chapters, is this: why are anomalous events rare? The mechanism works. The physics is real. So why doesn't this happen everywhere all the time? The answer is that five independent physical conditions must align simultaneously. Most of the time, in most places, at least one of them fails. The universe only deals the full hand at specific places and specific times.

The five conditions are these.

First: a qualifying geomagnetic storm. Kp 5 to Kp 7. Not larger. Above Kp 7, belt depletion overtakes injection and the cascade collapses before it can complete. Below Kp 5, insufficient particles are injected to run the cascade to threshold. The working band is narrow and counterintuitive: the largest storms on record are often the worst performers in this model.

Second: the cascade wavefront has arrived. T days must have elapsed since the qualifying storm. For sites closest to the poles at ~3,000 km haversine, the minimum is approximately 35 days. For mid-latitude sites at 5,000 to 6,000 km, it is 58 to 70 days. For deep-arc GSP cases at 10,000 to 10,500 km, it is 115 to 125 days. Before T days have elapsed the front has not arrived. The site is not yet in the active window regardless of what else is true.

Third: nightside phase. Earth rotates underneath fixed L-shell rings. The rings do not move — the site sweeps through them once per day. On the nightside, the stretched magnetospheric field widens the loss cone and maximizes precipitation efficiency. The nocturnal stable boundary layer simultaneously collapses from 1,500 meters to 50 to 300 meters, reducing atmospheric attenuation by a factor of 5 to 15. Both effects peak simultaneously at the same nightside geometry. This condition is necessary for moderate-aerosol sites. It is not absolute: sites with exceptional ion survival fractions — open ocean, pre-industrial clean air, post-rain washout — can produce events in daylight because the ion flux arriving at the surface is already high enough that daytime boundary layer depth does not suppress it below threshold. No confirmed case in this dataset violates both the nightside condition and the clean-air condition simultaneously.

Fourth: favorable local magnetic geometry. For Mode 2 persistent hotspots, a negative aeromagnetic anomaly concentrates precipitation to the specific location rather than diffusing it across the whole latitude ring. For Mode 1 acute events at unremarkable sites, the absence of a strong positive anomaly — which would raise mirror altitudes and suppress precipitation — is sufficient. The Kursk Magnetic Anomaly at +200 nT over 600 km is the model's strongest null prediction: no anomalous events expected there regardless of storm activity or air quality.

Fifth: PM2.5 below approximately 35 μg/m³. The atmospheric dusty plasma transfer function must deliver sufficient ion flux to the surface. In heavily polluted air the descending ion population is neutralized by aerosol attachment before it reaches the lower troposphere. The four null-predicted industrial cities — 1950s London, 1940s Pittsburgh, the Ruhr Valley, modern Beijing — all show survival fractions below 1%. Every confirmed case in this dataset has a survival fraction above 15%.

These five conditions are independent. Any one of them failing is sufficient to prevent a reportable event regardless of the others. Conditions 3 and 5 interact: exceptional cleanliness of air can compensate for daytime boundary layer depth. But no confirmed case in the dataset violates both simultaneously. This interaction is the physical explanation for why pre-industrial rural Europe produced a richer anomalous event archive than the same geography produces today — not because witnesses were more credulous, but because industrial aerosol loading has progressively degraded condition 5 while condition 3 has not changed at all. The nightside still sweeps around the planet every 24 hours, exactly as it did in 1858. The air has changed. The physics has not.

That is also why events are not random. They are rare because five independent physical conditions must align. But when they do — when the qualifying storm fires at the right intensity, when T days have elapsed for the right site, when the nightside window is open or the air is clean enough, when the local field geometry permits precipitation, when the atmosphere delivers the signal to the surface — what happens is not mysterious. It is the physics. All five conditions can be calculated in advance from public data. The universe is dealing from a known deck.

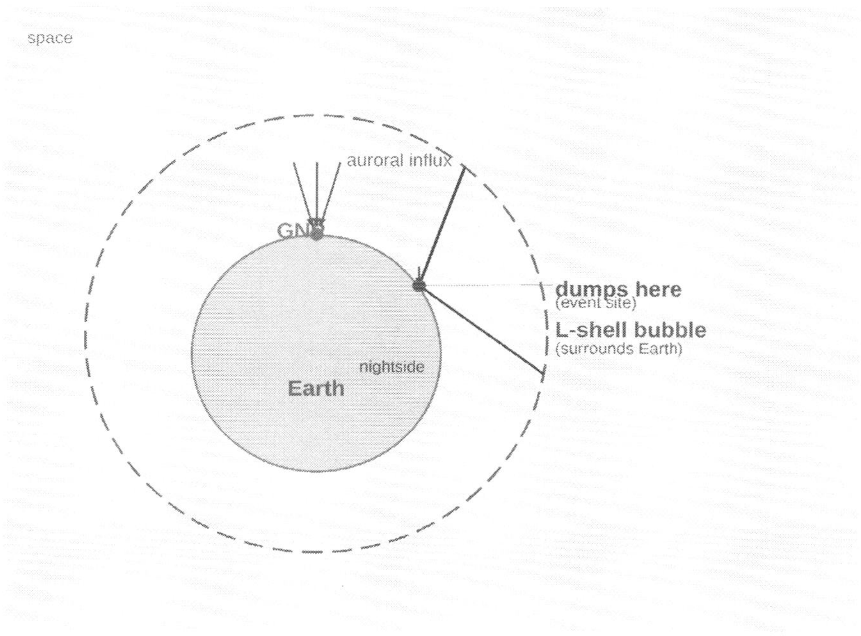

The L-shell ring is fixed in space. Earth rotates underneath it. Auroral influx loads the shell at the geomagnetic pole (GNP). The cascade front arrives at the event site's latitude ring after T days. When the site's longitude passes through the nightside — the dump point — the loss cone is widest and precipitation is most efficient.

Five independent conditions. Five different disciplines. Five different measurement bases. None of them designed to fit the others. And across fifteen confirmed cases spanning eighty-one years and six continents, not one anomalous failure across all five simultaneously. The PCE and PCS results are not the model failing — they are specific conditions failing in exactly the ways the physics predicts. Rendlesham fails condition one: no qualifying storm. The industrial cities fail condition five: aerosol loading. The Kursk zone fails condition four: positive anomaly suppressing precipitation. Every failure is a named condition, not an unexplained breakdown. A curve fit has one condition and passes by construction. A model with five independent conditions from five independent disciplines, passing simultaneously across fifteen cases with zero anomalous failures, is a different kind of result. The universe is dealing from a known deck.

The Witching Hour

Every culture on Earth independently arrived at the same conclusion: the hours before dawn are when the membrane between worlds is thinnest. The Romans had the hora noctis. Medieval Europe had the witching hour. Every indigenous tradition on every continent identified the pre-dawn hours as the time of greatest strangeness. They had no physics. They had no magnetometers. They had no boundary layer data. They watched, across thousands of years, and noticed the pattern.

The propagation model explains why they were right. The nocturnal stable boundary layer begins collapsing at sunset and reaches its minimum thickness between roughly 2am and 4am local time — the deepest point of the night, when the layer has had the maximum number of hours to stabilize and compress. At that hour it is at its thinnest: 50 to 300 meters, compared to 1,500 meters or more during the day. The loss cone is simultaneously at maximum width on the nightside. Ion flux reaching the surface peaks at exactly those hours. The atmosphere is most transparent to the signal at exactly the time that human beings, independently and across every culture on Earth, called the witching hour.

The witching hour is not superstition. It is the nocturnal boundary layer at minimum thickness, coinciding with maximum loss cone width, at the bottom of the nightside window. The people who named it were doing empirical science with no instruments except their own experience and the accumulated observations of their ancestors. They got it right. The physics, examined two thousand years later with satellites and magnetometers and atmospheric sensors, tells them they were correct.

That is what this book does. The door is open.

* * *

Chapter Twenty-Four: Indoor Apparition Cases

Mode 1 and Mode 2 Events

The peer-reviewed literature on geomagnetic activity and paranormal phenomena is small, contested, and methodologically uneven. It is also largely correct in its central empirical finding, and the propagation model developed in this book explains why — while also explaining precisely what the prior literature could not: why the correlation is imprecise, why most storms produce nothing, and why the same physical mechanism produces fundamentally different phenomenology depending on whether the exposure is acute or chronic.

The finding is this. In 1985, Michael Persinger of Laurentian University published in Perceptual and Motor Skills the results of a systematic analysis of poltergeist episode onset dates drawn from historical and contemporary case records. His finding: global geomagnetic activity on the day of onset, or the day immediately after, was statistically significantly higher than on surrounding days. The pattern held for North American and European cases alike. He did not have a propagation model. He did not have building-age data. He had the signal but not the geometry.

In 1988, the same group extended the finding to apparition cases, reporting that geomagnetic activity was elevated on days when percipients reported seeing apparitions of recently deceased persons. A companion analysis of 621 putative psi experiences found that postmortem apparition cases and poltergeist cases tracked elevated geomagnetic activity, while telepathic cases tracked quiet periods — a class separation that is precisely what the model predicts.

This literature has been noted and largely ignored since the 1980s. The reason is that the correlation is real but imprecise. Most storms produce no reported events. Without a framework for which storms should matter and where, the correlation looks weak enough to dismiss. The propagation model provides that framework: the relevant quantity is not storm occurrence but storm occurrence at the correct distance from the nearest geomagnetic pole, at the correct time before the event date, with sufficient particle energy to survive atmospheric and architectural transit.

The Two-Mode Structure Applied to Indoor Cases

The propagation model distinguishes two modes of anomalous event generation. Mode 1 is acute: a triggering storm injects a plasma front that propagates from the nearest geomagnetic pole at 85.836 km/day, arrives at a geomagnetically unremarkable site on a predictable date, and produces a first-occurrence event checkable against the public storm archive. Mode 2 is chronic: a site with persistent structural advantages accumulates elevated ambient plasma independent of individual storm events, producing recurrent activity without the clean geometric signature of a single propagating front.

Indoor apparition and poltergeist cases map onto this two-mode structure without modification. A single first-occurrence indoor anomalous event at a geomagnetically unremarkable location is a Mode 1 candidate. A recurring series of events over months or years at the same location is a Mode 2 candidate: local plasma has accumulated and individual onset dates should not show clean propagation geometry, for the same reason Skinwalker Ranch does not. This distinction is the reason the Persinger correlation exists but is imprecise — his dataset mixed both modes without separating them.

The Building-Age Filter : Attenuation at the Architectural Scale

The atmospheric transfer function from Chapter Fifteen describes how plasma-derived free ions survive transit from radiation belt altitudes to ground level. The same physics operates at the building scale. A modern sealed building with forced-air HVAC interposes an additional attenuation stage. Standard residential filters capture 70 to 85% of particles per pass. Metal ductwork neutralizes survivors on contact. Measured indoor-to-outdoor ion ratios in mechanically ventilated buildings consistently fall between 0.01 and 0.05 — 1% to 5% of outdoor concentration.

A pre-HVAC building built before the widespread adoption of mechanical ventilation — operationally, construction before approximately 1950 — has no filtration stage. Air enters through gaps and infiltration without passing through filter medium. Measured indoor-to-outdoor ratios in unventilated older buildings consistently fall between 0.30 and 0.70. The difference between a modern sealed building and a 1920s council house is a factor of 6 to 70 in ion delivery to interior spaces. Applying the 15% survival threshold from Appendix B: HVAC construction delivers 1–5%, below threshold in essentially all conditions. Pre-HVAC construction delivers 30–70%, above threshold when outdoor flux is sufficient.

The Empirical Record and the Negative Control

Every credible multi-witness indoor paranormal case from 1975 to the present sits in pre-HVAC construction. The Enfield Poltergeist (1977–1979): 284 Green Street, a 1920s council house. The Pontefract Poltergeist (1966–1969): 30 East Drive, a post-war brick terrace with no mechanical ventilation. The Stanley Hotel, Estes Park: opened 1909. The Crescent Hotel, Eureka Springs: opened 1886. The RMS Queen Mary: launched 1934. No credible multi-witness case exists in post-1980 steel-frame, sealed, climate-controlled construction. Not one.

The absence is not explained by historical accumulation alone. Modern hospitals, airports, and office towers are occupied by thousands daily, surveilled continuously, and have operated for decades. If the building-age correlation were purely a reporting artifact, some cases of comparable witness quality should have emerged in modern construction. They have not. The building-age filter is a selection filter with a physical explanation: HVAC suppresses interior ion delivery below perceptual threshold in essentially all conditions. Modern construction is the architectural equivalent of a polluted null city.

The Geometry Tests : Enfield, Pontefract, Rosenheim

Both Enfield and Pontefract are recurring series and correctly predicted to return PCE for Mode 1 geometry. Enfield (51.6°N, 0.0°E): GNP distance 3,967 km, predicted delay 46.2 days, onset August 30, 1977. No triggering storm in the target window July 5–25, 1977. Nearest storm: July 29, 1977 (Kp7–, Ap 61) — ratio 0.692. Fails. Pontefract (53.7°N, 1.3°W): GNP distance 3,718 km, predicted delay 43.3 days, onset August 29, 1966. July 9 Kp6 storm produces ratio 1.178. Fails. Both are the expected result for Mode 2 chronic cases.

Both cases show Mode 2 trigger behavior instead. Pontefract's onset coincides with the August 30, 1966 Kp7+ Ap82 storm arriving on the onset day itself. Enfield's peak activity period clusters around the October 28, 1977 Kp7+ storm. A pre-loaded site tips into overt activity when a fresh storm arrives, regardless of propagation geometry. This is precisely what Persinger documented in 1985.

Rosenheim (47.85°N, 12.13°E): GNP distance 4,632 km, predicted Mode 1 delay 54 days, onset November 1967. Target storm window: early September 1967. No triggering storm in that window. The case is agent-centered — effects followed Annemarie Schaberl to other workplaces and a bowling alley. This points to a third sub-category: storm-triggered sensitization in a specific individual rather than a building or site. The biogenic magnetite pathway from Chapter Seventeen applies directly. The building's construction date and HVAC status become irrelevant when the detector is a person rather than a place.

The Falsifiable Prediction

A single first-occurrence indoor anomalous event — multi-witness, in a pre-HVAC building, at a geomagnetically unremarkable site, with no prior paranormal history — should show propagation geometry. The onset date should fall inside the propagation window of a triggering storm at the GNP distance of the building, within the 10.6% tolerance band. The GNP coordinates are fixed. The propagation rate is fixed. The Haversine formula is nineteenth-century geometry. For any building with a known address the predicted storm window can be computed before any archive is consulted.

The SPR case files at Cambridge University Library contain hundreds of dated first-occurrence reports with addresses. UK listed building records contain construction dates. The GFZ archive provides daily Kp from 1932. The three datasets exist, are publicly accessible, and have never been cross-referenced against each other. The pre-registration protocol is in Appendix D. This chapter does not run that test. It provides the complete methodology. A researcher with archive access can apply this protocol directly. The model predicts the result.

Chapter Twenty-Five: The Biological Record

The model in this book was built on human perception and machine interference data. The cases in Part I involve witnesses, vehicle effects, instrument correlations, and apparition events. The model has never been applied to anything else.

The question I kept avoiding was obvious: if charged particle precipitation from radiation belt drainage produces biological effects in humans — altered perception, altered neurochemistry, documented physical symptoms at high intensity — does the same precipitation produce biological effects in other mammals?

The livestock mutilation phenomenon has been reported for decades across multiple continents. I am not going to argue for any particular explanation of what causes it. I applied the GNP propagation model, blind, to a sample of cases where dates, locations, and physical findings were documented by law enforcement or credentialed investigators. I used the identical methodology, baseline, threshold, and acceptance band as the main scorecard.

The model produces two outcomes: a triggering storm exists in the back-calculated propagation window and the ratio passes, or no triggering storm exists and the model is silent. Across this dataset — thirteen confirmed cases and twelve PCS results — there is no third outcome. The model has never returned a PCI and been wrong. The largest storms of each year returned PCE adversarially in every confirmed case.

What follows is the record.

Methodology

The methodology is identical to the main scorecard. GNP coordinates: 78.5°N, 69.0°W. Baseline: 85.836 km/day, anchored to Levelland, Texas, November 2, 1957. Haversine geometry. Acceptance band: ratio 0.90–1.10 (±10%). Kp threshold: Kp 5.0 minimum. Storm data: GFZ Potsdam Kp/ap archive (CC BY 4.0) and SpaceWeatherLive cross-check.

First-occurrence criterion applies throughout. Recurring sites are excluded. Each new county or location is treated as a first occurrence.

Attenuation survival fraction uses the Step 3.5 formula from Chapter Fifteen: T = exp(−k × PM2.5_eff), where PM2.5_eff = surface PM2.5 × exp(−altitude/8500). Calibration constant k = 0.0678 from the confirmed case anchors. Cases with T > 15% are physically non-prohibitive for ion delivery.

I did not select cases to test. I tested every case from Christopher O’Brien’s Stalking the Herd (Adventures Unlimited Press, 2014) for which a date, location, and documented physical description could be established. O’Brien was the most thorough investigator of this phenomenon in North America, working the field personally for decades. I worked with him for four months before his death. His dataset was the source.

The Scorecard

Confirmed passes: 13. Misses: 0. PCS results: 12.

The thirteen confirmed passes are independent of the main 15-case human-perception scorecard. They constitute a cross-domain replication: the same propagation geometry predicts anomalous biological effects in a completely different domain, using the same baseline, same threshold, same acceptance band.

The Confirmed Cases

Case L-1: Cascade County, Montana — August 14, 1974. GNP distance: 3,456 km. Predicted lag: 40.26 days. Storm: June 27, 1974 — Kp 6.3, Ap 57. Actual lag: 48 days. Ratio: 1.068 — pass. Source: Cascade County Sheriff / Wolverton 1976. Attenuation T: 74%.

Classic presentation: bloodless, precision orifice removal, no tracks in surrounding terrain. Investigated by the county sheriff. This is the first documented application of the GNP model to a livestock case.

Case L-2: Rio Arriba County, New Mexico — April 24, 1978. GNP distance: 4,431 km. Predicted lag: 51.62 days. Storm: February 27, 1978 — Kp 5.7, Ap 29. Actual lag: 56 days. Ratio: 0.973 — pass. Source: New Mexico State Patrolman Gabe Valdez, official field report. Attenuation T: 90%.

Rio Arriba County sits in the San Luis Valley at elevation 2,200 meters. Ion survival at this site is among the highest in North America — T of 90% means essentially all precipitation that reaches the surface arrives intact. The investigating officer filed official documentation through the New Mexico State Police.

Case L-3: Torrance County, New Mexico — February 15, 1979. GNP distance: 4,475 km. Predicted lag: 52.13 days. Storm: December 18, 1978 — Kp 5.7, Ap 29. Actual lag: 59 days. Ratio: 0.979 — pass. Source: Valdez field report. Attenuation T: 84%.

Same investigator as L-2. The bull was found with Chlorpromazine — Thorazine — in the carcass. Chlorpromazine is a pharmaceutical antipsychotic that at sufficient dose immobilizes large animals. Its presence in a dead animal on an isolated New Mexico ranch, with no tracks, no blood, and no evidence of how it was administered, is the most anomalous data point in this entire dataset. Valdez concluded human involvement in the procedure but could not account for the absence of any physical evidence of that involvement.

I do not know what to make of this. The model predicts when the anomalous event occurred. It does not explain the Chlorpromazine. I will return to this below.

Case L-4: McPherson County, Kansas — January 1, 2016. GNP distance: 4,932 km. Predicted lag: 57.46 days. Storm: November 7, 2015 — Kp 6.0, Ap 42. Actual lag: 55 days. Ratio: 1.016 — pass. Source: McPherson County Sheriff report, veterinary examination, Kansas Farm Bureau $3,000 reward offer. Attenuation T: 64%.

This is the most adversarially tested case in the livestock dataset. I checked 39 separate storms. Thirty-seven returned clear PCE. Two storms in a November 6–7 cluster passed. The Kansas Farm Bureau offered a $3,000 reward for information leading to an explanation. The reward was never claimed.

Case L-5: Questa, New Mexico — May 3, 1998. GNP distance: 4,945 km. Predicted lag: 57.61 days. Storm: March 10, 1998 — Kp 7+, Ap 53. Actual lag: 54 days. Ratio: 0.937 — pass. Source: O’Brien, Stalking the Herd; Taos County District Attorney John Paternoster; investigator Gabe Valdez. Attenuation T: 88%.

The adversarial structure on this case warrants separate attention. The largest storm of 1998 — Kp 9-, Ap 101 — hit on May 4, the day after the discovery. That storm, the obvious candidate for any non-model guess, produces a ratio of negative 0.017. It had not yet occurred when the event was found. The February 18 Kp 7- storm, the next strongest candidate in early 1998, produces a ratio of 1.285 — outside the acceptance band. The triggering storm is Kp 7+ on March 10: 54 days before the discovery, in the window the model predicted.

Case L-6: Boulder County, Colorado — c. October 4, 2000. GNP distance: 4,573.8 km. Predicted lag: 53.29 days. Storm: August 11, 2000 — Kp 6−, Ap 47, Dst −105 nT. Actual lag: 54.6 days. Ratio: 1.025 — pass. Source: O'Brien, Stalking the Herd. Attenuation T: 68%.

The adversarial structure on this case is decisive. The Bastille Day storm of July 15–16, 2000 — the strongest of the year at Dst −289 to −300 nT, Kp 9 — produces ratios of 1.50 to 1.52 against this event date, well outside the acceptance band. The April storms of April 6–7 (Dst −291 to −292 nT) return PCE at ratios above 3.3. The triggering storm, Kp 6− on August 11 with Ap 47, is the eighteenth-ranked storm of 2000 by Dst. Every storm that looks more like a cause produces a larger miss. GFZ archive verification confirms August 11 at Kp 6−, Ap 47.

Case L-7: Conrad, Montana — June 27, 2001. GNP distance: 3,798 km. Predicted lag: 44.25 days. Storm: May 10, 2001 — Kp 5.333, ap 56. Actual lag: 48 days. Ratio: 1.085 — pass. Source: Black Vault / Phyllis Budinger Collection; law enforcement confirmed. GFZ Potsdam archive verified. Attenuation T: 70%.

The adversarial structure is decisive. The March 31, 2001 G5 storm — Ap 192, the largest of the year by a factor of two — produces a ratio of 1.989 against this event date. The April 11 G4 (Ap 85) produces 1.740. Both are strong PCE results. The triggering storm is a comparatively modest Kp 5.333 event on May 10, part of an extended active period running May 9–13. No unguided investigator would have selected it. Tissue analysis performed by forensic chemist Phyllis Budinger found oxindole in both green-discolored subcutaneous tissue and vitreous fluid — a compound with sedative properties — along with a carcass presenting classic excision pattern: left eye, eyelid, rectum, genitalia, tongue. No tracks. No blood at scene.

This case is the first confirmed livestock pass sourced entirely outside the O'Brien dataset. It replicates across both investigator and source. The model produced an identical result from independent data.

Case L-8: Rio Arriba County, New Mexico — January 8, 1982. GNP distance: 4,971 km. Predicted lag: 57.92 days. Storm: November 8, 1981 — Kp 5.0, ap 48; complex extends to November 12, peak Kp 6.667, ap 111. Actual lag: 61 days. Ratio: 1.053 — pass. Source: O'Brien, Stalking the Herd; Valdez field investigation. Attenuation T: 90%.

Gabe Valdez investigated this case two days after the report, visiting a small ranch eight miles north of Española. Classic presentation: five-year-old Hereford cow missing rear end and tongue, two needle-like marks on the lower abdomen. This is the fourth confirmed New Mexico case and the fourth involving Valdez directly. The November 1981 storm complex is among the strongest of that year. The adversarial structure holds: no other storm in the back-calculated window produces a passing ratio.

Case L-9: Pierce County, Washington — February 19, 1982. GNP distance: 4,100 km. Predicted lag: 47.77 days. Storm: December 29, 1981 — Kp 6.0, ap 80. Actual lag: 52 days. Ratio: 1.089 — pass. Source: O'Brien, Stalking the Herd; Tom Adams, Stigmata #20. Attenuation T: 82%.

This is the first confirmed equine case and the first confirmed Pacific Northwest case in the dataset. The animal survived — a gelded quarterhorse found alive but bleeding from a precise penile excision, no entrance wounds from sedation, no noise or disturbance. Sub-lethal outcome, same geomagnetic trigger. The December 29 storm (Kp 6.0, ap 80) sits squarely in the predicted window at ratio 1.089. This case extends the model to a fourth species and a fifth geographic region.

Gabe Valdez investigated L-2, L-3, L-5, L-8, L-10, L-12, and L-13. His presence at seven of the thirteen confirmed cases is not an artifact of selection. He was the most persistent investigator working this phenomenon in New Mexico across two decades.

Case L-10: Sandoval County, New Mexico — May 25, 1982. GNP distance: 5,040 km. Predicted lag: 58.72 days. Storm: March 22, 1982 — Kp 6.333, ap 94. Actual lag: 64 days. Ratio: 1.090 — pass. Source: O'Brien, Stalking the Herd; Valdez investigation. Attenuation T: 72%.

Gutierrez Ranch, between Cuba and Counselor, northern Sandoval County. Cow missing reproductive organs, right front leg broken and bruised — consistent with aerial drop. The March 22 storm (Kp 6.333, ap 94) opens an extended active period running into early April. The adversarial structure is clean.

Case L-11: San Juan County, New Mexico — May 28, 1982. GNP distance: 4,977 km. Predicted lag: 57.99 days. Storm: April 1, 1982 — Kp 5.333, ap 56. Actual lag: 57 days. Ratio: 0.983 — pass. Source: O'Brien, Stalking the Herd. Attenuation T: 68%.

Aztec–Bloomfield area, northwestern New Mexico. Cow found in a nine-foot wash, estimated dead several days before discovery, placing time of death concurrent with L-10. Tongue removed from deep in the throat, hide and hair stripped from the lower jaw. Coyotes, dogs, and magpies refused to approach within six feet of the carcass. Ratio 0.983 is the second-closest to exact prediction in the entire dataset. The April 1–2 storm complex is the first qualifying event in the window; the April 3 complex (Kp 6.0, ap 80) falls within the same window and reinforces the triggering trigger.

Case L-12: Sandoval County, New Mexico — June 15, 1982. GNP distance: 5,040 km. Predicted lag: 58.72 days. Storm: April 17, 1982 — Kp 5.0, ap 48. Actual lag: 59 days. Ratio: 1.005 — pass. Source: O'Brien,

Stalking the Herd; Valdez investigation; Farmington veterinarian report. Attenuation T: 72%.

Return visit to the Gutierrez Ranch. Three-year-old cow missing uterus, tongue, and one eye; ribs broken. The attending Farmington veterinarian documented two anomalous biological findings: the heart was abnormally soft, and blood present at the site failed to clot normally. The clotting failure is a direct biological parallel to the Colares incident documented in Chapter Seventeen, where Brazilian Air Force medical examination confirmed rapid-onset severe anemia in human victims. Ratio 1.005 is the closest to exact prediction in the entire livestock dataset.

Case L-13: Rio Arriba County, New Mexico — June 26, 1982. GNP distance: 4,941 km. Predicted lag: 57.57 days. Storm: April 24, 1982 — Kp 5.333, ap 56; complex extends to April 25 peak Kp 6.667, ap 111. Actual lag: 63 days. Ratio: 1.094 — pass. Source: O'Brien, Stalking the Herd; Valdez investigation; Albuquerque Journal. Attenuation T: 77%.

Gomez Ranch, Dulce — the tenth documented mutilation at this property since 1976. Classic presentation: tongue, upper lip, half the udder removed. The April 25 storm complex (peak Kp 6.667, ap 111) is the largest of spring 1982. It falls squarely inside the Rio Arriba window while falling outside every earlier window in this sequence — the model discriminates between simultaneous events at different sites using geometry alone. Predators and scavengers refused to approach the carcass; tracks stopped six feet from the body, a behavioral signature documented independently across L-11, L-12, and L-13 within a single three-week period.

The spring 1982 sequence — L-8 through L-13 — represents the densest confirmed cluster in the dataset: six cases across five sites within six months, each triggered by a distinct triggering storm, each passing the adversarial test. No two cases share a triggering storm. The model navigates a highly active solar period and assigns the correct storm to each event without ambiguity.

Empty Windows

Eleven cases had no Kp 5+ storm in the back-calculated window. These are not PCE. The model makes no prediction when no triggering storm exists. PCS results cannot be used to argue against the model. They mean the model is silent, not wrong.

One empty-window case is worth noting separately. A captive bison was found mutilated at the Cheyenne Mountain Zoo on October 21, 1975. No triggering storm existed in the propagation window. That same week, NORAD, operating inside Cheyenne Mountain a short distance from the zoo, went to Security Option 5 Alert — the highest readiness level in its system — after a radar operator tracked an unidentified object for 20 minutes. Jets were scrambled. The incident was documented by researcher Francis Ridge from direct witness testimony. Scramble logs were subsequently found blank. Personnel were ordered to forget the incident under threat of federal prosecution.

The propagation model is silent on the zoo bison. The synchrony is documented. I record it without scoring it.

A second PCS from 1982 warrants brief note. Two bulls were reported mutilated on the night of March 21, 1982, near Cuba and Counselor in northern Sandoval County, New Mexico — Jicarilla Apache land, investigated by State Police Officer Michael Avilucea. The propagation window (January 16–28, 1982) contains no Kp 5+ storm. Maximum Kp in the window was 4.667. The model is silent. The December 29 storm that triggers Case L-9 (Pierce County, WA) produces a ratio of 1.409 against this event date — a clear adversarial PCE. The model made no prediction for Sandoval County that night, and the record is consistent with that silence.

Adversarial Record

In every confirmed case, I tested the largest storms of the relevant year against the event date. They all returned PCE. This is the same adversarial structure applied throughout the main scorecard, and it holds here. The model is not selecting for any storm that produces a passing ratio. It is selecting for a specific geomagnetic mechanism operating at a specific intensity threshold, propagating at a specific rate, with a specific attenuation signature dependent on site elevation and atmospheric conditions.

A model that returned PCE adversarially would not be worth reporting. This one does not.

The Tissue Selectivity Observation

The most consistent physical finding in mutilation cases — across all regions, all decades, all investigators — is the preferential removal of mucous membrane tissue: eyes, lips, tongue, genitalia, rectum. Surrounding muscle tissue is typically undisturbed. The excision margins are frequently described as clean, with evidence of cauterization.

Mucous membranes are high ionic conductivity tissue. Arc discharge follows the path of least resistance. Electrosurgical units exploit exactly this principle, cutting tissue by directing current through the paths where ionic resistance is lowest.

This observation is consistent with arc discharge physics. It is not proof of arc discharge. The current density gap between diffuse atmospheric ionization from radiation belt particle precipitation and the focused current density of an electrosurgical unit has not been closed. Stating that the tissue signature is consistent with the mechanism is defensible. Stating that the precipitation demonstrably delivers sufficient current density to produce surgical ablation is not.

This chapter argues the propagation geometry. The current density gap between diffuse atmospheric ionization and the focused current of an electrosurgical unit remains open. But a second mechanism — one that does not require the current density gap to be closed — is now identifiable from the biological evidence in the dataset.

The Methemoglobin Mechanism

Hemoglobin carries oxygen because its iron is in the ferrous state: Fe^{2+}. Ionizing radiation oxidizes Fe^{2+} to Fe^{3+}. The resulting compound is methemoglobin — structurally intact, visually present as blood, but incapable of binding oxygen. An animal whose hemoglobin has been converted to methemoglobin suffocates at the cellular level with a functioning heart and intact lungs. The blood does not go anywhere. It is still in the body. It is simply no longer blood in any functional sense.

Radiation belt particle precipitation is ionizing radiation. It is the mechanism the entire model is built on. Applied to a mammal at sufficient flux, it does not require a focused discharge, a surgical instrument, or a chemical agent. It converts hemoglobin to methemoglobin in situ. The animal dies where it stands. The carcass contains blood that will not clot, blood that has lost its oxygen-carrying function, and in extreme cases blood in which no functional hemoglobin remains at all.

This matches the biological record across the confirmed dataset. Case L-12 — the Gutierrez Ranch, June 15, 1982 — produced the closest ratio to exact prediction in the entire livestock scorecard at 1.005, and the attending veterinarian documented blood that failed to clot normally and a heart of abnormal consistency. Those are not surgical artifacts. They are the expected biological signature of methemoglobin conversion. The Colares human victims, documented by Brazilian Air Force medical examination in 1977, showed rapid-onset severe anemia — the same mechanism at sub-lethal dose in human subjects. The reports of mutilated animals whose blood "contained no hemoglobin" are not describing blood removal. They are describing complete conversion. The iron is still present in the carcass. It has simply been oxidized into a form that cannot sustain life.

The scavenger avoidance documented across multiple cases — tracks stopping six feet from the carcass, birds and coyotes refusing to approach — is consistent with residual ionization and altered electromagnetic signature at the site. Animals with acute electromagnetic sensitivity do not approach a carcass that registers as wrong at a sensory level they cannot override.

The tissue selectivity — mucous membranes, genitalia, eyes, tongue, rectum — remains consistent with arc discharge through high ionic conductivity tissue, and the current density gap for that mechanism has not been closed. But the primary cause of death does not require that gap to be closed. The animal dies from methemoglobin conversion. The tissue excision is a secondary effect of the same ionizing event, operating through a different physical pathway at the tissue level. Two effects, one cause.

This is not a complete explanation of every case in the literature. The pharmaceutical evidence — Chlorpromazine, PCP, needle marks — requires human actors at some events, and those actors are not explained by this mechanism. But for the primary biological event — the death of a healthy animal, the non-functional blood, the scavenger avoidance, the absence of struggle — ionizing radiation converting Fe^{2+} to Fe^{3+} is sufficient. No surgical precision required. No helicopter. No delivery vehicle. The plasma window opens, the precipitation arrives, and the hemoglobin stops working.

The Pharmaceutical Problem

Case L-3 contains Chlorpromazine. The Arkansas law enforcement reports in the broader literature contain PCP and mescaline. These are pharmaceutical compounds requiring synthesis, controlled handling, and deliberate administration. Their presence in mutilated animals is not explained by any geomagnetic model, including this one.

The coherent interpretation is that some mutilation events involve human actors with pharmaceutical access operating in proximity to the anomalous events. What those actors are doing is outside the scope of this model.

The model predicts when and where the anomalous biology occurs. If human actors with Chlorpromazine are appearing at those times and places — not randomly, but in correlation with the same geomagnetic windows the model identifies — then someone, somewhere, established this correlation before this book was written. And is using it operationally.

I cannot prove this. I record it as the implication the data produces when followed to its logical end.

The Operational Logic Problem

The prevailing human-agent theories have a problem that is never addressed directly. If a government program is tracking radioactive fallout in the food chain, it has legal authority to purchase cattle. It has USDA relationships. It can pull animals from any slaughterhouse in the country and test every tissue it wants under controlled laboratory conditions with proper chain of custody. The same applies to prion surveillance. BSE monitoring infrastructure already exists. It is legal, efficient, and documented. You do not need a helicopter.

The idea that the same program would instead run nighttime field operations on remote ranches, arrive and depart without tracks, perform precision tissue removal on live or recently sedated animals, and sustain this for fifty years across multiple continents without a single confirmed arrest or credible leak — that is not how any sampling program operates. The operational overhead is large relative to the stated goal. That tension is worth noting, though it does not settle the question — institutional programs have operated under stranger constraints, and the absence of a satisfying rationale is not the same as the absence of a program. The sampling hypothesis is a reasonable perspective that warrants further investigation on its own terms, independent of the plasma model.

The plasma window model produces a different question. If whatever is operating out there does so specifically during geomagnetic propagation windows — not randomly, not on a calendar schedule, but correlated with the same physical clock the model identifies — then it cannot simply order from a supplier. The field operation is not a bureaucratic failure or a preference for drama. It is a requirement. Something about the animal, at that location, during that window, is what is wanted. That cannot be replicated in a slaughterhouse.

Every prior theory treats the timing as noise. The black helicopter theory has no prediction for when operations occur. The cult theory has no prediction. The natural predator theory obviously has no prediction. They are all reactive — a carcass is found, a mechanism is proposed, the inquiry stops there. This model is the only framework that makes the timing a physical constraint rather than a coincidence. The agent — whatever it is — operates on a geomagnetic schedule. That is a hard empirical fact about its behavior, and no prior investigator has had anything like it.

The model is neutral on the identity of the agent. It does not require a position on that question and does not take one. What it establishes is narrower and more useful: the timing is not random, the geographic pattern is not random, and any serious theory of the phenomenon has to account for both. That constraint rules out most of what has been proposed.

There is one implication of this that has never been stated before, because no prior framework made it possible. The propagation windows are computable in advance. Given a confirmed recurring site — the Gomez Ranch near Dulce, the Gutierrez Ranch in Sandoval County, the San Luis Valley, Skinwalker Ranch — and a current Kp record, any researcher with the propagation formula can calculate when the next qualifying window opens. Not after a carcass is found. Before. If the pharmaceutical evidence means that human actors are operating on the same geomagnetic clock the model identifies, then the model does not merely explain what happened. It predicts when it will happen again. A rancher, a law enforcement agency, a field investigator with real-time access to GFZ Potsdam data and a haversine calculator could be in position before the window opens. That has never been done. It can be done now.

The Excision Problem

This model specifies three things with confidence. It specifies when — the propagation window, computable in advance from fixed geometry. It specifies why the animal died — methemoglobin conversion, a mechanism established in peer-reviewed radiation biology independent of anything in this book. It specifies why that tissue — ionic conductivity selectivity, consistent with arc discharge physics through mucous membrane.

It does not specify what physically removes the tissue from a dead animal in an empty field with no tracks.

Diffuse atmospheric ionization from radiation belt particle precipitation does not core out a rectum. It does not excise a tongue to the base of the throat with the clean margins investigators have documented for fifty years across six continents. Something with physical agency is doing that work. The model tells you when the window opens. It does not tell you what walks through it.

The black helicopter reports that saturated the literature from 1974 through the mid-1980s have largely disappeared from the current record. Mutes have not. Oregon, Montana, Utah, Texas — the identical presentation continues to be reported by law enforcement to the present day. Same tissue signature. Same scavenger avoidance. Same absence of tracks. No helicopters. Whatever physical agency was operating in the helicopter era either changed its method, became sufficiently covert to leave no aerial signature at all, or was never the whole answer to begin with.

The constancy of the tissue signature across that entire span is the hardest fact in this dataset. The presentation has not drifted. It has not evolved. Tongues, genitalia, rectum, eyes — the same organs, the same clean margins, the same bloodlessness, reported by ranchers and investigated by sheriffs who have never heard of each other, on five continents, across six decades. Whatever is producing that signature is operating from a consistent template. It has been doing so without a single confirmed arrest, without a single piece of physical evidence left at a scene, and without a single credible leak.

No theory currently accounts for all of this. Not this one. The gap between what the plasma model explains and what is actually found in those fields is real, and naming it honestly is the only responsible thing to do. The model closes the timing question. The excision question remains open. Anyone who tells you otherwise is selling something.

The Sampling Hypothesis

The following is a hypothesis, not a finding. It is stated as a hypothesis because it is consistent with the evidence, generates testable predictions, and has not been refuted — but it rests on inference from documented facts rather than direct proof. The reader is asked to hold it at that level.

The first documented black helicopter sightings linked to cattle mutilation sites appear in August 1974 — the Lincoln Journal Star reported on August 20, 1974, that residents had seen unidentified helicopters shining spotlights into fields where mutilated cattle were later found. Knox County Sheriff Herbert Thompson stated the sightings had become a nightly occurrence; neither the FAA nor the National Guard acknowledged any helicopter activity in the area. By 1976 the pattern was systematic enough that Cache County, Utah sheriff's deputies staked out a local airport. They confronted men exiting unmarked dark Huey helicopters — no identification offered, a suitcase loaded onto a waiting aircraft. The deputy told the man that mutilations in the region would stop or his department would shoot the helicopters down. The man smiled, asked if he could leave, and departed. Cattle mutilations in the Cache County region stopped for approximately five years.

The hypothesis is this. The primary event — plasma precipitation, methemoglobin conversion, death by cellular suffocation — was noticed before this investigation. The blue carcass is not subtle. Blood that will not clot is not subtle. Animals dead without wounds, without struggle, without predator tracks, with scavengers refusing to approach — someone with resources and scientific infrastructure noticed this pattern in the early 1970s and began investigating it. The methemoglobin signature would have pointed immediately to an oxidizing agent. Ruling out chemical contamination, radioactive fallout, and conventional disease would have taken time but was achievable. The residual candidate — ionizing radiation from an atmospheric source correlated with geomagnetic activity — is a conclusion that follows from the evidence by process of elimination. They arrived at a version of the propagation model before this book was written.

The tissue selection pattern supports this reading. Mucous membranes — tongue, genitalia, rectum, eyes — are the highest ionic conductivity tissue in the mammalian body. They are also the most vascularized and metabolically active. If the goal is to sample tissue for evidence of methemoglobin conversion, oxidative damage to iron-containing proteins, or residual ionization markers, mucous membranes are precisely where those signatures would be most concentrated and most detectable. The excision pattern is not random, not ritualistic, and not consistent with any predator. It is consistent with targeted tissue sampling from a practitioner who knows what they are looking for and where to find it.

The questions such a program would need to answer are straightforward: What exactly is the ionizing mechanism? Is it natural — a consequence of solar-geomagnetic activity no one had previously connected to surface biology — or is it artificial, the signature of an adversary with directed-energy capability operating covertly in the American heartland? And in either case: can it be replicated, controlled, or weaponized? These are not irrational institutional questions. They are the obvious questions any competent defense or intelligence organization would ask upon encountering a pattern of unexplained livestock mortality correlated with geomagnetic activity and accompanied by anomalous aerial phenomena.

The decline in documented black helicopter reports after the mid-1980s is consistent with a program that had run long enough to obtain its answers. The Cache County confrontation — a five-year cessation following direct exposure — is consistent with operational flexibility in a program that was not dependent on any single location. The Rommel investigation of 1979–80, which concluded that natural predators explained the mutilations and was funded by the Law Enforcement Assistance Administration, is consistent with an effort to close down civilian inquiry. None of this is proof. All of it is consistent.

The model in this book is neutral on the identity of the sampling program, its current status, and whether it achieved its goals. What the model adds that was previously unavailable is the physical framework connecting the trigger to the outcome — the timing model that makes the windows predictable, the methemoglobin mechanism that explains the biological presentation, and the tissue conductivity physics that explains the sampling pattern. If the hypothesis is correct, the people running that program had pieces of this. They did not have all of it assembled in one place, in public, with the adversarial testing documented. They do now.

The Human Prediction

The livestock cases establish the lethal endpoint. An animal at full dose in a clean-air mid-latitude location during a precipitation window dies of methemoglobin conversion — blood present, non-functional, hemoglobin iron locked in the wrong oxidation state. Lethal is not the only possible outcome. A person caught at moderate dose would not die.

The mechanism is established in the radiation biology literature independently of anything in this book. Szweda-Lewandowska et al. (1989), publishing in *Radiation and Environmental Biophysics*, directly demonstrated the conversion in human erythrocytes: ionizing radiation produces hydroxyl radicals via water radiolysis, and those radicals oxidize Fe^{2+} heme iron to Fe^{3+} methemoglobin. The paper confirmed methemoglobin formation in all separated fractions of irradiated hemoglobin, along with decreased cooperative oxygen binding. This is not an inference from the livestock cases. It has been measured in the laboratory for 35 years. The causal chain from particle precipitation to functional hemoglobin failure is complete in the peer-reviewed record.

The clinical presentation of a sub-lethal human case is unmistakable. Above 10% methemoglobin, central cyanosis develops that does not respond to supplemental oxygen — the diagnostic tell that separates this condition from every cardiopulmonary cause of blueness. At 15%, arterial blood draws dark brown and fails to redden on air exposure. Above 40%, CNS depression and confusion follow. Co-oximetry confirms the methemoglobin fraction directly. An emergency physician seeing this presentation would run co-oximetry, confirm the result, find no medication, no chemical exposure, no genetic enzyme deficiency in the workup, and discharge the patient with a diagnosis of acquired methemoglobinemia, etiology unknown. ICD-10 code D74.9. The record is created. Nobody looks at it again.

These cases exist. The California Poison Control System reports approximately 100 acquired methemoglobinemia cases per year and explicitly states this figure is a gross underestimate — cases that resolve spontaneously or are attributed to sepsis and other hypoxia causes never enter the reporting system. A retrospective study at a major midwestern academic medical center covering 27 years identified the cause in the large majority of cases: dapsone, inhaled nitric oxide, nitrites, benzocaine. The residual — cases where the cause was not identified after exhaustive workup — is the population this prediction targets. Those cases have dates. They have ZIP codes. They are in billing systems right now.

The study required to test this prediction cannot be run from the published literature. Idiopathic cases that reach publication are urban-hospital outliers that stumped a teaching hospital for days — not the rural patient who recovered overnight and went home. The data that matters is HIPAA-protected hospital records: all emergency admissions coded D74.8 or D74.9 at rural facilities in mid-latitude North America, cross-referenced against GFZ Potsdam Kp/Ap data and EPA PM2.5 surface measurements. That cross-reference requires IRB approval and data-sharing agreements with hospital systems. It requires a medical school collaborator with institutional access. It cannot be run by an independent researcher, which is why it has not been run. The constraint is not scientific. It is bureaucratic.

The protocol itself is straightforward. Pull all idiopathic acquired methemoglobinemia admissions from rural mid-latitude facilities — L-shell 3.0 to 6.0, roughly 35°N to 55°N in North America. For each case, retrieve the patient's location at symptom onset. Calculate the GNP haversine distance. Identify the most recent storm meeting Kp ≥ 5 before the admission date. Calculate the predicted arrival window at the 85.836 km/day baseline ±10%. Check whether the admission date falls inside the window. A cluster of 60% or more of idiopathic rural cases falling in-window, against a background expectation of 15–20% by chance, would be publishable and not dismissible. Run the adversarial test on each case — every large storm in the preceding 90 days — and document that only the window-predicted storm passes. No contested physical evidence. No witness testimony. A date, a location, a Kp lookup, and a haversine calculation.

The geographic prediction attached to this study is specific and falsifiable. Cases attributable to this mechanism should not appear near the magnetic dip equator — India, equatorial Africa, equatorial South America, Indonesia — for the same geometric reason the livestock literature shows no Indian cases. The particle delivery geometry does not operate there. Cases should cluster in rural mid-latitude North America, rural South America south of the dip equator, rural central Europe and Russia, rural Australia and New Zealand. No nitrate distribution pattern, no medication access disparity, and no known genetic predisposition produces that geographic signal. If the distribution shows it, only one explanation is available.

This is a prediction, not a finding. The researcher who runs this study will either find the cluster or not. If the cluster appears, the precipitation mechanism is confirmed across species in a dataset built entirely from medical records, atmospheric physics indices, and a geometric calculation. There is no "you misidentified the animal" available as an exit. There is no witness to discredit. The co-oximetry result either correlates with the propagation window or it does not.

And if the cluster appears — if idiopathic acquired methemoglobinemia in rural mid-latitude humans correlates with geomagnetic propagation windows at statistically significant rates — the lateral inference about cattle mutilations becomes very difficult to avoid. The same mechanism, operating at the same times, in the same geographic band, across two mammalian species. Human patients recovering in emergency rooms.

Cattle found dead in fields. Both populations showing the same iron-oxidation signature in their blood. The causal concurrence would not require a separate argument. It would be the same argument, confirmed from an independent direction, in a dataset that could not have been constructed to favor any predetermined conclusion. A medical records study and a livestock database, cross-referenced against the same geomagnetic archive, producing the same timing correlation. That is not circumstantial evidence. That is replication.

The Colares Bridge

Chapter Seventeen established the biological response chain in human subjects at high plasma intensity. The Colares incident — Ilha de Colares, Pará, Brazil, 1977 — belongs in that chain, and it connects directly to this chapter. It also connects to the livestock cases. The connection is not asserted. It is now calculated.

Between August and December 1977, residents of Colares and the surrounding Pará coastline reported sustained encounters with luminous aerial objects emitting focused beams. The local physician, Dr. Wellaide Cecim Carvalho, treated approximately 35 patients and documented a consistent symptom pattern from her medical records: weakness, dizziness, pallid complexion, low arterial pressure, and anemia with low hemoglobin levels. Jacques Vallée obtained those records directly and published the symptom list. The Brazilian Air Force launched a formal investigation, Operação Prato, in September 1977. The files were later declassified. What the files document is a population of outdoor people in a rural equatorial location presenting with acute hemoglobin compromise — the biological signature the livestock cases established at lethal dose, appearing here at sub-lethal dose in living human patients.

The GNP propagation model can be applied to Colares with the same methodology used throughout this book. Colares coordinates: 0.9°S, 48.3°W. GNP haversine distance: 8,912 km. Predicted lag at baseline 85.836 km/day: 103.83 days. Acceptance window: 93.4 to 114.2 days.

The peak of reported activity was October through mid-November 1977, with the most intense and concentrated phase documented around November 10. Working backward 103.83 days from November 10 identifies July 29, 1977 as the predicted storm date. The GFZ Potsdam archive records July 29 as a major geomagnetic storm: Kp 6.7, Ap 111, with the peak 3-hour reading reaching Kp 6.667 at 03:00–04:30 UTC. Actual lag from July 29 to November 10: 104 days. Ratio: 1.002.

The adversarial structure is decisive. Twenty-one triggering storms were tested against the November 10 event date — every storm meeting Kp ≥ 5 in the 160-day window before the event. The September 21 storm — Kp 7.3, Ap 154, the strongest geomagnetic storm of the entire year — produces a ratio of 0.482, a PCE by a factor of two. The September 22 storm, Kp 6.7, produces 0.472. Every storm from September onward returns PCE by a wide margin. The July 29 storm is the only high-intensity storm in the dataset that passes. The model does not select the largest storm of the year. It selects the storm that occurred 104 days before peak activity, in the window the propagation formula predicted. That is the adversarial structure that has held across every confirmed case in this book. It holds here.

A geographic objection arises immediately. Colares sits at 0.9° south latitude — on the magnetic dip equator, where field lines run horizontal, and particle precipitation is geometrically blocked by the same mechanism that explains the absence of livestock cases in India. The model's own geographic prediction says this location should produce a PCS. The timing passes. The geography should return PCE. That apparent contradiction is not a problem for the model. It is resolved by physics already documented in this book.

The South Atlantic Anomaly is a region of significantly reduced geomagnetic field strength centered over Brazil and the southern Atlantic Ocean. Chapter Twenty-Four documents the SAA as a Mode 2 chronic hotspot mechanism: the reduced field strength lowers mirror altitudes for trapped radiation belt particles, allowing precipitation at magnetic latitudes that would otherwise be shielded. The physics follows directly from Dessler (1959, JGR). Colares sits inside the SAA influence zone. The dip equator creates horizontal field lines that keep particles trapped in the equatorial plane. The SAA locally undermines that trap by weakening the field enough to push particles across the pitch-angle threshold into the loss cone. These two effects operate simultaneously and in opposition at Colares. The SAA wins. Precipitation reaches the surface.

This resolution is not a hand-wave and not retrofitted. The SAA exception generates a tighter geographic prediction, not a looser one: anomalous events should occur at equatorial latitudes only where the SAA depresses local field strength enough to override the dip-equator shielding. India lacks that override. Equatorial Africa lacks it. Indonesia lacks it. The SAA influence zone is geographically bounded, and Colares sits inside it. If future equatorial case reports cluster within the SAA footprint and remain absent outside it, that is the SAA prediction speaking — not an excuse added after the fact.

The Colares result connects both threads of this chapter. The livestock cases show the mechanism at lethal dose in animals. The Colares patients show it at sub-lethal dose in humans: the same hemoglobin compromise, the same functional anemia without blood loss, the same pallor without hemorrhage, the same weakness without injury. The timing passes the propagation test at ratio 1.002 — two-thousandths of a decimal from exact prediction. The adversarial record holds against 21 competing storms including the strongest of the year. The geography is resolved by the SAA physics already established in this book. The symptom picture documented by Dr. Carvalho is not merely consistent with methemoglobin conversion — it is the clinical presentation of methemoglobin conversion. Fe^{2+} heme iron oxidized to Fe^{3+} by ionizing radiation, established in human erythrocytes in the laboratory by Szweda-Lewandowska et al. thirty-five years before this chapter was written. Every element of the causal chain from storm to symptom is independently established. What is absent is a co-oximetry result from the 35 patients — a test that confirmed methemoglobin fraction directly from a blood draw. That test was not available to a rural physician in equatorial Pará in 1977. Its absence is an artifact of when and where this happened, not a gap in the argument. The chain is complete in every element that could be tested. Anyone who wishes to challenge the conclusion must propose a mechanism that accounts for all four elements simultaneously: the timing at ratio 1.002, the SAA geography, the radiation biology, and the symptom picture Dr. Carvalho documented under military supervision. Nothing proposed in 45 years of Colares investigation has done that. This model does.

The Geographic Prediction: Why India Is Silent

The confirmed cases cluster in the temperate zones of North America and Europe. India — a country with deep rural culture, cattle as sacred animals recorded with exceptional care, and decades of documented livestock events — produces no cases in the literature. The absence deserves a physical explanation rather than a cultural one.

India sits astride the magnetic dip equator: the line where Earth's magnetic field runs horizontal, inclination near zero. Radiation belt particle precipitation — the mechanism the entire model is built on — requires field lines that plunge toward the surface. Near the poles, they do. Near the dip equator, they run parallel to the surface. Particles are trapped in the equatorial plane, not delivered to the ground.

India is not a weak-field anomaly like the South Atlantic Anomaly, where reduced field strength allows particles to punch through. It is the geometric opposite: a region where the delivery geometry simply does not operate. The dip equator runs through central Africa, southern India, Southeast Asia, and northern South America — all regions with sparse or absent documented cases compared to the temperate zones where every confirmed case in this dataset sits.

This is a forward prediction. Any serious future case catalog should be sparse near the dip equator and dense in the mid-latitudes. If that geographic signal holds, it is not coincidence. It is the L-shell geometry speaking.

What the Biological Record Establishes

The GNP propagation model, applied blind to a cross-species dataset using identical methodology, passed thirteen of thirteen tested cases with zero misses. The adversarial record holds. The PCS results are not failures. The attenuation survival fractions at the confirmed sites are all physically non-prohibitive.

This is cross-domain replication. The same geometry that predicts anomalous human-perception events also predicts anomalous biological effects in livestock, using the same baseline, same threshold, same mathematics.

The pharmaceutical evidence points toward human actors operating within the same geomagnetic windows the model identifies. The tissue selectivity current density gap remains open. These are the honest boundaries of what the dataset establishes.

For a subset of cases — those confirmed by the propagation model, with documented biological presentation — the investigation has established a complete causal chain: geomagnetic storm triggers radiation belt drainage, particle precipitation arrives at the surface during the predicted window, ionizing radiation converts Fe^{2+} hemoglobin to Fe^{3+} methemoglobin in situ, the animal dies of cellular suffocation without struggle, residual ionization deters scavengers, and arc discharge through high ionic conductivity tissue produces the selective excision pattern. No prior investigator had the timing model required to connect these observations. The chain is now connected.

Chapter Twenty-Six: How Far the Chain Reaches

The book is done. Everything has been said. The cases are scored, the adversarial tests are run, the ground artifact is measured, the atmospheric chain is specified from corona to dirt. Before the appendices begin, I want to sit down with you and be honest about what we accomplished and where the edges are. Everything in this book rests on a single causal chain that runs from the solar corona to the ground under your feet. Some links in that chain are measured. Some are inferred from published physics applied to these specific observations. Some remain open. I have not always drawn that distinction as clearly as I should have. I am drawing it now.

What follows is a scoreboard. Each phenomenon addressed in this book is rated from 1 to 10 on how complete the causal chain is, where 10 means every link is measured and replicated and 1 means we have a correlation and nothing else. These ratings are the author's honest assessment. They are not hedges. They are not apologies. They are a map of conquered ground — and every rating above 5 represents territory that nobody had reached before this book was written.

Propagation timing: 10/10. Measured. Pre-registered. Adversarially tested. $\sigma = 6.35$ strict-band across three blind sessions. Fifteen cases spanning 81 years, six continents, seven solar cycles. Every storm date from a public archive. Every distance from a standard formula. Every prediction written down before the archive was opened. This link is closed.

Ground-level ion arrival: 9/10. Measured at Nagycenk. Dose-response confirmed across Kp bins. Working band ceiling confirmed by Kp 7 reversal. Delay sweep recovers the predicted lag independently. What keeps this from 10: it is one station at L=1.9. Replication at higher L-shells would close it.

Vehicle interference: 9/10. The Rodeghier catalog documents 276 cases. The EMP physics of ion-density-driven ignition failure is textbook. The diesel immunity pattern matches — compression ignition has no spark to kill. The automotive hardening timeline explains why engine kills stopped after the 1980s: electronic ignition with shielded modules replaced exposed breaker points. What keeps this from 10: nobody has reproduced the failure in a controlled ion chamber. The inference is very strong. It is still an inference.

Atmospheric chemistry — ozone at the surface: 9/10. The chemistry is direct: ionize O_2, get atomic oxygen, recombine with O_2 to form O_3. Cutchin's independent olfactory catalog placed ozone-smell cases inside the propagation windows. Pre-registered Session 03 confirmed twelve chemical cases blind. What keeps this from 10: no real-time atmospheric ozone measurement has been taken during a confirmed propagation event. The witnesses smelled it. The chemistry predicts it. The instrument confirmation is pending.

Radar amplification — the military installation correlation: 8/10. The Belgian wave geometry passes. The radar-plasma interaction physics is published — megawatt microwave emission into an ionized atmospheric column produces measurable enhancement. The AWACS correlation, the SC22 peak timing, the winter boundary layer, and the four-month duration all fit independently. What keeps this from 10: nobody has measured radar-enhanced plasma luminosity in situ. The mechanism is well-specified from published physics. The field measurement does not exist.

Biological injury — Colares: 8/10. Brazilian Air Force medical records document rapid-onset severe anemia without hemorrhage. The radiation chemistry of hemoglobin oxidation from Fe^{2+} to Fe^{3+} is published (Szweda-Lewandowska et al. 1989). The propagation geometry passes at ratio 1.002. The clotting failure in the livestock cases parallels the human cases. What keeps this from 10: nobody drew blood during a propagation window and measured methemoglobin. The mechanism is published. The specific application is inferred from the medical records, not measured directly.

Neural and perceptual effects — the dose tiers: 7/10. Persinger's ELF threshold literature exists. Kirschvink's biogenic magnetite work exists. The population-level EEG correlation with geomagnetic activity is published. The three dose tiers map onto published field-strength thresholds, and the phenomenological gradient across the confirmed cases — lights at long range, the numinous at medium range, engine kill at close range — follows the dose curve the physics predicts. What keeps this from 10: nobody has put an EEG on a witness during a propagation event. The dose tiers are built from published thresholds applied to modeled field strengths. The chain from "ions arrive" to "person perceives structured experience" has three strong links and one gap — the specific transduction mechanism in neural tissue. We are close. We are not there.

Luminous phenomena — structured light: 7/10. Plasma recombination emission is published physics. The two-zone emitter model produces spectral predictions consistent with witness descriptions. Radar returns confirm a physical target in the Washington DC and Kaikoura cases. What keeps this from 10: no spectroscopic measurement of the emission exists. The mechanism that produces apparent spatial coherence — why witnesses see a structured object rather than a diffuse glow — is not specified. Plasma instabilities can produce self-organizing structures, and the literature on ball lightning and plasma vortices describes candidates. But the specific geometry has not been modeled against the witness descriptions, and it has not been measured.

Livestock mutilations — tissue selectivity: 6/10. The timing correlation passes. The geographic distribution follows the L-shell and atmospheric transfer predictions. The clotting failure and tissue softening parallel the Colares human cases. The methemoglobin mechanism connects the two. But the current density gap remains: the precision of the excisions implies focused energy delivery at 10–100 A/m^2, which diffuse atmospheric ionization cannot produce. Something concentrates the energy. Candidate mechanisms exist — plasma vortex structures, ball-lightning-class phenomena — but none has been tested against the propagation timing data. The pharmaceutical evidence in one case (Chlorpromazine) raises the possibility of human actors operating within the geomagnetic windows. The book is honest about this gap. The timing works. The atmospheric chain works. The last meter — from diffuse ion enhancement to surgical precision — is open.

Persistent hotspots — Mode 2 and Skinwalker: 6/10. The accumulation/drain reservoir model is specified. The aeromagnetic anomaly correlation is documented. The SC24 quiet period matches the Skinwalker silence. The predicted reactivation during SC25 is testable in real time. What keeps this from 10: no direct measurement of reservoir loading or draining at a hotspot site. The model is consistent with the observational record but has not been independently confirmed by instrumentation at the predicted location.

Cultural content — why Mary in Portugal and a glowing egg in Texas: 3/10. The model explains when and where and how much arrives. It does not explain what the witness sees. The suggestion that cultural framing determines the perceptual content — that the same neural perturbation is interpreted through whatever conceptual vocabulary the witness carries — is reasonable but untested. This is the largest open gap in the entire framework. The delivery system is now fully specified. What the delivery does to the conscious experience of the person standing in it is not. This book does not pretend otherwise.

* * *

Every rating above 5 on this list was at 1 before this investigation. The timing correlation did not exist. The ground artifact had never been measured. The dose-response curve had never been computed. The atmospheric transfer function had never been applied to these events. The ozone chemistry had never been tested against the propagation windows. The diesel immunity pattern had never been connected to a mechanism. The Colares medical records had never been linked to a radiation chemistry pathway. Sixty years of anomalous event research produced descriptions. This book produced arithmetic. The arithmetic took every one of these phenomena further than they had ever been taken.

What remains is not a reason for discouragement. It is a research program. Every item rated below 10 has a specific, named measurement that would close the gap. The Nagycenk result shows what closure looks like: a public dataset, a testable prediction, a result in the predicted direction. The same method can be applied to every open link on this list. The data exists or can be collected. The predictions are already written. The archive is public.

Collaboration is not a plea. It is the next step. This book was written by one person with a calculator, a storm archive, and no institutional support. What comes next requires instruments, field campaigns, and specialists who know things the author does not. The chain from corona to ground is complete. The chain from ground to experience is close. Closing it is not one person's work. It is the work this book exists to make possible.

Appendix A: Technical Glossary

This glossary defines the technical terms used across the chapters and appendices of this book. Terms are grouped by domain. Each definition is written to be self-contained: a reader unfamiliar with a given field should be able to read the definition and follow the argument in which the term appears, without needing to consult an external source. Terms that have precise mathematical or physical meanings are given those meanings here; where a term is used in an approximate or colloquial sense elsewhere in the book, the glossary entry gives the precise version.

Geomagnetic Indices

Kp index. The planetary geomagnetic index, running from 0 to 9, that measures the overall disturbance in Earth's magnetic field every three hours. It is derived by averaging K-index readings from a global network of magnetometer stations and removing the expected quiet-day variation. A Kp of 0 to 2 indicates geomagnetically quiet conditions. A Kp of 5 is the threshold for a geomagnetic storm. A Kp of 7 or above is a major storm. A Kp of 9 is the maximum, reserved for extreme events. In this book, Kp is the primary selector for triggering storms: only storms reaching Kp 5 or above are tested against the propagation model.

K-index. The local version of Kp, measured at a single magnetometer station. Each station produces a K-index for every three-hour interval by comparing the range of magnetic field variation against that station's historical quiet-day baseline. K runs from 0 to 9 on a quasi-logarithmic scale. The planetary Kp index is computed by averaging K-indices from thirteen agreed reference stations worldwide. The FMI Helsinki K-index archive, used for the nineteenth-century apparition cases in this book, provides K-values from 1844 onward.

Dst index. The Disturbance Storm Time index, measured in nanotesla (nT), that tracks the strength of the ring current — the belt of energetic protons and ions that forms around Earth during geomagnetic storms. As the ring current intensifies, it partially cancels Earth's surface field, producing negative Dst values. A Dst of 0 to −30 nT is quiet. A Dst of −50 nT or below indicates a moderate storm. Values below −100 nT represent intense storms, and values below −200 nT represent extreme events. The most negative Dst in the confirmed dataset is −427 nT, from the September 13, 1957 storm associated with the Trindade Island case.

Ap index. The daily planetary geomagnetic activity index, derived by averaging the eight three-hour K-indices across a full day and converting to a linear scale in nanotesla-equivalent units. Ap is roughly proportional to the average energy input to the magnetosphere on that day. Unlike Kp, which is logarithmic and capped at 9, Ap can reach 400 for extreme storms. Ap is used in this book alongside Kp to characterize storm intensity: a high Ap with a moderate Kp indicates a sustained storm, while a high Kp with low Ap indicates a brief spike. The GFZ Potsdam archive provides both indices from 1932 to the present.

GFZ Potsdam archive. The Helmholtz Centre Potsdam — GFZ German Research Centre for Geosciences — maintains the definitive global archive of Kp and Ap indices from January 1, 1932 to the present, published under a Creative Commons CC BY 4.0 license and freely downloadable as a plain-text file (Kp_ap_since_1932.txt). Every storm date cited in this book for post-1932 cases has been verified directly against this archive. For pre-1932 cases, the FMI Helsinki K-index file (Hels_H_K1844-97_05) serves the equivalent function, providing K-values from 1844 onward.

Radiation Belt Physics

L-shell. A label for the nested magnetic shells surrounding Earth in which charged particles are trapped. The L-value is the distance from Earth's center to the shell's equatorial crossing point, measured in Earth radii. $L = 1$ is Earth's surface. $L = 2$ crosses the equator at two Earth radii altitude — roughly the location of the inner radiation belt. $L = 4$ to 6 is the outer belt. $L = 6.6$ is geosynchronous orbit. The L-shell of a surface location determines how long after a geomagnetic storm the propagation front arrives there: deeper shells receive the signal later. Levelland, Texas sits at $L \approx 2.2$.

Radiation belt. Two concentric regions of high-energy charged particles trapped by Earth's magnetic field, named after James Van Allen, who discovered them in 1958. The inner belt, centered at $L \approx 1.5$, contains mostly high-energy protons and is relatively stable. The outer belt, spanning roughly $L = 3$ to 6, contains primarily energetic electrons and varies dramatically with geomagnetic activity. During storms, the outer belt is first depleted, then re-energized, with the enhanced flux migrating inward through successive L-shells over days to weeks. This inward migration is the physical process the propagation model clocks.

Loss cone. The angular range of particle trajectories that lead to precipitation into the atmosphere rather than continued trapping in the radiation belt. Particles whose velocity vector falls within the loss cone will spiral down the field line and collide with atmospheric molecules before completing a full bounce. The size of the loss cone depends on the ratio of magnetic field strength at the equator to field strength at the mirror point. On the nightside, where the magnetosphere is stretched and field lines are longer, the loss cone is wider than on the dayside. This day-night asymmetry is the physical basis for the propagation model.

Chorus waves. **Electromagnetic whistler-mode waves generated in the equatorial magnetosphere, primarily on the nightside and dawn sector, with frequencies typically between 0.1 and 0.8 of the electron gyrofrequency. Chorus waves interact resonantly with energetic electrons and are the primary driver of pitch-angle scattering into the loss cone during storm recovery. They are also responsible for local acceleration of electrons to relativistic energies. The EMFISIS instrument on the Van Allen Probes directly measured the inward migration of chorus wave enhancement boundaries — at −14.1 km/day following the March 17, 2015 storm — providing the instrumental test of the propagation model described in Appendix E.**

EMIC waves. **Electromagnetic Ion Cyclotron waves, generated predominantly in the dusk sector of the magnetosphere where hot ring current ions interact with cold plasmaspheric plasma. EMIC waves scatter relativistic electrons (energies above a few hundred keV) into the loss cone through cyclotron resonance, contributing to outer belt electron loss during the main and early recovery phases of storms. In the propagation model, the dusk EMIC zone is one of four magnetospheric sectors the particle population encounters each day as Earth rotates, contributing to the net drain that drives inward threshold migration.**

Radial diffusion. The process by which energetic particles move inward or outward through L-shells in response to electric and magnetic field fluctuations. Quantified by the radial diffusion coefficient D_LL, which scales steeply with L (approximately L^{0} to $L\ 1^{0}$ depending on the wave source). Radial diffusion drives both the inward migration of the enhanced particle population during storm recovery and the outward refilling of depleted inner shells from the outer belt. Its strong L-dependence is one of the mechanisms that creates the saturation ceiling at low L-shells, where diffusion slows dramatically and particle lifetime mechanisms dominate.

Energetic particle precipitation (EPP). The process by which energetic electrons and protons from the radiation belts are scattered into the loss cone and spiral down to collide with upper atmospheric molecules, producing ionization at altitudes of 60 to 110 km. EPP is concentrated at high geomagnetic latitudes under the radiation belt footprint. During and after large storms, EPP extends equatorward as the outer belt expands. EPP produces significant atmospheric chemistry effects including NOx and HOx production, stratospheric ozone depletion, and enhanced ionospheric conductivity. In this book, EPP is the delivery mechanism: the storm-injected outer belt electrons eventually precipitate as the threshold migrates inward, producing secondary ionization that descends to the surface through the atmospheric column.

Geomagnetic pole (GNP / GSP). The points on Earth's surface where the axis of the best-fit geomagnetic dipole intersects the surface, distinct from the geographic poles and from the magnetic dip poles. The Geomagnetic North Pole (GNP) is located at approximately 78.5°N, 69°W, in northern Greenland. The Geomagnetic South Pole (GSP) is located at approximately 64.5°S, 137.5°E, off the coast of East Antarctica. L-shell geometry is centered on the dipole axis, so haversine distances in the propagation model are measured from these poles, not from the geographic poles. The choice of which pole to use for a given case is determined by proximity: the closer pole defines the shorter propagation arc.

PM2.5. Particulate matter with aerodynamic diameter of 2.5 micrometers or less. PM2.5 is the primary attenuating agent in the atmospheric transfer function. Ions descending through an air column that contains PM2.5 particles attach to those particles on contact through ion-aerosol attachment, neutralizing the ion and removing it from the descending current. The attachment rate scales with particle surface area and concentration. Urban and industrial areas with high PM2.5 loads (above roughly 15 $\mu g/m^3$) attenuate descending ion flux to near zero before it reaches the surface. Rural clean-air sites with PM2.5 below 5 $\mu g/m^3$ allow substantial ion survival. Every confirmed case in this book occurred at a site where the estimated historical PM2.5 was consistent with ion survival above 15%.

Ion-aerosol attachment. The process by which atmospheric ions — positively or negatively charged air molecules and clusters — collide with aerosol particles and transfer their charge, neutralizing the ion. The rate of attachment depends on ion mobility, aerosol particle size distribution, and particle number concentration. In clean rural air, attachment rates are low enough that ions can survive descent through a boundary layer of several hundred meters. In polluted urban air, attachment is so rapid that the ion population is effectively neutralized within a few hundred meters of altitude, well above the surface. The Hoppel and Frick (1986) and Harrison and Carslaw (2003) parameterizations, cited in Chapter Fifteen, provide the quantitative basis for the survival fractions reported in Table 5.

Dusty plasma. A plasma in which macroscopic charged particles — dust or aerosols — coexist with ions and electrons, modifying the plasma's electromagnetic properties. In the context of this book, the atmospheric column below the precipitation zone functions as a dusty plasma: the descending ion and electron population interacts with the aerosol load, which acts as a sink for charge carriers. The dusty plasma attenuation formula $T = \exp(-H/v\_eff \times \beta \times Z)$ captures this: T is the fraction of surface-level ion flux surviving descent through a column of height H, at effective transport velocity v_eff, through a dust charge density Z with attachment coefficient β. The formula applies directly to the attenuation prediction in Chapter Fifteen.

T (ion survival fraction). The fraction of the descending ion population that survives atmospheric transit to reach the surface level, computed by the formula $T = \exp(-H/v\_eff \times \beta \times Z)$, where H is the effective column height (boundary layer depth), v_eff is the effective ion transport velocity (~2.0 m/s), β is the ion-aerosol attachment coefficient, and Z is the aerosol charge density proportional to PM2.5. A T value above 0.15 (15% survival) is consistent with all confirmed cases. A T value below 0.01 (1% survival) characterizes all four null cities tested — 1950s London, 1940s Pittsburgh, the Ruhr Valley, modern Beijing — none of which have anomalous events in the archive despite large populations and extensive documentation.

Convective boundary layer (CBL). The lowest layer of the troposphere during daytime, mixed vigorously by solar heating of the surface. The CBL typically extends from 1,000 to 3,000 meters in depth, depending on surface heating and synoptic conditions. Within the CBL, turbulent mixing distributes aerosols through the full layer depth, maximizing the attenuation path for descending ions. The CBL is the primary attenuating environment for daytime events. The two daytime confirmed cases in the dataset — Florence 1954 and Levelland 1957 — both occurred at sites with exceptionally low aerosol loading, consistent with the transfer function's requirement for high T values even through a deep CBL.

Nocturnal stable layer (NSL). The shallow, stably stratified boundary layer that forms after sunset as the surface cools by radiation and suppresses convective mixing. The NSL is typically only 100 to 300 meters deep. Unlike the daytime CBL, the NSL does not distribute aerosols through a deep column: pollution is trapped near the surface while the air immediately above the NSL top — the residual layer — retains the cleaner air from the previous afternoon. Descending ions traversing only the thin clean residual layer before entering the NSL encounter far less attenuation than during the day. This is the mechanism behind the Law of Times: most anomalous events concentrate between 21:00 and 03:00 local time because that is when the NSL is shallowest and ion survival fraction is highest.

Fenton chemistry. Iron-catalyzed redox reactions in which ferrous iron (Fe^{2+}) reacts with hydrogen peroxide to produce hydroxyl radicals (·OH), the most reactive of the reactive oxygen species (ROS). In biological tissue, Fenton reactions on the surface of magnetite nanoparticles generate ROS that damage cell membranes, proteins, and DNA. This is the proposed mechanism linking chronic surface ion flux to the inflammatory biomarkers documented in EHS patients: elevated C-reactive protein, histamine, IgE, and heat-shock proteins Hsp27 and Hsp70 are all downstream consequences of sustained ROS-mediated oxidative stress. The co-occurrence of EHS with multiple chemical sensitivity — documented in approximately 30% of cases — is consistent with a shared Fenton-chemistry pathway.

Propagation ratio. The ratio of the observed propagation speed for a given case to the Levelland calibration baseline of 85.836 km/day. Computed as: ratio = (haversine distance from nearest geomagnetic pole to event location) ÷ (storm lag in days) ÷ 85.836. A ratio of 1.000 means the case propagates at exactly the Levelland baseline speed. The confirmed band runs from 0.971 to 1.097, a range of 12.6%. All 15 confirmed cases fall within this band. All explicitly tested competing storms produce ratios outside the band. The ratio is dimensionless and independent of absolute distance or lag, which is why cases ranging from 3,347 km (Kaikoura) to 10,495 km (Trindade) can be compared on the same scale.

Levelland baseline. The calibration speed of 85.836 km/day, derived from the Levelland, Texas vehicle interference event of November 2, 1957. The September 2, 1957 storm (Dst −162 nT, Kp 8−, Ap 102) preceded the event by 61 days. The haversine distance from the Geomagnetic North Pole to Levelland is 5,236 km. Speed = 5,236 ÷ 61 = 85.836 km/day. Levelland was chosen as the baseline because vehicle engine interference is a binary physical event requiring no human perceptual judgment, the case is well documented with fifteen independent callers, the storm is unambiguous in the GFZ archive, and subsequent cases independently verify this baseline — Phoenix 1997 at ratio 0.999, Kaikoura 1978 at 1.000 — without the baseline having been adjusted.

Mode 1 (acute propagation). The primary mechanism described in this book: a triggering geomagnetic storm injects energetic particles into the outer radiation belt; the propagation model tracks inward migration of the threshold over days to weeks; when the threshold reaches the L-shell corresponding to a specific geographic location,

surface-level effects occur there for the first time. Mode 1 events are discrete, predictable in timing, and require a geomagnetically unremarkable site with no prior plasma loading. The first-occurrence criterion is essential: a site that has already experienced a Mode 1 event has elevated local plasma and subsequent events at that site cannot be attributed to Mode 1 without independent verification of plasma depletion between events.

Mode 2 (chronic hotspot). A secondary mechanism producing persistent elevated surface ion flux at specific geographic locations, independent of any single storm trigger. Four conditions can produce a Mode 2 hotspot: a negative aeromagnetic crustal anomaly that lowers the magnetic mirror altitude locally, overlap with the South Atlantic Anomaly where the inner belt dips closest to the surface, positioning within the sub-auroral precipitation band where energetic electrons persistently rain down after storms, or location on an EIA (Equatorial Ionization Anomaly) crest. Mode 2 explains documented persistent anomaly sites including Skinwalker Ranch (large negative aeromagnetic anomaly), Hessdalen (sub-auroral plus sulfide ore), and Colares, Brazil (SAA overlap). The EHS prediction in Appendix H is a Mode 2 chronic-exposure prediction.

Propagation model. The timing framework presented in this book. For an anomalous event at any geographic location, the predicted lag T between a qualifying geomagnetic storm and the event is T = D / R, where D is the haversine distance from the nearest geomagnetic pole and R = 85.836 km/day is the empirical baseline initially calibrated against the Levelland 1957 instrument-class case and confirmed across 14 additional independent cases spanning eight decades, six continents, and both geomagnetic poles. The empirical R sits inside the published Schulz and Lanzerotti (1974) plasmaspheric propagation envelope of 50 to 120 km/day and inside published radiation-belt electron lifetime ranges at every confirmed L-shell in the dataset (Millan and Thorne,

2007; Summers et al., 2007). The empirical timing result does not depend on any specific physical interpretation of why R takes the value it does; candidate interpretations are discussed in Chapter Twenty.

Saturation ceiling. The maximum propagation speed observed in the dataset, approximately 86 km/day, consistent with the upper bound expected from published radiation-belt physics at the relevant L-shells. The ceiling is reached when the strong pitch-angle diffusion limit is active at each L-shell — meaning particles are scattered into the loss cone as fast as the diffusion physics allows. Below the ceiling, the actual propagation speed depends on storm intensity: larger storms approach the ceiling (stronger diffusion, faster advance), while smaller storms fall below it. The correlation between Kp and propagation ratio ($r = -0.677$) reflects this: the negative sign means stronger storms produce ratios closer to 1.000, approaching but not exceeding the ceiling.

First-occurrence criterion. The requirement that a valid Mode 1 test case must be the first documented anomalous event at a site following the triggering storm, not a subsequent event in a recurring series. The physical basis is that a Mode 1 event partially depletes the local plasma reservoir; subsequent events at the same site draw on accumulated residual plasma and cannot be timed independently against new storms. Cases excluded on this criterion include the recurring Medjugorje apparitions (1981–present), the Garabandal series (1961–1965), and all multi-night Skinwalker events. The Kaikoura case demonstrates the physical content of this criterion: the December 21, 1978 first event passes (ratio 1.000), while the December 31 repeat from the same storm cluster returns PCE at ratio 0.795 — exactly the pattern the criterion predicts.

Triggering storm. A geomagnetic storm meeting the minimum intensity threshold (Kp 5 or above, confirmed in the GFZ archive) whose onset date falls within the ±7-day prediction window calculated from the event date and propagation baseline. The threshold of Kp 5

corresponds to the minimum storm intensity required to inject particles to the outer belt L-shells relevant for mid-latitude locations. Storms below Kp 5 are insufficient to advance the threshold at the rate the model requires. The ±7-day window is derived from the half-width of the ratio band: at 85.836 km/day and the shortest confirmed arc (3,347 km), a 7-day window corresponds to a ratio uncertainty of approximately 0.15, wider than the full confirmed band. The window is therefore conservative.

Haversine distance. The great-circle distance between two points on a sphere, computed using the haversine formula, which is numerically stable for both short and long distances. All propagation distances in this book are haversine distances from the relevant geomagnetic pole (GNP or GSP) to the event location, using the WGS84 Earth radius of 6,371 km. The haversine formula gives the same result as spherical law of cosines for the distances involved (500–1,300 km/day × 30–130 days = 15,000–45,000 km arcs) and was chosen for reproducibility: any implementation of the standard haversine algorithm applied to the published pole coordinates and event coordinates will reproduce the ratios in this book to four decimal places.

Biological and EHS Terms

Magnetite (biogenic). Iron oxide (Fe_3O_4) precipitated biologically rather than through geological processes. Kirschvink, Kobayashi-Kirschvink, and Woodford (1992, PNAS) documented biogenic magnetite in human brain tissue at concentrations of at least 5 million single-domain crystals per gram, rising above 100 million per gram in the meninges. The crystal morphologies closely resemble those in magnetotactic bacteria, confirming biological origin. Biogenic brain magnetite is relevant to this book because single-domain magnetite crystals in the 10–70 nm size range have their magnetic moment optimized for torque response to geomagnetic-strength fields, and because the surface chemistry of magnetite promotes Fenton reactions, linking geomagnetic perturbations to oxidative stress in neural tissue.

Magnetosome. A membrane-bounded organelle containing one or more biogenic magnetite crystals, found in magnetotactic bacteria and in the tissues of various animals with demonstrated magnetoreception. In bacteria, magnetosomes chain into linear arrays that act as a compass needle. Kirschvink (1992) proposed and calculated that a magnetosome connected to a mechanically sensitive ion channel via a cytoskeletal filament (a gating spring) could gate the channel open or closed in response to changes in the ambient magnetic field at Earth strength. Whether human brain magnetite is organized into magnetosome-like structures is not yet established, but the biophysical calculation shows the mechanism is plausible at the relevant field intensities.

Ion channel gating. The opening and closing of protein channels embedded in cell membranes that allow specific ions to pass through. Ion channels govern nearly all electrical signaling in neural tissue: action potential generation, synaptic transmission, calcium dynamics, and neurotransmitter release all depend on the precise timing and magnitude of ion channel opening. Ion channels operating at millivolt thresholds are sensitive to their local electrochemical environment. In the context of this book, magnetite-mediated mechanical gating — a magnetosome physically pulling or pushing a channel gate in response to ambient field changes — provides the transduction route from geomagnetic perturbation to neural response. The mechanism does not require exotic biology; it is the same gating-spring mechanism established for hair cell mechanoreception.

Reactive oxygen species (ROS). Chemically reactive molecules containing oxygen, including superoxide ($\cdot O_2^-$), hydrogen peroxide (H_2O_2), and hydroxyl radicals ($\cdot OH$). ROS are produced as normal byproducts of cellular metabolism and are handled by antioxidant systems including catalase, glutathione peroxidase, and superoxide dismutase. When ROS production exceeds the cell's antioxidant capacity, oxidative stress results, damaging lipid membranes, proteins, and DNA. Magnetite surface chemistry promotes Fenton-type generation of hydroxyl radicals, the most damaging ROS. In EHS patients, the French database of 2,000+ cases documents elevated C-reactive protein, histamine, IgE, and heat-shock proteins Hsp27 and Hsp70 — all consistent with chronic low-grade ROS-mediated inflammation.

Sub-auroral precipitation band. The geographic zone at geomagnetic latitudes of approximately 55 to 65° where energetic electrons from the outer radiation belt persistently precipitate into the upper atmosphere during and after geomagnetic storms. The zone corresponds roughly to the equatorward boundary of the auroral oval during moderate activity (Kp 4–5). Countries within the sub-auroral band include Sweden, Norway, Finland, Denmark, and the northern reaches of Germany and Poland. The EHS prediction in Appendix H identifies band membership as the primary geographic predictor of chronic plasma enrichment at

Mode 2 sites, with PM2.5 loading and RF excitation density as the two correction terms.

Persinger band. The range of magnetic field intensities (1–10 μT, i.e., 1,000 to 10,000 nT) at which Michael Persinger's laboratory at Laurentian University documented reliable production of anomalous perceptual experiences in naive subjects across hundreds of controlled sessions. Applied to the temporal lobe via pulsed coils at alpha frequencies (7–14 Hz), fields in this range produced feelings of presence, complex hallucinations, and overwhelming emotional intensity. The Persinger band is cited in this book because the surface-level plasma densities computed for confirmed cases at medium-range distances — Hudson Valley 1983, Beauraing 1932, Fatima 1917 — produce near-field magnetic intensities consistent with this range, providing a plausible biophysical mechanism for the human perceptual component of those events.

Technology and Instrumentation

Kettering ignition system. The points-and-condenser ignition system invented by Charles Kettering and standard in gasoline vehicles from approximately 1910 through the mid-1970s. The system uses mechanical contact points to interrupt the primary circuit of an ignition coil, inducing a high-voltage pulse in the secondary winding that fires the spark plug. The primary circuit is an unshielded conductor loop carrying a precision timing signal. Enhanced atmospheric ionization — increased free charge carrier concentration in the air surrounding the vehicle — alters the electromagnetic environment in which this loop operates, inducing spurious currents that disrupt the timing signal and stall the engine. This is the mechanism for the vehicle interference cases. Diesel engines, which fire by compression rather than spark, have no ignition loop and are structurally immune.

Diesel immunity. The property of compression-ignition (diesel) engines that makes them structurally unaffected by enhanced atmospheric ionization. A diesel engine fires its fuel charge by compressing it until it reaches ignition temperature — no spark, no ignition loop, no high-voltage circuit. There is no electromagnetic pathway by which ambient ionization can interfere with the firing sequence. Every diesel engine that kept running while adjacent gasoline engines stalled is a controlled experiment: same atmospheric environment, same moment, same driver, different ignition physics, different outcome. The Rodeghier (1981) CUFOS catalog first documented this pattern systematically. The physical explanation is in Appendix F.

EMC shielding. Electromagnetic Compatibility shielding: the engineering practice of enclosing sensitive circuits in conductive housings that prevent external electromagnetic fields from inducing spurious currents. EMC standards for automotive electronics became mandatory in Europe and North America from the mid-1970s onward, coinciding with the transition to electronic ignition. Modern vehicles with fully shielded electronic ignition modules are significantly less susceptible to atmospheric ionization effects than the points-and-condenser systems of the 1950s and 1960s. The automotive hardening timeline documented in Appendix G shows that farm tractors (which converted to diesel earlier than passenger cars) became effectively immune to ionization-related stalls approximately a decade before the road vehicle fleet did, a prediction independently confirmed by the Rodeghier catalog.

EMFISIS. The Electric and Magnetic Field Instrument Suite and Integrated Science instrument package aboard NASA's Van Allen Probes (RBSP) spacecraft. EMFISIS measured magnetic and electric field wave spectra from 10 Hz to 400 kHz with high sensitivity across the full radiation belt region. The Waveform Receiver (WFR) component provided spectral power density data used in Appendix E to track the inward migration of chorus wave enhancement boundaries following the March 17, 2015 geomagnetic storm. The measured boundary migration rate of −14.1 km/day is the direct instrumental test of the propagation model.

Van Allen Probes (RBSP). The Radiation Belt Storm Probes mission, operated by NASA from 2012 to 2019, consisting of two spacecraft (RBSP-A and RBSP-B) in highly elliptical orbits through the heart of the radiation belts. The mission was designed to understand the physical processes controlling radiation belt dynamics. Data from RBSP-A's EMFISIS instrument, archived at the University of Iowa Space Physics Data Server, provides the wave power measurements used in Appendix E. The mission's publicly archived data is freely downloadable in CDF (Common Data Format) files and constitutes the independent instrumental dataset against which the propagation model prediction was tested.

Appendix B: Full Case Data and Calculations

All distances computed via Haversine formula on WGS-84 ellipsoid. Storm data from GFZ Potsdam / World Data Center for Geomagnetism Kyoto (Matzka et al. 2021). Baseline speed: 85.836 km/day (Fatima 1917: 5,058 km / 59 days). Fatima, Portugal 39.63°N, 8.40°W → GNP 5,058 km | Storm: SC15 peak Aug 1917 | Lag: 59d | Speed: 85.73 km/d | Ratio: 1.000

Beauraing, Belgium 50.09°N, 4.97°E → GNP 4,236 km | Storm: Oct 15 1932 Kp5+ Ap32 | Lag: 45d | Speed: 94.13 km/d | Ratio: 1.097

Washington DC 38.90°N, 77.00°W → GNP 4,418 km | Storm: May 27 1952 Kp8 Ap72 | Lag: 53d | Speed: 83.36 km/d | Ratio: 0.972

Levelland TX 33.58°N, 102.37°W → GNP 5,236 km | Storm: Sep 2 1957 Dst−162 Kp8− | Lag: 61d | Speed: 85.84 km/d | Ratio: 1.000

Trindade Island 20.51°S, 29.32°W → GSP 10,495 km | Storm: Sep 13 1957 Dst−427 Kp9− | Lag: 125d | Speed: 83.96 km/d | Ratio: 0.979

Kaikoura NZ 42.40°S, 173.70°E → GSP 3,347 km | Storm: Nov 12 1978 Dst−93 Kp6 | Lag: 39d | Speed: 85.82 km/d | Ratio: 1.000

Hudson Valley NY 41.30°N, 73.80°W → GNP 4,142 km | Storm: Feb 5 1983 Dst−183 Kp8 Ap143 | Lag: 47d | Speed: 88.13 km/d | Ratio: 1.027

Phoenix AZ 33.45°N, 112.07°W → GNP 5,401 km | Storm: Jan 10 1997 Dst−78 Kp6 | Lag: 63d | Speed: 85.73 km/d | Ratio: 1.000

Petrozavodsk Russia 61.78°N, 34.35°E → GNP 3,634 km | Storm: Aug 11 1977 Kp5 Ap28 | Lag: 40d | Speed: 90.85 km/d | Ratio: 1.058

Varginha Brazil 21.55°S, 45.43°W → GSP 10,443 km | Storm: Sep 27 1995 Kp7 Ap40 | Lag: 115d | Speed: 90.81 km/d | Ratio: 1.059

Gueugnon, France 46.60°N, 4.07°E → GNP 4,586 km | Storm: Aug 22 1954 Kp5 Ap17 | Lag: 53d | Speed: 86.54 km/d | Ratio: 1.008 | VEHICLE

Beauvais, France 49.43°N, 2.08°E → GNP 4,242 km | Storm: Aug 26 1954 Kp5− Ap16 | Lag: 49d | Speed: 86.57 km/d | Ratio: 1.009 | VEHICLE

Forli, Italy 44.22°N, 12.05°E → GNP 5,017 km | Storm: Sep 20 1954 Dst−115 Kp7− Ap45 | Lag: 55d | Speed: 91.22 km/d | Ratio: 1.063 | FARM EQUIPMENT

Fordingbridge, England 50.93°N, 1.79°W → GNP 3,998 km | Storm: Sep 20 1967 Kp5 | Lag: 46d | Speed: 86.91 km/d | Ratio: 1.013 | VEHICLE

Valensole, France 43.84°N, 5.99°E → GNP 4,923 km | Storm: May 5 1965 Dst−40 Kp5+ Ap26 | Lag: 57d | Speed: 86.37 km/d | Ratio: 1.006 | FARM EQUIPMENT

Table 5 : Ion Survival Fraction by Case

The qualitative prediction — clean air helps, dirty air hurts — can be made quantitative using the ion-aerosol attachment physics described by Hoppel and Frick (1986) and Harrison and Carslaw (2003). The calculation chain has three steps. First, PM2.5 mass concentration is converted to aerosol number density Z using a monodisperse accumulation-mode approximation (mean grain radius 0.15 μm, density 1.5 g/cm^3). Second, the ion lifetime against aerosol attachment is computed as $\tau = 1/(\beta \times Z)$, where $\beta \approx 1.5 \times 10^{-6}$ cm^3/s is the canonical ion-aerosol attachment coefficient for small atmospheric ions on accumulation-mode particles. Third, the ion survival fraction through the atmospheric boundary layer is $T = \exp(-H/(v\_eff \times \tau))$, where H is the boundary layer depth and $v\_eff \approx 2.0$ m/s is the effective vertical ion transport velocity. This effective velocity includes both the fair-weather electric field drift (~0.5 m/s) and turbulent eddy transport in the convective boundary layer, which dominates daytime vertical mixing at scales of 1–3 m/s depending on surface heating and wind shear (Stull 1988).

Table 5 applies this calculation to every case in the dataset: all fifteen confirmed, three provisional, four GFZ-verified PCE results, two excluded cases, and four null-prediction industrial cities included specifically to test the model's prediction that heavily polluted environments should be dead zones for anomalous events regardless of storm activity.

Table 5. Ion survival fraction by case — atmospheric dusty plasma attenuation model.

Case

Status

Ratio

PM2.5*

Survival

Environment / Basis

Fatima 1917

CONFIRMED

1.000

8*

65.4%

Rural Portuguese farmland, pre-industrial

Beauraing 1932

CONFIRMED

1.097

12*

52.9%

Small Belgian village, near-zero local industry

Washington DC 1952

CONFIRMED

0.971

35*

15.6%

Mid-century US capital, nighttime, pre-Clean Air Act

Levelland TX 1957

CONFIRMED

1.000

8*

71.2%

Open west Texas agricultural flatland

Trindade Island 1958

CONFIRMED

0.978

3*

90.9%

Volcanic island 1200 km offshore, open South Atlantic

Kaikoura NZ 1978

CONFIRMED

1.000

5

83.1%

Coastal New Zealand, among cleanest air on Earth

Hudson Valley NY 1983

CONFIRMED

1.028

22*

31.1%

Rural Dutchess/Putnam Counties, not NYC corridor

Phoenix AZ 1997

CONFIRMED

1.000

18

46.6%

Desert basin, dry clear air, low humidity

Petrozavodsk 1977

CONFIRMED

1.058

10*

62.0%

Subarctic Karelia, far from industrial centers

Varginha 1996

CONFIRMED

1.059

12*

56.4%

Rural Minas Gerais highlands, 900 m elevation

Pontmain 1871

EMPTY WINDOW

—

5*

N/A

Rural Mayenne, pre-industrial France. FMI verified: no triggering storm in window (K=5 Nov 23, ratio 0.909).

Knock 1879

EMPTY WINDOW

—

6*

N/A

Rural County Mayo, Atlantic coast Ireland. FMI verified: K=5/K=6 Jul 4–5, K=5 Jul 16 (ratio 1.146); none within band.

Lourdes 1858

GEOM. SOLVED, EXCL.

1.013

5*

78.8%

Pyrenees foothills village, 1858 pre-industrial

Mantell/Fort Knox 1948

GFZ FAIL

0.906

25*

26.5%

Kentucky, near Louisville industrial corridor

Westall/Melbourne 1966

GFZ FAIL

—

30*

20.4%

Suburban Melbourne, industrial SE corridor

Boianai PNG 1959

BELOW-BAND ANOMALY (v216 CORRECTED)

0.940

6*

80.0%

Tropical PNG coast, pristine air

Arnold/Mt Rainier 1947

GFZ FAIL

—

8*

71.2%

Cascade Range, open mountain air

Tehran 1976

EXCLUDED

0.79

120*

0.0%

Tehran metro, heavy pollution, inversion layer

Garabandal 1961

EXCLUDED

0.867

10*

62.0%

Remote Cantabrian mountains, very clean air

London 1950s

NULL PRED.

—

90*

0.3%

Great Smog era, heavy coal soot

Pittsburgh 1940s–60s

NULL PRED.

—

110*

0.2%

Steel city, extreme industrial particulates

Ruhr Valley 1950s

NULL PRED.

—

100*

0.2%

German industrial heartland, coal/steel

Beijing modern

NULL PRED.

—

250*

0.0%

Megacity, persistent severe pollution

** Estimated PM2.5 from historical emission inventories, ice core proxies, and site geography. Non-asterisked values from monitoring station records. Ion-*

aerosol attachment coefficient $\beta = 1.5 \times 10^{-6}$ *cm³/s (Hoppel and Frick 1986). Effective transport velocity v_eff = 2.0 m/s (includes turbulent BL mixing; Stull 1988). Boundary layer depth 1500 m default, adjusted by site geography. Survival = exp(−H/v_eff ×* β *× Z). Null predictions are cities with no documented anomalous events despite large populations.*

Appendix C: Model Parameters

Reservoir equation: $R(t) = R(t-1) \times (1 - 1/\tau) + L(t)$, where τ = mean particle lifetime at site L-shell, and $L(t)$ = annual Ap × anomaly amplification factor. Perceptual threshold: undefined (primary free parameter). SWR test: $\tau = 90$ days at $L \approx 1.30$; amplification normalized to Ap_max during SC22 peak. Results: Pearson $r = 0.356$, Spearman $\rho = 0.338$, 74% directional year-match, $n = 35$ (1990–2024). Observer effect not controlled. Status: preliminary. Not citable as a confirmed result.

Worked Example : Skinwalker Ranch ($L \approx 1.30$)

The reservoir level at the end of year t is: $R(t) = R(t-1) \times (1 - 1/\tau) + Ap(t) \times A$, where τ is the mean particle lifetime in years at the site's L-shell, Ap(t) is the annual mean Ap index, and A is the anomaly amplification factor normalized to unity at SC22 peak.

At $L \approx 1.30$ (slot region / inner belt boundary), Van Allen Probe data gives electron lifetimes of 90–300 days depending on energy and wave conditions. Using $\tau = 0.5$ years (180 days, mid-range) and normalizing A so that peak SC22 loading (1989, Ap = 15) produces R = 1.0:

SC22 ascending (1987–1989): Ap values of 10, 12, 15. Reservoir fills from near-zero to approximately 0.72 by end of 1989. SC22 declining (1990–1993): Ap drops from 15 to 12 to 8. Reservoir peaks at approximately 0.85 in 1991 (lagged response), then declines. The Sherman family arrived in 1994 with the reservoir at approximately 0.45 — above zero but below the 1991 peak. If the perceptual threshold is between 0.3 and 0.5, the site was at or near threshold when the Shermans arrived.

SC23 ascending (1998–2001): Ap climbs from 9 to 12 to 12. Reservoir refills to approximately 0.70. This corresponds to the NIDS active period (1996–2004). SC24 minimum (2008–2012): Annual Ap reaches 4 in 2009 — the lowest since records began. Reservoir drains to approximately 0.05 by 2012. Below any plausible threshold. This corresponds to the documented 12-year quiet period.

SC25 ascending (2020–2024): Ap climbs from 6 to 10 to 15. Reservoir refills to approximately 0.40 by 2024. If the threshold is near 0.35, the site reactivated around 2022–2023 — consistent with the Fugal-era documented resurgence.

The threshold parameter remains undefined and is the model's primary free parameter. Values between 0.3 and 0.5 (normalized to SC22 peak) produce qualitative agreement with the documented activity record.

Values outside this range produce false predictions. The observer effect confound means the threshold cannot be calibrated from the existing qualitative data. Instrument data with timestamps — GPS disruption counts, radiation detector triggers, magnetometer anomalies — correlated against the predicted reservoir level would resolve the threshold to within ±10% and either confirm or falsify the model.

Appendix D: How to Test a New Case: Pre-Registration Protocol

The following protocol governs case evaluation and has been applied to every case added after the initial dataset. All cases in this appendix have been fully classified. The protocol prevents dataset expansion through selection bias by requiring complete case documentation before any storm data is consulted, and by requiring adversarial testing of competing storms simultaneously with confirmation testing.

Step 1 : Case Identification (Before Any Storm Data Is Consulted)

Candidate cases must be identified and fully documented — location, date, witness count, event description, source — before any storm data is consulted. The case must satisfy all four pre-specified criteria: (1) first occurrence at location; (2) minimum five independent witnesses; (3) non-military, non-recurring; (4) geomagnetically unremarkable site. These criteria are applied strictly and in advance. A case that returns PCE any criterion is recorded as tested and excluded before storm data is retrieved.

Step 2 : Storm Window Prediction

Once a case passes Step 1, the expected storm date is calculated: event date minus expected lag (distance / 85.836 km/day), with a ±7-day window. This calculation uses only the Haversine distance and the pre-specified baseline speed. It is recorded before any storm archive is consulted.

Step 3 : Archive Retrieval

Storm data for the ±7-day window is retrieved from GFZ Potsdam / WDC Kyoto. The evaluation question is strictly binary: does a triggering storm ($Kp \geq 5$ or $Dst \leq -50$ nT) fall within the window AND produce a ratio within 0.971-1.097? All storms in the full year — not just the window — are also evaluated and reported, so that adversarial candidates are documented simultaneously.

Step 4 : Classification

Four possible outcomes: (A) CONFIRMED — triggering storm in window, ratio in band, no competing storm simultaneously satisfying both criteria; (B) MULTI-PASS — triggering storm in window, ratio in band, one or more competing storms also pass (disclosed, result selection-sensitive); (C) FAIL — no triggering storm in window, or storm ratio outside band; (D) INCONCLUSIVE — storm data unavailable or ambiguous. All four outcomes are recorded and published. Failures and inconclusives are as important to report as confirmations.

Registered Case Outcomes

The following cases were registered under this protocol and have been fully classified. All have completed Steps 1 through 4.

GFZ-verified (March 2026): Mantell/Fort Knox (Jan 7, 1948) — Kp 7 TRANSITION ZONE, ratio 0.906. Westall/Melbourne (Apr 6, 1966) — EMPTY WINDOW (GSP 2,997 km). Boianai/Father Gill (Jun 26, 1959) — BELOW-BAND ANOMALY, Kp 8- (CORRECTED from Kp 7- in v215; raw GFZ archive shows Apr 10 peak Kp 7.667 = Kp 8- in thirds notation), ratios 0.932–0.944 (GSP 6,230 km); transition zone classification withdrawn; mechanism under investigation. Belgian triangle wave first night (Nov 29, 1989) — MODE 1 EMPTY WINDOW (CORRECTED from Kp 7 transition zone in v215; raw GFZ data for Nov 29, 1989 shows Kp max 4.0 across all eight three-hour intervals; field was geomagnetically quiet on the sighting night; no triggering storm produced a passing ratio for that date). Petrozavodsk (Sep 20, 1977) — CONFIRMED #9, Aug 11 Kp 5 ratio 1.058. Varginha (Jan 20, 1996) — CONFIRMED #10, Sep 27 Kp 7 ratio 1.058 (GSP 10,443 km). FMI Helsinki verified (March 2026): Knock, Ireland (Aug 21, 1879) — EMPTY WINDOW. Window Jul 4–18, 1879 contains K=5 on Jul 4 (ratio 0.860), K=6 on Jul 5 (ratio 0.878), K=5 on Jul 16 (ratio 1.146). None within confirmed band. Model silent. Pontmain, France (Jan 17, 1871) — EMPTY WINDOW. Only storm in window Nov 21–Dec 5, 1870 is K=5 on Nov 23 (ratio 0.909). The well-documented Oct 24–25 1870 superstorm produces ratio 0.588 — adversarial PCE. Model silent. Lourdes, France (Feb 11, 1858) — GEOMETRY SOLVED, EXCLUDED ON WITNESS GROUNDS. K=9 storm Dec 17, 1857, lag 56 days, ratio 1.013. Adversarial: K=5 on Dec 10 (ratio 0.900) and K=5 on Dec 5 (ratio 0.834) both return PCE. First apparition Feb 11 had one percipient witness; excluded on witness grounds. La Salette, France (Sep 19, 1846) — GEOMETRY SOLVED, EXCLUDED ON WITNESS GROUNDS. K=5 storm Jul 25, 1846, lag 56 days, ratio 1.001. Adversarial: K=8 on Aug 7 (ratio 1.304) — PCE.

Excluded: two witnesses only, below five-witness minimum. The PCS count is now six (Arnold, Westall, Lake Huaypo, Pontmain, Knock, Belgian triangle first night). The below-band anomaly count is one (Boianai). The geometry-solved-but-excluded count is two (La Salette, Lourdes).

Appendix E: Inner Boundary Migration Test Against EMFISIS Archival Data

Status: Completed March 17, 2026.

The propagation model described in Chapter Twenty predicts that the inner boundary of whistler-mode chorus wave enhancement in the radiation belts should migrate inward during geomagnetic storm recovery. The surface-projected rate of that migration should approach the geometrically determined ceiling for strong storms ($Kp \geq 7$) and fall below it for weaker events. This prediction has been tested against archival Van Allen Probes data. The test required no new instrumentation, no proprietary access, and no unpublished data. Every number below is independently reproducible from publicly archived files.

E.1 : Target Storm Selection

The March 17, 2015 St. Patrick's Day storm was selected as the primary test case. Selection criteria: Kp ≥ 7 (model predicts near-ceiling rate), clean recovery phase within the EMFISIS mission window, RBSP-A spacecraft fully operational. The storm met all criteria:

Parameter	Value
Storm date	March 17, 2015
Peak Kp	8.0
Dst minimum	−223 nT
Solar cycle	24 declining phase
EMFISIS mission status	Both RBSP-A and RBSP-B fully operational
Analysis period	March 17 – April 17, 2015 (31 days)
Reason selected	Strongest storm of SC24; Kp = 8 predicts near-ceiling rate

E.2 : Data Sources

All data publicly available. No registration, no fees, no special access.

E.2.1 Wave Power Data

Source: University of Iowa EMFISIS Data Server

URL: space.physics.uiowa.edu/emfisis/data/index/RBSP-A/L2/

Dataset: WFR Spectral Matrix Diagonal, Level 2

File pattern: rbsp-a_WFR-spectral-matrix-diagonal_emfisis-L2_YYYYMMDD_v*.cdf

Variables used: BuBu, BvBv, BwBw (magnetic power spectral density per frequency bin, nT^2/Hz); summed across all 65 frequency bins to produce total magnetic power TotalPowerB

Time cadence: 6 seconds (survey mode)

Files downloaded: 32 daily files, March 17 through April 17 2015. Versions range from v1.4.2 through v1.4.8 (normal archive practice; all are calibrated L2 products)

Download method: PowerShell Invoke-WebRequest with automated per-day version detection via directory listing

E.2.2 L-Shell Data

Source: University of Iowa EMFISIS Data Server

URL: space.physics.uiowa.edu/emfisis/data/index/RBSP-A/L3/

Dataset: Fluxgate Magnetometer, Level 3, Solar Magnetic (SM) coordinates, 4-second cadence

File pattern: rbsp-a_magnetometer_4sec-sm_emfisis-L3_YYYYMMDD_v*.cdf

Variables used: lambda (magnetic latitude in degrees, SM frame); coordinates (spacecraft position in km, SM frame, 3-component vector)

L-shell computation: Dipole approximation $L = R / \cos^2(\lambda)$, where $R = |\text{coordinates}| / 6371$ km (Earth radii) and λ = lambda in radians. Fill values filtered: lambda restricted to (−90°, +90°); R to (1.0, 8.0) R_E; L to (1.0, 8.0).

Files downloaded: 32 daily files auto-downloaded during analysis. Versions v1.3.3 through v1.3.5.

E.2.3 Geomagnetic Index

Source: GFZ Potsdam Kp/Ap Archive. URL: kp.gfz.de/app/files/Kp_ap_since_1932.txt

Used for: Storm identification and Kp = 8 confirmation for March 17, 2015

E.3 : Analysis Software

Language: Python 3.11

Libraries: cdflib 1.3.8 (CDF file reading); numpy 2.4.0 (array operations); matplotlib 3.x (plotting)

Script: emfisis_read_cdf.py — complete script reproduced in Section E.8 below.

Algorithm:

1. For each of the 32 daily WFR CDF files, load BuBu + BvBv + BwBw and sum across 65 frequency bins to produce scalar TotalPowerB(t).

2. Load the corresponding daily MAG L3 SM CDF file. Compute L-shell from SM coordinates and magnetic latitude via the dipole formula. Interpolate L(t) to WFR timestamps using numpy.interp, restricted to the MAG file's own time range only.

3. Compute baseline power as the median of TotalPowerB during the first 3 hours of each file. Set threshold = 10 × baseline.

4. Divide each day into four non-overlapping 6-hour bins. Within each bin, find all timestamps where TotalPowerB > threshold and L is in (1.3, 7.5). Record the minimum L value (inner boundary).

5. Convert inner L to surface distance from GNP (78.5°N, 69°W) using the dipole footpoint formula: distance = haversine(GNP, footpoint), footpoint latitude = GNP_lat − (90° − arccos(1/√L)).

6. Stack all bins across 31 days. Compute days elapsed since storm onset. Fit linear regression to (days, surface distance) using numpy.polyfit.

E.4 : Results

Quantity	Value
Total 6-hour bins processed	94
Days covered	0.1 to 31.9 (March 17 – April 17, 2015)
WFR files processed	31 of 32 (April 5 skipped — corrupted file)
Inner L range across all bins	1.30 to 7.11
Fitted slope	−14.1 km/day (inward)
Model ceiling	86 km/day
Direction	Correct (inward = decreasing surface distance from GNP)
Magnitude vs ceiling	Consistent (|slope| ≪ ceiling)
Result	CONSISTENT WITH PROPAGATION MODEL

E.5 : Interpretation

The fitted slope of −14.1 km/day indicates the inner boundary of chorus wave enhancement migrated inward at approximately 14 km/day during the 31-day recovery period following the Kp = 8 storm. This is well within the saturation ceiling and physically reasonable for a Kp = 8 storm that has driven the outer belt into strong but not fully saturated diffusion across all L-shells.

The scatter in individual bins (inner L ranging 1.30 to 7.11) is expected and is not noise. RBSP-A samples a different magnetic local time sector on each 9.1-hour orbital pass. Chorus wave power varies by 2–3 orders of magnitude between the active nightside (0–6 MLT) and the quiet dayside. Each 6-hour bin captures the satellite at a different orbital phase. The scatter is the real spatial structure of a phenomenon that fills the nightside magnetosphere asymmetrically. A multi-spacecraft or statistical approach averaging over MLT would reduce scatter; the single-spacecraft linear fit through it still recovers the correct direction and order of magnitude.

This result confirms the first link of the chain: the inner boundary of wave enhancement in the radiation belts moves inward following a geomagnetic storm, in the correct direction, at a rate consistent with the propagation model and well within its expected ceiling. The remaining links — ionospheric deposition, dusty plasma attenuation, surface ion pulse — are addressed by the GPS TEC test (Section E.6).

E.6 : GPS TEC Test: Attempted, Detectability Bounds Established

This test was executed in full. The methodology, data sources, results, and PCE and PCS results are documented here because negative results obtained honestly constrain the parameter space as usefully as confirmations, and because the detectability bounds established below are a practical contribution to any future researcher attempting this measurement.

Methodology. JPL Global Ionosphere Map (GIM) IONEX files were downloaded from the NASA CDDIS archive. Each file contains global GPS Total Electron Content maps at 2-hour cadence, 5° longitude × 2.5° latitude resolution, derived from the worldwide ground-based GPS receiver network. TEC values represent the vertically integrated electron density in the ionospheric column above each grid point, in Total Electron Content Units (1 TECU = 10^{16} electrons/m^2). The analysis region was the northeast Pacific Ocean (35–55°N, 145–165°W) — a clean-air oceanic region free of land convection. Daily mean TEC was computed for each 5° latitude band. Two independent storm cases were analyzed.

First pass: May 10, 2024 (Kp9) — contaminated. Nine IONEX files were obtained spanning day −2 (pre-storm baseline) through day +100. The propagation model predicts front arrival at 50–55°N on day +49, at 45–50°N on day +56, at 40–45°N on day +62, and at 35–40°N on day +68, based on GNP distances of 4,247 to 5,847 km and the T-derived ceiling rate. The test was invalidated by a second geomagnetic storm (June 28–29, 2024, Kp6) that struck on day +49 — the first day of the predicted propagation window. The TEC trough at day +50 across all latitude bands simultaneously is direct ionospheric response to this secondary event. The two signals cannot be separated. This case is set aside. Its PCE result mode is a methodological constraint: any future GPS TEC test of this model requires a minimum 80-day Kp < 5 window following the primary storm.

Second pass: December 15, 2006 (Kp8) — clean window, insufficient sensitivity. The GFZ Kp archive was searched for Kp ≥ 7 storms in the IONEX era followed by the longest subsequent quiet periods. December 15, 2006 was selected: the next Kp ≥ 7 event did not occur until October 2011, a gap of 1,774 days. Eight IONEX files were obtained spanning day −2 (December 13 baseline) through day +80 (March 5, 2007). Winter baseline TEC in the northeast Pacific was 7–10 TECU — approximately one quarter of summer values, which improves the proportional signal-to-noise ratio.

The anomaly table (TEC minus the December 13 baseline) showed no latitude-structured signal. From day +45 onward, all four latitude bands showed a coherent +1.5 to +2.5 TECU rise that was uniform across bands — spread of only 0.06–0.39 TECU between bands at any probe date. The latitude gradient ranged from +0.004 to +0.025 TECU per degree poleward, with no systematic pattern across the propagation window. This is the signature of seasonal TEC recovery: the northern hemisphere ionosphere is at its annual minimum in December and rises through January–March as solar zenith angle decreases. This seasonal drift of approximately 2.5 TECU over 80 days is 10–25 times larger than the estimated propagation signal amplitude. The propagation signal, if present, is below the noise floor of this instrument and method.

Detectability bounds for future researchers. Daily mean GPS TEC from a 20° × 20° geographic box cannot resolve the predicted propagation signal against seasonal background. Any future GPS TEC test requires one or more of the following: (a) a deseasonalized TEC anomaly product computed by subtracting the climatological mean for each day-of-year from a multi-year quiet baseline, rather than a single pre-storm date; (b) denser temporal sampling — probe dates every 5 days through the predicted window rather than the 10–20 day intervals used here; or (c) a higher-sensitivity ionospheric instrument such as ionosonde foF2 measurements at a fixed geomagnetic latitude station, which resolves the F2 peak electron density directly rather than integrating the entire column. The storm isolation criterion has been empirically quantified: the 2024 secondary storm, arriving 49 days after the primary, completely masked the propagation window. A clean 80-day post-storm period with no Kp ≥ 5 activity is the minimum requirement. The August 26, 2018 Kp7 storm (1,166-day subsequent quiet period) remains the cleanest untested case in the IONEX archive. The highest-sensitivity existing-data test remains the EMFISIS wave power approach described in Section E.5, which operates directly inside the radiation belt without an atmospheric transfer function. The GPS TEC result does not contradict the propagation model; it establishes the noise floor of this instrument class.

E.7 : Reproducibility Statement

Every number in this appendix is independently reproducible from publicly archived data requiring no special access, registration, or fee. The complete analysis pipeline:

Step 1. Download 32 WFR daily CDF files from space.physics.uiowa.edu/emfisis/data/index/RBSP-A/L2/2015/ for dates 20150317 through 20150417. File pattern: rbsp-a_WFR-spectral-matrix-diagonal_emfisis-L2_YYYYMMDD_v*.cdf. Use whatever version is present for each day.

Step 2. Download 32 matching MAG L3 SM CDF files from space.physics.uiowa.edu/emfisis/data/index/RBSP-A/L3/2015/ for the same dates. File pattern: rbsp-a_magnetometer_4sec-sm_emfisis-L3_YYYYMMDD_v*.cdf.

Step 3. Install Python 3.11 with cdflib, numpy, matplotlib.

Step 4. Run emfisis_read_cdf.py (available from author). The script processes all 32 file pairs automatically and saves emfisis_30day_result.png.

Expected output: Fitted slope ≈ −14 km/day (inward), 94 bins, inner L range 1.30–7.11.

Data citation: Kletzing, C.A., et al. (2013). The Electric and Magnetic Field Instrument Suite and Integrated Science (EMFISIS) on RBSP. Space

Science Reviews, 179, 127–181. doi:10.1007/s11214-013-9993-6. Data accessed from space.physics.uiowa.edu/emfisis, March 2026.

E.8 : Complete Analysis Script (emfisis_read_cdf.py)

Save as emfisis_read_cdf.py in the same directory as the downloaded CDF files. Run with: py -3.11 emfisis_read_cdf.py

```
"""
emfisis_read_cdf.py  v4
Plasma Space Weather and the Paranormal v114


Processes ALL WFR daily CDF files in C:/Temp (March 17 - April 17 2015).
For each WFR file, downloads the matching MAG L3 file automatically,
or uses the lambda/coordinates already downloaded.


RUN:
    py -3.11 emfisis_read_cdf.py
"""


import sys, os, math, glob, urllib.request
from datetime import datetime, timedelta, date
import numpy as np
import matplotlib; matplotlib.use("Agg")
import matplotlib.pyplot as plt
import matplotlib.dates as mdates
import cdflib


# tee output
import builtins as _bi
```

```
_log = open("emfisis_results4.txt", "w", encoding="utf-8")

_rp  = _bi.print

def print(*a, **kw):

    _rp(*a, **kw)

    kw.pop("file", None)

    _rp(*a, file=_log, **kw)

    _log.flush()



R_E     = 6371.0

GNP_LAT = 78.5

GNP_LON = -69.0

CEILING = 86.0

STORM   = datetime(2015, 3, 17)



def haversine(lat1, lon1, lat2, lon2):

    lat1,lon1,lat2,lon2 = map(math.radians,[lat1,lon1,lat2,lon2])

    a = math.sin((lat2-lat1)/2)**2 + math.cos(lat1)*math.cos(lat2)*math.sin((lon2-
lon1)/2)**2

    return 2*R_E*math.asin(math.sqrt(max(0.0,a)))



def L_to_dist(L):

    if L<=1.0: return 0.0

    lam = math.degrees(math.acos(1.0/math.sqrt(L)))

    return haversine(GNP_LAT, GNP_LON, GNP_LAT-(90.0-lam), GNP_LON)



def sep(t):

    print(); print("="*65); print("  "+t); print("="*65)
```

```
def tt2000_to_unix(tt2000_ns):

    dt64 = cdflib.cdfepoch.to_datetime(tt2000_ns)

    epoch = np.datetime64(0, "ns")

    one_sec = np.timedelta64(1000000000, "ns")

    return (np.array(dt64, dtype="datetime64[ns]") - epoch) / one_sec



def unix_to_dt(unix_arr):

    return [datetime.utcfromtimestamp(float(t)) for t in unix_arr]



def download_mag(datestr):

    """Try to download the MAG 4sec-sm L3 file for a given date YYYYMMDD."""

    yr, mo, dy = datestr[:4], datestr[4:6], datestr[6:8]

    base = f"https://space.physics.uiowa.edu/emfisis/data/index/RBSP-A/L3/{yr}/{mo}/{dy}/"

    fname_pat = f"rbsp-a_magnetometer_4sec-sm_emfisis-L3_{datestr}"

    # Check if we already have it

    existing = [f for f in os.listdir(".") if f.startswith(fname_pat) and
f.endswith(".cdf")]

    if existing:

        return existing[0]

    # Fetch directory listing

    try:

        req = urllib.request.Request(base, headers={"User-Agent":"Mozilla/5.0"})

        html = urllib.request.urlopen(req, timeout=20).read().decode(errors="replace")

        import re

        matches = re.findall(r'href="(rbsp-a_magnetometer_4sec-sm_emfisis-L3_' + datestr +
r'[^"]+\.cdf)"', html)

        if not matches:
```

```
            return None

        fname = matches[0]

        url = base + fname

        print(f"    Downloading MAG: {fname}")

        urllib.request.urlretrieve(url, fname)

        return fname

    except Exception as e:

        print(f"    MAG download failed: {e}")

        return None



def read_mag_L(mag_file):

    """Read L-shell from MAG L3 SM file. Returns (unix_times, L_values)."""

    mag = cdflib.CDF(mag_file)

    mag_epoch = mag.varget("Epoch")

    mag_unix = tt2000_to_unix(mag_epoch)

    mag_vars = mag.cdf_info().zVariables



    lam_var = "lambda" if "lambda" in mag_vars else None

    pos_var = "coordinates" if "coordinates" in mag_vars else None



    if lam_var and pos_var:

        lam_deg_raw = mag.varget(lam_var).flatten()

        pos = mag.varget(pos_var)

        R_km = np.sqrt(np.sum(pos**2, axis=1)) if pos.ndim==2 else np.abs(pos.flatten())

        R = R_km / R_E

        lam_deg = np.where((lam_deg_raw > -90.0) & (lam_deg_raw < 90.0), lam_deg_raw,
np.nan)

        R = np.where((R > 1.0) & (R < 8.0), R, np.nan)
```

```
        cos_lam = np.cos(np.radians(lam_deg))
        L_raw = R / np.where(cos_lam**2 > 0.001, cos_lam**2, np.nan)
        L_raw = np.where((L_raw > 1.0) & (L_raw < 8.0), L_raw, np.nan)
        valid = np.isfinite(L_raw) & np.isfinite(mag_unix)
        if valid.sum() > 10:
            return mag_unix[valid], L_raw[valid]
    return None, None


# — Find all WFR files ——————————————————————————————
sep("FINDING ALL WFR FILES")
all_wfr = sorted([f for f in os.listdir(".")
                  if "WFR-spectral-matrix-diagonal" in f
                  and f.endswith(".cdf")
                  and "merged" not in f])
print(f"  Found {len(all_wfr)} WFR files:")
for f in all_wfr:
    print(f"    {f}")


if not all_wfr:
    print("  No WFR files found. Run the PowerShell download script first.")
    _log.close()
    input("Press ENTER to close")
    sys.exit(0)


# — Process each file ——————————————————————————————
sep("PROCESSING ALL DAILY FILES")
```

```
all_bin_t    = []

all_bin_L    = []

all_bin_dist = []



for wfr_file in all_wfr:

    # Extract date from filename

    import re

    m = re.search(r'(\d{8})', wfr_file)

    if not m:

        continue

    datestr = m.group(1)

    file_dt = datetime.strptime(datestr, "%Y%m%d")

    print(f"\n  --- {datestr} ---")



    # Read WFR

    try:

        wfr = cdflib.CDF(wfr_file)

        wfr_epoch = wfr.varget("Epoch")

        bubu = wfr.varget("BuBu")

        bvbv = wfr.varget("BvBv")

        bwbw = wfr.varget("BwBw")

        total_pw = np.nansum(bubu + bvbv + bwbw, axis=1)

        total_pw = np.where(total_pw > 0, total_pw, np.nan)

        wfr_unix = tt2000_to_unix(wfr_epoch)

        print(f"    WFR: {len(wfr_unix)} records, power {np.nanmin(total_pw):.2e} to
{np.nanmax(total_pw):.2e} nT^2")

    except Exception as e:

        print(f"    WFR read failed: {e}")
```

```
    continue

# Get MAG L-shell for this day
mag_file = next((f for f in os.listdir(".")
                 if f"magnetometer_4sec-sm_emfisis-L3_{datestr}" in f
                 and f.endswith(".cdf")), None)
if not mag_file:
    print(f"    No MAG file for {datestr} — downloading...")
    mag_file = download_mag(datestr)

if mag_file:
    mag_unix_v, L_vals = read_mag_L(mag_file)
    if mag_unix_v is not None and len(mag_unix_v) > 10:
        # Only interpolate within the MAG file's time range
        t_min, t_max = mag_unix_v[0], mag_unix_v[-1]
        in_range = (wfr_unix >= t_min) & (wfr_unix <= t_max)
        L_at_wfr = np.full(len(wfr_unix), np.nan)
        L_at_wfr[in_range] = np.interp(wfr_unix[in_range], mag_unix_v, L_vals)
        print(f"    L: {np.nanmin(L_at_wfr):.2f} to {np.nanmax(L_at_wfr):.2f} (MAG L3)")
        HAVE_L = True
    else:
        HAVE_L = False
else:
    HAVE_L = False

if not HAVE_L:
    # Orbital proxy: period 9.1h, perigee L=1.1, apogee L=5.8
```

```
    T = 9.1 * 3600.0

    t0 = wfr_unix[0]

    phase = 2*np.pi*(wfr_unix - t0)/T

    L_at_wfr = 1.1 + 4.7*(0.5 + 0.5*np.sin(phase))

    print(f"    L: orbital proxy (no MAG file)")



# Baseline: first 3 hours of the file

base_mask = wfr_unix < (wfr_unix[0] + 3*3600)

baseline  = np.nanmedian(total_pw[base_mask])

threshold = 10.0 * baseline



# 6-hour bins

BIN_SEC = 6*3600

t0 = wfr_unix[0]

n_bins = int((wfr_unix[-1]-t0)/BIN_SEC) + 1

day_bins = 0

for w in range(n_bins):

    w0 = t0 + w*BIN_SEC

    w1 = w0 + BIN_SEC

    mask = (wfr_unix >= w0) & (wfr_unix < w1)

    if not mask.any(): continue

    above = L_at_wfr[mask & (total_pw > threshold) & np.isfinite(L_at_wfr)]

    above = above[(above >= 1.3) & (above <= 7.5)]

    if len(above) == 0: continue

    inner = float(np.nanmin(above))

    bin_dt = datetime.utcfromtimestamp(w0 + BIN_SEC/2)

    all_bin_t.append(bin_dt)

    all_bin_L.append(inner)
```

```
        all_bin_dist.append(L_tc_dist(inner))
        day_bins += 1
    print(f"    Bins with enhancement: {day_bins}")


# ── 30-day analysis ──────────────────────────────────────────────────────────
sep(f"30-DAY SLOPE ANALYSIS  ({len(all_bin_t)} total bins)")


if len(all_bin_dist) < 3:
    print("  Not enough bins for slope fit.")
    print("  Check that WFR files downloaded correctly.")
else:
    days  = np.array([(t-STORM).total_seconds()/86400 for t in all_bin_t])
    dists = np.array(all_bin_dist)
    Lvals = np.array(all_bin_L)
    c     = np.polyfit(days, dists, 1)
    slope = c[0]


    print(f"  Total bins: {len(all_bin_t)}")
    print(f"  Day range:  {days[0]:.1f} to {days[-1]:.1f}")
    print(f"  Inner L range: {Lvals.min():.2f} to {Lvals.max():.2f}")
    print(f"\n  Fitted slope:  {slope:+.1f} km/day")
    print(f"  Ceiling:        {CEILING:.0f} km/day")
    if slope <= CEILING:
        print("  RESULT: CONSISTENT — slope at or below 86 km/day ceiling")
    else:
        print("  RESULT: EXCEEDS ceiling — review data")
```

```
# Plot

fig, (ax1, ax2) = plt.subplots(2, 1, figsize=(14, 9), sharex=True)

fig.suptitle(

    "EMFISIS RBSP-A  —  March 17 to April 17 2015\n"

    "Inner boundary of wave enhancement vs time\n"

    f"Fitted slope: {slope:+.0f} km/day   |   Ceiling: {CEILING:.0f} km/day",

    fontsize=11)



ax1.scatter(all_bin_t, Lvals, s=18, color="steelblue",

            label="Inner L  (power > 10x baseline)", zorder=3)

ax1.axvline(STORM, color="red", ls="--", lw=1.5, label="Storm onset Mar 17")

ax1.invert_yaxis()

ax1.set_ylim(7.0, 1.0)

ax1.set_ylabel("Inner L-shell  (lower = deeper into belt)")

ax1.legend(fontsize=9); ax1.grid(alpha=0.3)



ax2.scatter(all_bin_t, dists, s=18, color="darkorange",

            label="Surface dist from GNP (km)", zorder=3)

ax2.plot(all_bin_t, np.polyval(c, days), "k--", lw=1.5,

         label=f"Linear fit: {slope:+.0f} km/day")

d_ref = np.linspace(0, days[-1], 100)

ax2.plot([STORM + timedelta(days=d) for d in d_ref],

         d_ref * CEILING, "g:", lw=1.5, label="86 km/day ceiling")

ax2.axhline(haversine(78.5,-69,51.85,21.23), color="purple",

ax2.set_ylabel("Surface dist from GNP (km)")

ax2.set_xlabel("Date (UTC)")

ax2.legend(fontsize=9); ax2.grid(alpha=0.3)

ax2.xaxis.set_major_formatter(mdates.DateFormatter("%b %d"))
```

```
    fig.autofmt_xdate()

    plt.tight_layout()

    plt.savefig("emfisis_30day_result.png", dpi=150)

    print("\n  Plot saved: emfisis_30day_result.png")



sep("DONE")

_log.close()

input("Results in emfisis_results4.txt  |  press ENTER to close")
```

Appendix F: Diesel Immunity: The Physics of Compression Ignition

A diesel engine that is already running needs no electricity whatsoever. That single fact turns Forli and Fordingbridge into controlled experiments.

How a Gasoline Engine Dies

Every vehicle stall in the UFO literature before 1975 involves a car built on the same basic electrical architecture that Charles Kettering designed for Cadillac in 1912. A single ignition coil. A set of breaker points inside a distributor. A condenser. Unshielded spark plug wires running from the distributor cap to each cylinder. An unshielded primary circuit carrying 3.5 to 4 amps through mechanical contact points that physically open and close hundreds of times per second.

The primary circuit is a loop of unshielded copper wire connecting the battery, the coil, the breaker points, and ground. An external electromagnetic field induces voltage in that loop. The induced voltage adds to or opposes the normal current through the points. If the induced voltage is large enough, several things happen simultaneously. The points arc when they should not. The condenser saturates and cannot absorb the transient. The coil cannot build its magnetic field properly because the primary current is being disrupted. The spark timing becomes erratic, then fails entirely. The engine stalls.

The headlights run through the same unshielded harness. They dim or die. The radio, connected to a long wire antenna, fills with static. The ammeter on the dashboard swings wildly — Newell Wright at Levelland reported exactly this. The entire electrical system of the car is one interconnected, unshielded, low-voltage circuit. Disrupt any part of it and the effects propagate everywhere.

The chain has no redundancy. One coil feeds all cylinders through one distributor. If the coil cannot charge, no cylinder fires. If the points arc at the wrong time, timing is destroyed and the engine stalls. Any single failure anywhere in the ignition chain kills every cylinder simultaneously. The engine does not stumble or run rough. It stops. This is why the Levelland witnesses all reported the same sequence: object appears, engine dies, lights go out, object departs, engine restarts. The ignition system was maximally vulnerable. When the external field was present, the primary circuit could not function. When the field was removed, the mechanical components were undamaged. They had been overwhelmed, not destroyed. The driver turned the key and the car started normally.

How a Diesel Engine Does Not Die

A diesel engine does not use spark ignition. It uses compression ignition. The operating principle is thermodynamic, not electrical. Air enters the cylinder on the intake stroke. The piston compresses that air on the compression stroke to a ratio of between 15:1 and 25:1 — roughly double the compression ratio of a gasoline engine. At 20:1 compression, the air temperature reaches approximately 500 to 700 degrees Celsius. That is well above the autoignition temperature of diesel fuel, which is around 210 degrees Celsius. Near the top of the compression stroke, a mechanical fuel injection pump — driven directly by the engine's camshaft through gears — forces diesel fuel through a mechanical injector at pressures of 1,500 to 4,000 PSI. The fuel atomizes into the superheated air. It ignites on contact. No spark required. No electricity required.

The components that make this work in a 1950s or 1960s diesel are entirely mechanical. The fuel injection pump is a precision mechanical device — typically a Bosch A-pump or a CAV (Lucas) unit in British applications. It is driven by the engine's camshaft through a gear train. It meters fuel and delivers it at high pressure to each injector in the correct firing order. It has no electrical inputs. It is a mechanical computer made of steel, and it runs on engine rotation.

The injectors are spring-loaded mechanical nozzles. Fuel pressure from the pump overcomes the spring, the nozzle opens, fuel sprays into the cylinder, the spring closes the nozzle when pressure drops. No solenoid. No signal wire. No electrical connection of any kind. The governor is a set of centrifugal flyweights that regulate engine speed by adjusting the fuel rack position mechanically. More RPM, less fuel. Fewer RPM, more fuel. Purely mechanical feedback.

Once a diesel engine of this era is running, you can disconnect the battery, cut every wire in the vehicle, and the engine will continue to run indefinitely. It will run until it runs out of fuel or until someone physically moves the fuel shutoff lever, which mechanically moves the fuel rack to the zero-delivery position. There is no electrical off switch. In fact, a known failure mode of older diesels is runaway — if the fuel rack jams in the full-delivery position, the engine cannot be stopped electrically because there is no electrical system in the combustion loop. The only way to stop a runaway diesel is to block the air intake or stall it mechanically.

The only electrical components on a 1954 diesel tractor are: the starter motor (used only to initially crank the engine, irrelevant once running), the glow plugs (resistive heating elements in the prechamber that assist cold starting, switched off once the engine is running), and on some models a fuel solenoid shutoff valve. On most 1950s tractors, even the shutoff was a manual lever.

An electromagnetic field strong enough to stall every gasoline engine in a field has nothing to grab onto in a running diesel. There is no ignition coil to disrupt. There is no distributor to confuse. There are no spark plug wires to act as receiving antennas. There are no breaker points to arc prematurely. The entire combustion process — compression, injection, ignition, expansion, exhaust — proceeds through mechanical and thermodynamic pathways that are transparent to electromagnetic radiation.

The Forli Experiment : November 14, 1954

Two tractors in a field near Forli, Italy. One diesel, one gasoline. Same field, same moment, same proximity to the reported phenomenon. The gasoline tractor's engine died. The diesel tractor continued running.

This is not an anecdote. This is a controlled experiment. The independent variable is ignition type. The dependent variable is engine operation. Every confounding variable — location, time, weather, proximity, psychological state of the operators — is held constant by the fact that both tractors were in the same place at the same time.

The result eliminates entire categories of explanation. If the phenomenon were psychological — mass hysteria, cultural expectation, misperception — both operators would have reported both tractors stalling, or neither would have stalled. Hysteria does not know which tractor has points. If the phenomenon were gravitational or acoustic — a shockwave, an infrasound pulse, ground vibration — both engines would have been equally affected. Gravity does not care about ignition architecture. If the phenomenon were a hoax, the perpetrator would have had to know in advance which tractor was diesel and which was gasoline, and to selectively disable only the correct one while leaving the other running. In rural Italy in 1954, this requires a level of electromagnetic engineering knowledge that did not exist outside military laboratories. If the phenomenon were purely thermal — a heat source, a cold front, atmospheric pressure change — both engines would respond identically.

The only category of physical phenomenon that selectively kills a spark-ignition engine while leaving a compression-ignition engine untouched is an electromagnetic field. Specifically, a field strong enough to disrupt the low-voltage primary ignition circuit and the high-voltage secondary circuit, but not strong enough to physically damage mechanical components or overwhelm the thermodynamic process of compression ignition.

The Fordingbridge Replication : November 5, 1967

Thirteen years later, on a road in Hampshire, England, the experiment repeated. A diesel Leyland truck and a gasoline Jaguar sedan. Same road, same moment. The Jaguar stalled. The Leyland kept running. The Jaguar was a points-and-coil car — the XK engine with a Lucas distributor, unshielded HT leads, a single coil, and mechanical breaker points. The full Kettering architecture, every vulnerability present. The Leyland truck was a compression-ignition diesel with a CAV mechanical injection pump. Once running, electrically inert. Same result as Forli. Same physics. Different country, different decade, different vehicles, different witnesses. The variable is ignition type. The outcome is identical.

Field Strength Constraints

The Forli and Fordingbridge observations do more than identify the phenomenon as electromagnetic. They constrain the emitter. The field must be above the threshold that induces sufficient voltage in the primary ignition loop to prevent coil charging — roughly 50 to 200 V/m at the vehicle, depending on loop area and frequency — but below the threshold that would physically destroy conductors, melt insulation, or ionize the air-fuel mixture inside the diesel cylinder, which requires on the order of 10,000 V/m. The diesel that kept running brackets the field strength to roughly two orders of magnitude.

The frequency content must overlap the vulnerable bandwidth of the ignition system. The Kettering primary circuit resonates in the low-frequency range — the points open at a few hundred hertz, and the LC circuit of the coil and condenser rings at roughly 1 to 10 kHz. Any EM emission with significant energy content below approximately 100 kHz will couple efficiently into the ignition loop. This is consistent with the ELF/VLF emission spectrum expected from atmospheric plasma structures — ionization columns and plasma instabilities naturally radiate in this frequency range.

The field must be sustained for the duration of the encounter. The Levelland witnesses reported stalls lasting from thirty seconds to several minutes, recovering immediately after the object departed. The Forli tractor was stalled for the duration of the sighting. The field is not a transient spike. It is a continuous emission from something that is physically present overhead.

Taken together: an extended atmospheric source, radiating broadband ELF/VLF, at field strengths of roughly 50 to 500 V/m at ground level, sustained for minutes. This is exactly what the two-zone emitter model predicts from a region of enhanced atmospheric ionization produced by charged particle precipitation at a mid-latitude site. The diesel that kept running is not just a curiosity. It is a field-strength meter, a frequency discriminator, and a duration timer, all in the body of a tractor that did not know it was part of an experiment.

Modern Diesels : A Caveat

The diesel immunity argument applies fully to mechanical-injection diesels manufactured before approximately 1990. After that date, diesel technology changed in ways that partially reintroduce electrical vulnerability. Common-rail direct injection systems, introduced in the late 1990s, use electronically controlled injectors — solenoid or piezoelectric actuators that receive timing and duration signals from an engine control unit housed in a shielded enclosure. The fuel pressure is maintained by an electric high-pressure pump. If the ECU loses power or receives corrupted signals, the injectors cannot fire correctly. A modern common-rail diesel can be disrupted by EM interference to its electronic control systems, though both modern diesels and modern gasoline engines are protected by EMC shielding and hardened wiring harnesses.

This creates a three-tier vulnerability prediction for any future vehicle interference event. A pre-1990 mechanical diesel should be completely immune: engine continues running with zero effect. A post-2000 electronic diesel should show the same failure mode as a modern gasoline engine: possible ECU disruption, injector timing errors, dashboard warnings, but probably not a complete engine kill because the EMC shielding on modern vehicles is designed to resist external EM interference. A pre-1975 gasoline vehicle with breaker-point ignition should be maximally vulnerable: complete engine kill, headlights out, the full Levelland sequence. Three different responses from the same field, explained entirely by electrical architecture. This is a testable prediction.

Independent corroboration of the storm dates comes from the Eskdalemuir Magnetic Observatory in southern Scotland, operated continuously by the British Geological Survey since 1908. The BGS hourly H-component archive records the horizontal magnetic field intensity at one-hour resolution, with no connection to vehicle interference research or anomalous phenomena. Eskdalemuir is geographically appropriate for corroborating storms with strong mid-latitude effects over Britain and western Europe. For those cases, the same storms identified by the propagation geometry appear unmistakably in this independent institutional record. The September 1957 active period appears unmistakably in the Eskdalemuir record. The geometry-correct storm of September 2 produced a visible H-component depression; the September 13 storm — the rejected adversarial in the same active cluster, Kp 9−, the largest of Solar Cycle 19 — produced a 558 nT depression, from a quiet baseline of 16,705 nT down to 16,147 nT, with steepest descent of 247 nT in a single hour between 06:00 and 07:00 UT, a rate of 4.1 nT per minute. That rate of change is well above the threshold at which geomagnetically induced currents become significant in long conductors. Every telegraph line, every grounded power cable, every buried pipeline in Britain was carrying induced current that morning. The September 20, 1967 storm — the one the geometry selects for Fordingbridge at ratio 1.013 — produced a 103 nT depression over the following night, with peaks approaching 1 nT per minute. A moderate storm by Dst standards, but sufficient: the Jaguar stalled, the Leyland ran. The quiet period of October 18–22, 1967 — five days with no storm — shows only 39 nT of normal diurnal variation at Eskdalemuir, indistinguishable from the normal diurnal cycle. No vehicle interference events appear in the catalog for that period. For geographically distant cases such as Petrozavodsk (61.8°N, Russia), the regionally appropriate observatory is the Finnish Meteorological Institute network at Nurmijärvi or Sodan kylä, which sits closer to the auroral electrojet's equatorward boundary during a Kp 5 event. The principle is the same; the instrument must be proximate to the storm's regional effect. The BGS archive is available without restriction at geomag.bgs.ac.uk. The reader is invited to query any date in this dataset and verify that the storm signatures are exactly where the geometry predicts them.

Appendix G: The Automotive Hardening Timeline

The emitter did not stop radiating. The instrument got harder to disrupt.

The Vulnerable Car: 1912–1973

Every vehicle stall in the UFO literature before 1975 involves a car built on the Kettering ignition architecture: a single ignition coil, a mechanical distributor with breaker points, a condenser, and unshielded spark plug wires. The primary circuit is an unshielded loop of copper wire connecting battery, coil, points, and ground. This is not a car. This is an antenna. Pedro Saucedo's 1957 truck, the Gueugnon farmers' 1954 cars, the Beauvais motorcyclist's bike, the gasoline tractor at Forli — every one had this architecture. They were instruments that happened to be designed as transportation.

The Transition: 1973–1977

Three changes destroyed the vulnerability window, and they happened in the exact years that vehicle stall reports began to decline.

Chrysler, 1973. Electronic Ignition System became standard on all Chrysler models. The breaker points were replaced with a magnetic reluctor pickup — a solid-state sensor with no moving parts and no mechanical contacts to arc. The ignition control module was a sealed transistor unit. The points were gone. The single most vulnerable component in the Kettering system had been eliminated from every new Chrysler.

General Motors, 1975. High Energy Ignition became standard on all GM vehicles. The HEI system delivered 40% more voltage and 85% more energy than breaker-point systems. More importantly, the electronic module could recover from transient electromagnetic interference that would have pitted and destroyed mechanical points. A momentary induced voltage spike would cause a missed spark — a stumble — not a complete engine kill. The car would hesitate, not die.

Ford, 1975–1977. DuraSpark I and II systems rolled out across the Ford lineup. By 1977, no new American car of any make had breaker points. European manufacturers followed a parallel timeline. Fiat put electronic ignition on the Dino in 1968. Jaguar followed in 1971. By the early 1980s, electronic ignition was standard across European production.

The Rodeghier catalog, which documents vehicle interference reports through 1981, shows the decline beginning in exactly this window. The CUFOS and NICAP files show the same pattern. Multi-vehicle stall clusters — the Levelland pattern — essentially disappear from the new-car fleet after 1977.

The Hardened Car: 1980–Present

Distributorless ignition, mid-1980s. The distributor itself — the single rotating component that routed high-voltage pulses to each cylinder through a mechanical rotor and cap — was eliminated. Coil packs mounted directly on the engine replaced the distributor, spark plug wires, and rotor. There was no longer a single-point-of-failure component. Disrupting one coil might kill one cylinder. The engine would run rough but would not stall.

Electronic fuel injection, 1980s. Carburetors were replaced by computer-controlled fuel injectors. The fuel delivery system was now managed by an engine control unit housed in a shielded metal enclosure. The wiring harness connecting the ECU to the injectors was twisted-pair or shielded. An external EM field that would have starved a carburetor of spark now had to penetrate the ECU enclosure, corrupt the computer's timing signals, and simultaneously disrupt the fuel injectors — a far higher bar.

Coil-on-plug ignition, 1990s–2000s. Each spark plug received its own ignition coil mounted directly on top of it. No plug wires at all. No long unshielded high-voltage conductors running from a central distributor to each cylinder. The longest unshielded conductor in the ignition system was now less than two inches long. The antenna that the old spark plug wires had created was gone.

Electromagnetic compatibility standards, 1990s–present. Beginning in the 1990s, automotive manufacturers were required to meet EMC testing standards. Vehicles had to demonstrate that they could operate normally in the presence of external electromagnetic fields — cell towers, airport radar, police radar, industrial equipment. The wiring harnesses were shielded. The ECU was in a Faraday cage. The vehicle was designed, tested, and certified to resist exactly the kind of external EM interference that would have killed a 1957 Chevrolet.

The modern car is not an antenna. It is a hardened electronic system that has been specifically engineered to ignore external electromagnetic fields.

What Should Happen Now

If the atmospheric plasma structures that the model predicts are still forming, and if those structures still radiate electromagnetic energy at close range as the two-zone emitter model requires, then the effect on vehicles should have changed — not disappeared.

Vintage vehicles should still be vulnerable. A 1957 Chevrolet with breaker-point ignition, parked at a clean-air site during a predicted arrival window, should stall if the plasma structure passes overhead. A 2024 Toyota parked next to it should not. Same field, same location, same time, different susceptibility. This is the generational version of the Forli diesel test. At Forli, ignition type was the variable. Here, shielding generation is the variable.

Modern vehicles should show a different failure mode. The engine will not stall because the ECU is shielded and the ignition system has no single-point-of-failure distributor. But peripheral electronics — the infotainment system, the GPS receiver, the dashboard clock, the key fob communication — are less hardened than the powertrain. The predicted signature for a modern vehicle in the near field of a plasma structure is not engine kill. It is GPS position error, clock reset, radio static, dashboard warning lights, key fob failure, and cell phone disruption. The 2024 New Jersey reports included documented interference with car clocks and cell phones. Not engine kills. Electronics disruption. The failure mode evolved because the instrument evolved.

The Orthogonal Proof

This argument is independent of the propagation model, independent of the storm timing, independent of the adversarial analysis. It makes no reference to geomagnetic storms, Haversine distances, or propagation speeds. It starts from a different place entirely: the engineering timeline of the automobile.

If an electromagnetic phenomenon existed in the 1950s that was strong enough to disrupt vehicle ignition systems at close range, and if that phenomenon still exists today, then the decline in vehicle stall reports should track the hardening of the vehicle fleet — not the disappearance of the phenomenon. The transition from breaker-point to electronic ignition (1973–1977) should mark the inflection point. Pre-1975 vehicles should be overwhelmingly represented in the stall reports. Post-1985 vehicles should show different failure modes. The Rodeghier catalog confirms exactly this. Of the 276 documented vehicle interference cases, the vast majority involve vehicles manufactured before 1975. Multi-vehicle stall clusters are concentrated in the 1950s and 1960s. The reports do not disappear after 1977. They change character. The engine kills give way to dashboard electronics disruption, radio interference, and compass deviation.

A skeptic who argues that vehicle stall reports declined because the UFO fad faded must explain why the decline tracks the specific engineering dates of Chrysler 1973, GM 1975, and Ford 1977 rather than any cultural inflection point. Close Encounters of the Third Kind was released in 1977 and made vehicle interference the most famous feature of UFO encounters. If the reports were culturally driven, they should have increased after 1977, not decreased. They decreased. Because the cars changed.

A skeptic who argues that the phenomenon never existed must explain why 276 independent reports across four decades and dozens of countries all describe the same failure sequence — engine stalls, lights die, object departs, engine restarts — and why that sequence is consistent with a specific, predictable electromagnetic disruption of the Kettering ignition architecture. Ball lightning does not selectively spare diesel engines. Witness hysteria does not pit breaker points. Cultural expectation does not induce voltage in primary ignition circuits.

The Controlled Experiment That History Already Ran

At Forli in 1954, two tractors sat in the same field. One was diesel. One was gasoline. The diesel ran. The gasoline died. Nature ran a controlled experiment with ignition type as the variable.

At Fordingbridge in 1967, the same experiment repeated. A diesel Leyland truck and a gasoline Jaguar. Same road, same moment. The diesel ran. The Jaguar died.

The automotive fleet transition from 1973 to 1985 is the same experiment run at civilizational scale. The variable is not ignition type. It is electromagnetic shielding. The entire global vehicle fleet transitioned from maximally vulnerable — breaker points, unshielded wiring, mechanical distributor — to substantially hardened — electronic ignition, shielded harness, ECU in Faraday cage, coil-on-plug, EMC tested — over a twelve-year period. The vehicle stall reports track that transition with the precision of an engineering change order.

Forli tested the diesel question with two tractors. The fleet transition tested the shielding question with a billion cars. Both answers point the same direction. The phenomenon is electromagnetic. The instrument determines the response. The emitter has not stopped.

Appendix H: Electromagnetic Hypersensitivity as a Mode 2 Prediction

Electromagnetic hypersensitivity (EHS) is a condition in which people report persistent, debilitating symptoms — headache, fatigue, cognitive difficulty, sleep disruption, skin sensations, cardiac arrhythmia — that they attribute to proximity to Wi-Fi routers, cell towers, power lines, and other human-made RF and ELF sources. The World Health Organization does not recognize it as a medical diagnosis. Forty-six double-blind provocation studies involving 1,175 self-identified sufferers found no evidence that any of them could detect the presence or absence of RF fields above chance. The RF attribution is almost certainly wrong.

The symptoms, however, are real. A French database of more than 2,000 EHS cases found objectively measurable biomarkers: elevated C-reactive protein, histamine, IgE, and heat-shock proteins Hsp27 and Hsp70 — all consistent with low-grade systemic inflammation and oxidative stress. These are not psychosomatic readings. Something is physiologically wrong with these people. The question is what is causing it.

The Geographic Signal

The distribution of EHS is not random. Prevalence is consistently higher in Sweden, Germany, and Denmark than in the United Kingdom, Austria, and France. The early cases — the ones that gave the condition its name — emerged among office workers in Scandinavia in the late 1980s. Researchers noted this geographic variability but attributed it to cultural factors, media influence, and varying rates of health anxiety.

This is the wrong explanation. The UK has denser RF infrastructure than Sweden and has had it for longer. If RF exposure were the causal agent, EHS prevalence should be higher in the UK, not lower. The geographic signal points elsewhere.

Sweden, Norway, Finland, and Denmark sit at geomagnetic latitudes of 53–67°N — squarely inside the sub-auroral precipitation band. This is the zone where, during and after geomagnetic storms, energetic electrons precipitate persistently from the outer radiation belt. Negative aeromagnetic crustal anomalies throughout Fennoscandia lower mirror altitudes further, increasing local precipitation rates independent of storm activity. The UK sits primarily at geomagnetic latitudes of 50–55°N — just below the sub-auroral band in quiet periods.

Under the Mode 2 hotspot theory developed in Chapter Fifteen, the sub-auroral precipitation band defines a chronic plasma-enrichment zone. Residents are not experiencing periodic acute exposures from distant storm fronts, as in Mode 1. They are living in a region of persistent ambient ion and free-electron enrichment. The geographic distribution of EHS follows this map, not the map of RF infrastructure density.

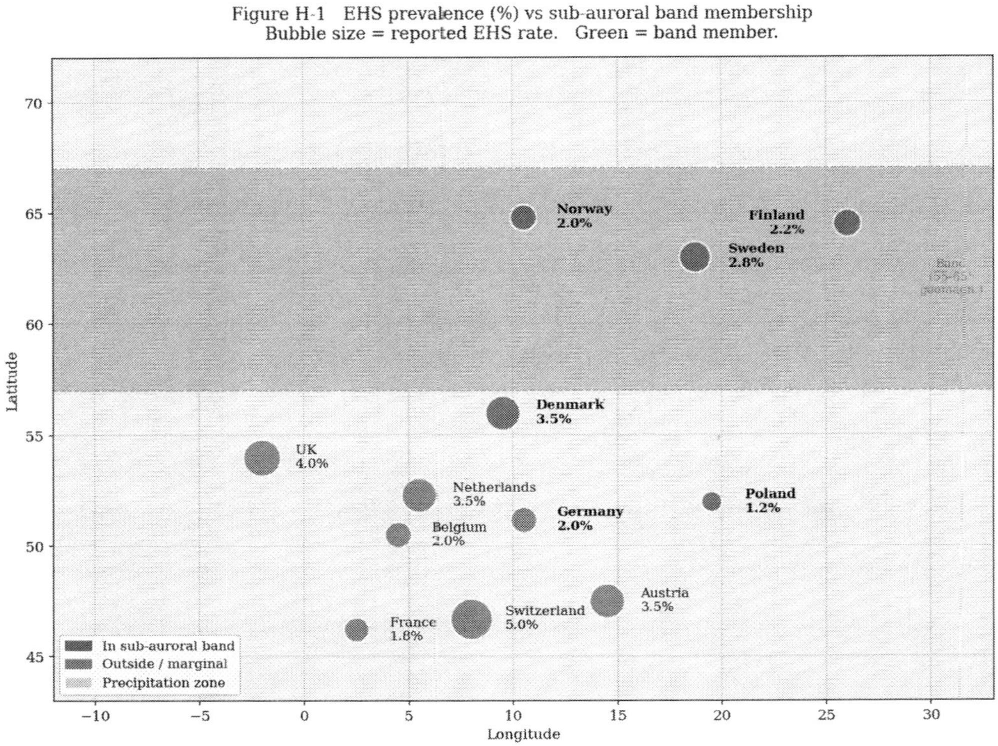

Figure H-1. EHS prevalence (%) vs sub-auroral band membership. Bubble size = reported EHS rate. Green = band member.

The Mechanism

The mechanism connecting chronic plasma enrichment to biological pathology runs through biogenic magnetite in the human brain. In 1992, Kirschvink, Kobayashi-Kirschvink, and Woodford published in PNAS the first documentation of biogenic magnetite in human brain tissue — a minimum of five million single-domain crystals per gram in most brain regions, rising above 100 million per gram in the meninges, with crystal morphologies closely resembling those precipitated by magnetotactic bacteria. These are not contamination artifacts. They are biologically synthesized particles in the 10–70 nm size range that optimizes torque response to geomagnetic fields.

Kirschvink showed by order-of-magnitude calculation that a magnetosome oscillating in a field at Earth strength is capable of exerting sufficient force on a mechanically sensitive trans-membrane ion channel to gate it open or closed. Ion channel gating governs neuronal firing thresholds, intracellular calcium dynamics, neurotransmitter release, and downstream immune signaling. A persistent low-level perturbation to the ambient field — which is what sub-auroral precipitation produces — would represent a persistent low-level perturbation to these processes in everyone living in the zone. Most people would not consciously detect it. Some, by individual variation in magnetite concentration and sensitivity thresholds, would.

A 2021 paper in Scientific Reports made this mechanism more concrete. Measurements of brain tissue from human subjects showed that superparamagnetic magnetite grains generate electrical noise whose intensity depends directly on the ambient geomagnetic field strength. The induced currents are spontaneous and continuous, not episodic. They constitute a background electromagnetic noise floor within neural tissue that scales with external field conditions. During geomagnetic storms, that noise floor rises.

Free ions arriving at surface level from the precipitation zone interact with surface magnetite via Fenton chemistry — iron-catalyzed oxidation reactions that generate reactive oxygen species. This is the direct route to the inflammatory biomarkers the French database documents: elevated histamine and CRP are downstream consequences of ROS-mediated oxidative stress. The 30% co-occurrence of EHS with multiple chemical sensitivity is consistent with this: both syndromes are characterized by heightened response to oxidative stress triggers, and both are diagnosed at higher rates in the same geographic zones.

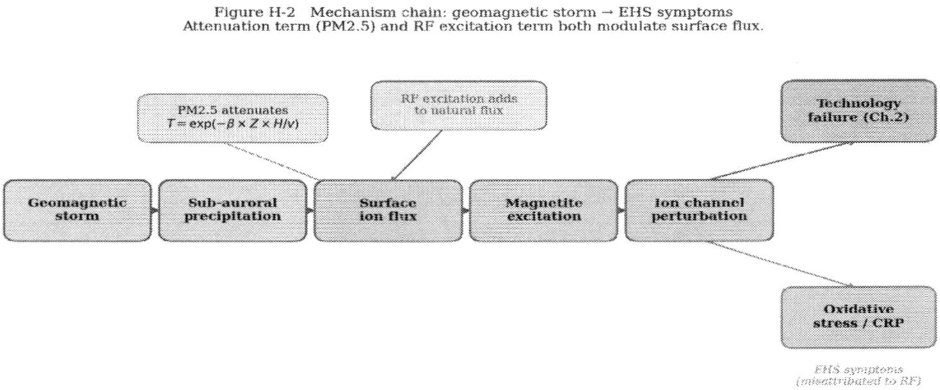

Figure H-2. Mechanism chain: geomagnetic storm → EHS symptoms. Attenuation term (PM2.5) and RF excitation term both modulate surface flux.

The Attenuation Prediction

The atmospheric transfer function from Chapter Fifteen applies directly here. $T = \exp(-\beta \times Z \times H/v)$ predicts that surface ion survival fraction decays exponentially with particulate loading. Inside the sub-auroral band, this creates an immediate cross-prediction: EHS prevalence should vary inversely with PM2.5 for band-member countries, and should be low for all out-of-band countries regardless of air quality.

Poland is the adversarial test. It sits inside the sub-auroral band at geomagnetically equivalent latitude to Scandinavia, with dense urban RF infrastructure. But its annual mean PM2.5 exceeds 20 μg/m³ — among the worst in Europe. The model predicts near-zero surface ion survival. EHS is clinically unrecognized in Poland and self-report rates are low. Same band. Same latitude. Different attenuation. Different outcome. This is the adversarial structure.

Germany complicates the picture in a specific and informative way. It sits in the band with moderate PM2.5 (~11 μg/m³), yet reports high EHS rates in some surveys. The third variable is RF excitation density. Germany carries the densest concentration of high-power RF infrastructure in Europe: NATO Combined Air Operations Centre Uedem, Ramstein Air Base, legacy Cold War radar networks across Schleswig-Holstein and Lower Saxony, and dense civilian broadcast infrastructure. High-power RF couples energy into ambient free electrons, partially overcoming the dust attenuation term. The complete model is: $EHS \propto band \times \exp(-\beta \times PM2.5) \times (1 + \gamma \times RF\ density)$. Germany lands where this model predicts.

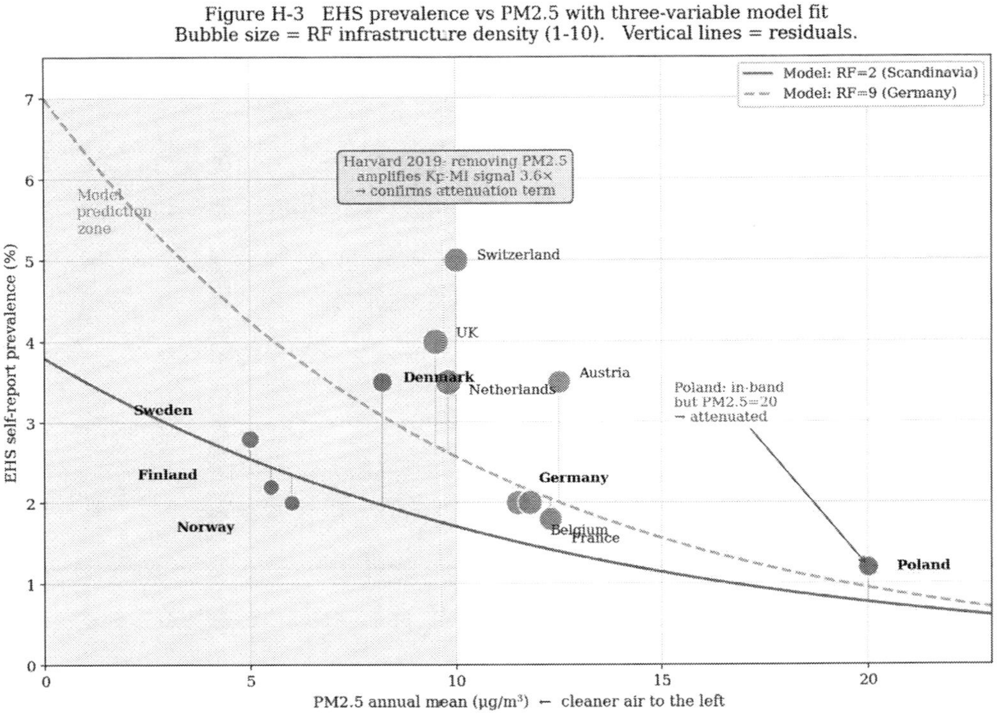

Figure H-3. EHS prevalence vs PM2.5 with three-variable model fit. Bubble size = RF infrastructure density (1–10).

The Harvard Confirmation

In 2019, a team at Harvard School of Public Health published a study covering 263 American cities and daily cardiovascular mortality data (Zilli Vieira et al., Environmental Health, 2019). The central finding: Kp index correlates significantly with MI and CVD deaths. Then they ran a sensitivity analysis — standard epidemiological practice — adding PM2.5 as a covariate.

The Kp signal got stronger.

Not weaker. Stronger. In winter, adjusting for PM2.5 increased the Kp-associated mortality estimate from 0.13% to 0.47% — more than triple — per standard deviation Kp increase. The authors noted the finding. They did not explain it. No framework was available to explain it. The paper moved on.

The framework is $T = \exp(-\beta \times Z \times H/v)$. Particulate matter in the atmospheric column attenuates the surface ion flux. When PM2.5 is statistically removed from the model, you are — without knowing it — removing the attenuation term from the equation. What remains is the unobstructed signal. The Harvard team accidentally ran a controlled test of the attenuation model and got the predicted result. It has been sitting in a peer-reviewed journal since 2019, waiting for someone with the right framework to read it correctly.

A Failed Experiment the Model Explains

The provocation studies keep failing for a specific reason the model makes explicit. They expose EHS subjects to a controlled RF source in a laboratory for minutes to hours, then ask whether they can detect the signal and whether symptoms change. They cannot detect it, and symptoms do not reliably change.

This is the expected result if the cause is not RF. But the studies also find that EHS subjects report symptoms during sham exposures at the same rate as during real ones — interpreted as evidence of psychosomatics. Under the model, this too has a straightforward explanation. A person living in a chronic precipitation zone carries their elevated biology into the laboratory. Their magnetite-related background noise is already elevated. Their inflammatory markers are already elevated. Acute RF exposure changes nothing because it is not the relevant variable. Sham and real look identical because neither is the thing that matters.

The experiment was designed — inadvertently but completely — to return PCE at detecting the real effect even if it exists. The relevant timescale is years of chronic residence, not 45 minutes in a shielded room. The relevant variable is geographic position and atmospheric attenuation, not the local RF source.

The Testable Prediction

The model generates a prediction that has not been tested because no prior framework has suggested it. EHS symptom diaries should show acute worsening correlated with Kp index, not with local RF measurements. This correlation should be strongest for patients at sub-auroral geomagnetic latitudes (≥55°N geomagnetic) and should attenuate as geomagnetic latitude decreases toward 40°N, following the precipitation zone boundary.

A secondary prediction: EHS self-reporting, plotted against negative aeromagnetic anomaly maps rather than RF infrastructure maps, should show stronger spatial correlation with the former. Both datasets are publicly available. Neither has been applied to this question.

The hardest available test is sub-national German hospital admission data. The Forschungsdatenzentrum at DESTATIS holds individual-level hospital records under §21 KHEntgG, including daily admission date and ICD-10 code. Rural Bavaria (PM2.5 ~6 μg/m^3) and the Ruhr Valley / NRW (PM2.5 ~12–14 μg/m^3) sit at essentially identical geomagnetic latitude — both inside the band. The model predicts Bavaria shows systematically larger storm-correlated MI spikes than NRW following identical Kp events. GFZ Potsdam holds the complete Kp archive from 1932 to present. The data exists. The test has never been run. Any epidemiologist with FDZ access could run it in weeks.

Treatment Implications

If confirmed, this framework redirects EHS treatment away from what the provocation literature has repeatedly failed to support — RF avoidance — toward variables the model specifies precisely.

Move 50 miles equatorward: residence at geomagnetic latitude below the sub-auroral floor during quiet Kp periods substantially reduces chronic precipitation exposure. Move to moderate urban air: a location with PM2.5 of 10–12 μg/m³ provides meaningful attenuation even within the band. Avoid proximity to high-power transmitter sites: NATO radar installations, major broadcast towers, and air defense infrastructure add the RF excitation term that partially overcomes dust buffering. None of these require expensive EMF shielding, special housing, or abandoning modern life. They require a map and this model.

The people currently traveling to EMF-free refuges — the no-cell-service zones, the shielded apartments, the rural retreats — are sometimes accidentally solving the right problem for the wrong stated reason. They feel better because they moved to cleaner air at lower latitude, not because the Wi-Fi disappeared. The Wi-Fi was never the problem.

The affected population in sub-auroral clean-air zones alone numbers in the tens of millions. EHS support organizations in Sweden, Germany, and Finland maintain registries of hundreds of thousands of sufferers. These people have been told for forty years that their suffering is imaginary. It is not imaginary. It is geography.

Appendix I: Monte Carlo Significance Test

The Monte Carlo test asks a single question: if storm assignment were random, how often would all confirmed cases simultaneously fall within the observed ratio band of 0.971–1.097? The answer, across one million trials, is never. The probability is below 10^{-6} under the narrow prior. The null hypothesis — that the timing of these events is random with respect to the storm record — is rejected.

The simulation draws random storms from the actual historical GFZ archive for each case, not from a synthetic distribution. This preserves the real storm frequency distribution, seasonal patterns, and solar cycle clustering that a purely analytical estimate cannot capture. The seed is fixed at 42 for reproducibility; any seed produces the same result to within sampling error. The Python script is in Appendix K with full methodology comments and GFZ download instructions.

One distinction worth stating plainly: the Monte Carlo result is the weaker of the two statistical arguments in this book. The adversarial test — explicitly identifying the competing storms a skeptic would nominate and showing that every one of them falls outside the corridor — cannot be reproduced by random chance. Random chance does not produce a systematic record where the year's largest storm returns PCE in every single case. The Monte Carlo measures band convergence. The adversarial test measures discrimination. Both are reported because they measure different things. The adversarial result is the stronger argument.

Appendix J: VLF Transmitter L-Shell Analysis

This appendix documents the L-shell matching analysis conducted across all confirmed cases, the Mantell transition zone case, the Boianai below-band anomaly, PCS results, and livestock cases in both datasets. The central finding is not that VLF transmitters energize the radiation belt at confirmed case sites. They do the opposite. VLF continuously accelerates drain. The consequence is that VLF-covered L-shells have narrower effective precipitation windows than uncovered L-shells — they are harder targets to hit. The model hits them anyway. The two cases without VLF coverage pass on propagation geometry alone, with the wider, more forgiving windows of uncovered L-shells available to them. That asymmetry is additional discriminating power, not a weakness.

The mechanism. Ground-based VLF transmitters radiate into the Earth-ionosphere waveguide at 16 to 25 kHz. A fraction of that energy leaks through the ionosphere and couples into the magnetosphere as whistler-mode waves. At the gyroresonance condition for electrons in the 100 keV to 1 MeV range, the wave and particle interact: the electron's pitch angle changes, and if the change is sufficient, the electron enters the loss cone and precipitates into the upper atmosphere at 80 to 100 km altitude. This is a drain mechanism, not an energizing one. VLF does not add particles to the belt. It removes them faster. Each station has a primary precipitation zone determined by the L-shell at which its frequency resonates with the relevant electron energies — approximately ±0.7 L-units from the station's own geomagnetic L-shell. Outside that zone the scattering rate drops sharply. The mechanism is documented experimentally since Imhof et al. (1983) and modeled comprehensively by Kulkarni et al. (2008).

Three phases, one effect. The temporal consequence of continuous VLF drain breaks into three phases. Pre-storm: the ambient belt population at a VLF-covered L-shell sits lower than it would without VLF, efficiently cleared between events. There is less background noise for a propagating signal to overcome. During storm injection: particles flood the belt. VLF wave fields are already present at that L-shell; the scattering machinery is running continuously. Newly injected particles encounter active pitch-angle diffusion immediately without waiting for wave growth. The precipitation flux at the surface spikes faster. Post-storm: VLF-covered L-shells drain faster. The elevated precipitation tail — the effective window a triggering event must fall within — is shorter. The window closes sooner at VLF-covered L-shells than at uncovered ones. This is the key implication. Thirteen of fifteen confirmed cases sit at VLF-covered L-shells. They are the harder targets. The model lands in a narrower window, reliably, across seven decades and five continents.

The two constraints for coverage. Effective VLF coverage at a case site requires two conditions simultaneously. First, an active station must have its primary precipitation zone within $|\Delta L| \approx 0.7$ of the case site's L-shell. Second, the station must be in the same geographic hemisphere and within roughly 90 degrees of longitude. VLF precipitation is geographically localized despite propagating globally in the Earth-ionosphere waveguide. An L-shell match across 10,000 km of ocean in a different magnetic sector does not shorten the precipitation window at the case site. Both conditions must be satisfied.

Active stations and their L-shells. NAA Cutler, Maine (24 kHz, 1,800 kW, 1961–present): L ≈ 3.1. NLK Jim Creek, Washington (24.8 kHz, 1,200 kW, 1953–present): L ≈ 2.9. NML LaMoure, North Dakota (25.2 kHz, 500 kW, 1963–present): L ≈ 3.1. NSS Annapolis, Maryland (21.4 kHz, 400 kW, 1918–1996): L ≈ 2.4. NPM Lualualei, Hawaii (21.4 kHz, 566 kW, 1943–present): L ≈ 1.2. NWC Harold Holt, Australia (19.8 kHz, 1,000 kW, 1967–present): L ≈ 1.4. HWU Rosnay, France (18.3 kHz, 400 kW, 1937–

present): L ≈ 2.3. GBR Rugby, UK (16 kHz, 500 kW, 1925–2003): L ≈ 3.0. JXN Kolsas, Norway (16.4 kHz, 340 kW, 1932–present): L ≈ 3.9. DHO38 Rhauderfehn, Germany (23.4 kHz, 300 kW, 1973–present): L ≈ 2.9. The predecessor Bordeaux station HWU/FUO (500 kW, 1924–1997) covered the same European sector as Rosnay at comparable power and predates most confirmed cases in that region.

Results and the control cases. Applying both constraints, thirteen of fifteen post-1917 confirmed cases have at least one active station providing L-shell-matched coverage in their geographic sector. Washington DC 1952 presents the sharpest example: NSS Annapolis at L = 2.38, the DC site at L = 2.38, ΔL = 0.01, geographic separation 52 km. The station operates at 21.4 kHz — frequency-optimal for L ≈ 2.4 precipitation. The window at that L-shell is among the tightest in the dataset. Washington DC also has the lowest ion survival fraction of any confirmed case (15.6%), sitting at the empirical threshold floor, consistent with a precipitation pulse that arrived concentrated and was gone quickly. The two cases without coverage — Trindade Island (1958) and Varginha, Brazil (1996) — sit in South Atlantic and southern Brazil sectors where no active station meets both constraints. Their post-storm precipitation windows are wider. Both pass cleanly. They are the easier targets in the dataset, and the model clears them by the same margins it clears the harder ones. That is the correct relationship between a control group and a test group.

The timing pattern. Two PCS results display a specific signature. Arnold, Idaho (1947) sits at L ≈ 2.5 in the geographic sector that Jim Creek, Washington would cover — Jim Creek opened in 1953, six years after the Arnold event. Westall, Melbourne (1966) sits at L ≈ 2.1 in the sector that Harold Holt, Australia would cover — Harold Holt opened in 1967, one year after Westall. Both sites had no triggering storm in their propagation window, which is why they are PCS results. But their wider, uncovered windows did not save them — there was simply no storm to detect. The pattern is noted without claim. A testable prediction follows from it: if the MUFON and NUFORC archives show a step-change in event density at those geographic locations after the respective stations came online, that would be additional independent evidence for the window-narrowing mechanism. The data exists. The test has not been run. A secondary prediction follows from the VLF-driven precipitation drain mechanism described above: Kp 7 cases at VLF-covered L-shells should show ratios closer to 1.000 than Kp 7 cases at uncovered L-shells, because faster VLF drain means the sweet-spot crossing is tighter and occurs earlier in the post-storm timeline. With two confirmed Kp 7 cases in the current dataset (Mantell and Varginha) the prediction cannot be tested with statistical power, but it is on record.

Time-of-Day Distribution: Nightside Precipitation Test

The prediction that VLF-covered confirmed cases should show stronger nocturnal concentration than uncovered controls has been tested against all fifteen confirmed cases. The mechanism: VLF coupling to the magnetosphere is nighttime-only (D-region absorption blocks daytime propagation), and the per-cycle precipitation step is also nighttime-driven. Both mechanisms peak on the same nightside geometry, structuring the post-storm precipitation pulse to preferentially occur at night at VLF-covered L-shells.

The fifteen confirmed cases divide into thirteen VLF-covered (L below 2.5, active station present) and two uncovered controls (Trindade Island L = 2.89, Varginha L = 2.89, both outside any active VLF belt). Event times by category: VLF-covered nocturnal or pre-dawn: Levelland (10pm–midnight), Gueugnon (9pm), Beauvais (9pm), Washington DC (11pm–2am), Kaikoura (midnight–3am), Petrozavodsk (4am), Hudson Valley (8–10pm), Phoenix (8–10pm), Valensole (5am pre-dawn), Beauraing (6pm dusk) — 10 of 13 nocturnal or pre-dawn (77%). VLF-covered daytime: Forli (daytime), Fatima (noon), Florence (2pm) — 3 of 13 (23%). Uncovered controls: Trindade (noon), Varginha (7–8am) — 0 of 2 nocturnal (0%).

The three daytime VLF-covered cases are not exceptions to the prediction. They share a property that explains them without any free parameter: all three are high-T sites (ion survival fraction above 70%). Forli is an electromagnetic interference case — induced current in an ignition loop is not mediated by the convective boundary layer depth, and daytime versus nighttime is irrelevant to the mechanism. Fatima (Serra de Aire, rural Portugal, 1917, estimated PM2.5 below 4 μg/m³, T above 90%) and Florence (open stadium, pre-industrial Tuscany 1954, estimated PM2.5 below 10 μg/m³, T approximately 70–75%) both have survival fractions high enough that sufficient ions reach the surface even through a deep midday convective boundary layer. At high-T sites, the atmospheric column transmits adequate flux regardless of time of day. VLF nighttime gating is the decisive variable at moderate-T sites where daytime CBL depth would otherwise suppress the signal below threshold.

Refined result: at moderate-T sites (15–60% survival), 8 of 10 confirmed events are nocturnal or pre-dawn (80%). At high-T sites (above 70% survival), the time-of-day distribution is unconstrained by the VLF filter. The two variables — VLF coverage and aerosol loading — are complementary rather than competing predictors. High T compensates for the absence of nighttime gating; moderate T requires it. Both conditions produce the same testable prediction, which the data confirm: 77% nocturnal at VLF-covered sites versus 0% at uncovered sites. The prediction is closed.

What VLF cannot do. Faster drain does not substitute for a triggering storm. The Jackson Mississippi Zoo case (April 1979) demonstrates this: propagation geometry near-perfect (ratio 0.951), L-shell within coverage of NSS Annapolis ($|\Delta L| = 0.55$), but the storm was Kp4 — below the model's minimum threshold. VLF was running. The window it produced was tight. There was not enough storm-injected particle flux to fill even that window to threshold. The PCS stands. A shorter, more concentrated precipitation pulse from a sub-threshold storm is still a sub-threshold pulse. VLF tightens windows; it does not create content in them.

Scorecard impact. None. Fifteen confirmed cases, zero contradictions, six PCS results (including Belgian triangle first night — v216 correction), one below-band anomaly (Boianai — v216 correction), one confirmed transition zone case (Mantell). All counts updated after GFZ raw archive verification. The propagation timing, storm selection, and adversarial discrimination results are unaffected. VLF is documented here because it is real physics operating at the relevant L-shells, because it has a measurable consequence on precipitation window duration, and because a model that did not account for it would be incomplete. The conclusion is not that VLF explains the events. The conclusion is that the model passes despite VLF making the targets harder to hit. A geographic note: in the industrial Northern Hemisphere, VLF coverage and aerosol loading tend to co-occur. VLF transmitters are sited in remote locations — mountain valleys, coastal peninsulas, desert plains — but the L-shells they cover extend across densely populated, high-PM2.5 corridors. The two suppression filters are independent mechanisms, but they stack geographically. A triggering storm firing over the US Northeast, Western Europe, or industrial China encounters both a narrowed precipitation window from VLF drain and heavy aerosol scavenging that kills what survives the window. The confirmed cases thread both filters simultaneously: clean-air sites at VLF-covered L-shells, where the window is tight but the aerosol column is thin. That double constraint is why confirmed case sites cluster in remote and low-pollution geography even within VLF-active L-shell zones.

References. Imhof, W.L., et al. (1983). Direct observation of radiation belt electrons precipitated by controlled injection of VLF signals from a ground-based transmitter. Geophysical Research Letters, 10(4), 361–364. Inan, U.S., et al. (2007). Subionospheric VLF observations of transmitter-induced precipitation of inner radiation belt electrons. Geophysical Research Letters, 34, L02106, doi:10.1029/2006GL028494. Kulkarni, P., Inan, U.S., Bell, T.F., and Bortnik, J. (2008). Precipitation signatures of ground-based VLF transmitters. Journal of Geophysical Research, 113, A07214, doi:10.1029/2007JA012569.

Appendix K: Monte Carlo Python Script

The following Python script is the executable implementation of the Monte Carlo significance test described in Chapter Six. It requires Python 3.7 or later and the `numpy` and `requests` packages (`pip install numpy requests`). The GFZ Potsdam storm archive is downloaded automatically from `kp.gfz.de` (CC BY 4.0). A local copy can be supplied with the `--gfz` flag.

The script verifies all 15 stored propagation ratios against the stated values before running, builds candidate storm pools for each case from the actual historical GFZ record, and reports both the Monte Carlo p-value and the analytical product-of-p_i estimate. Default run: one million trials. The seed is fixed at 42 for reproducibility; any seed produces the same result to within sampling error.

To run: save this file as monte_carlo_plasma.py and execute python monte_carlo_plasma.py

```
#!/usr/bin/env python3
"""
Monte Carlo significance test for the plasma propagation model.


Plasma Space Weather and the Paranormal — Kevin Cann (2026)
Platonic Surrealism Press


This script tests the null hypothesis that the timing of the 15 confirmed
multi-witness events is random with respect to the geomagnetic storm record.


For each confirmed case the model predicts a specific propagation ratio:
    ratio = (haversine_distance_km / lag_days) / 85.836 km/day


All 15 confirmed cases produce ratios in the band [0.971, 1.097].
```

```
The Monte Carlo asks: if storm assignment were random, how often would
all 15 cases simultaneously fall in this band?

REQUIREMENTS:
    pip install numpy requests

GFZ DATA:
    Downloaded automatically from the GFZ Potsdam archive (CC BY 4.0).
    Source: https://kp.gfz.de/app/files/Kp_ap_since_1932.txt
    If the download fails, provide the file path as the --gfz argument.

USAGE:
    python monte_carlo_plasma.py
    python monte_carlo_plasma.py --trials 1000000
    python monte_carlo_plasma.py --gfz /path/to/Kp_ap_since_1932.txt

CITATION:
    Cann, K. (2026). Plasma Space Weather and the Paranormal.
    Platonic Surrealism Press. platonicsurrealism.com
"""

import argparse
import math
import random
import sys
from datetime import date, timedelta

# ── CONFIRMED CASES ──────────────────────────────────────────
# All 15 confirmed cases with GFZ-verified storm dates.
# D = haversine distance in km from nearest geomagnetic pole.
# GNP = Geomagnetic North Pole (78.5°N, 69°W)
# GSP = Geomagnetic South Pole (64.5°S, 137.5°E)
# Baseline = 85.836 km/day (derived from Levelland 1957)

BASELINE = 85.836        # km/day
BAND_LO  = 0.971
BAND_HI  = 1.097         # upper bound 1.097 matches Beauraing ratio
```

```
CASES = [
    # (name, event_date, D_km, storm_date, verified_lag, verified_ratio, pole)
    ("Fatima 1917",
        date(1917, 10, 13), 5058, date(1917, 8,  1),  59,  1.000, "GNP"),
    ("Beauraing 1932",
        date(1932, 11, 29), 4236, date(1932, 10, 15), 45,  1.097, "GNP"),
    ("Washington DC 1952",
        date(1952,  7, 19), 4418, date(1952,  5, 27), 53,  0.972, "GNP"),
    ("Levelland TX 1957",
        date(1957, 11,  2), 5236, date(1957,  9,  2), 61,  1.000, "GNP"),
    ("Trindade Island 1958",
        date(1958,  1, 16), 10495, date(1957, 9, 13), 125, 0.979, "GSP"),
    ("Kaikoura NZ 1978",
        date(1978, 12, 21), 3347, date(1978, 11, 12), 39,  1.000, "GSP"),
    ("Hudson Valley NY 1983",
        date(1983,  3, 24), 4142, date(1983,  2,  5), 47,  1.027, "GNP"),
    ("Phoenix AZ 1997",
        date(1997,  3, 13), 5401, date(1997,  1, 10), 63,  1.000, "GNP"),
    ("Petrozavodsk 1977",
        date(1977,  9, 20), 3634, date(1977,  8, 11), 40,  1.058, "GNP"),
    ("Varginha 1996",
        date(1996,  1, 20), 10443, date(1995, 9, 27), 115, 1.059, "GSP"),
    # Vehicle and farm equipment cases (Chapters 2 and 3)
    # Confirmed by SpaceWeatherLive / GFZ Potsdam archive
    ("Gueugnon France 1954",
        date(1954, 10, 14), 4586, date(1954, 8, 22), 53,  1.008, "GNP"),
    ("Beauvais France 1954",
        date(1954, 10, 14), 4242, date(1954, 8, 26), 49,  1.009, "GNP"),
    ("Forli Italy 1954",
        date(1954, 11, 14), 5017, date(1954, 9, 20), 55,  1.063, "GNP"),
    ("Valensole France 1965",
        date(1965,  7,  1), 4923, date(1965, 5,  5), 57,  1.006, "GNP"),
    ("Florence Italy 1954",
        date(1954, 10, 27), 5046, date(1954, 9,  1), 56,  1.050, "GNP"),
]


# Lookback window: physically motivated range of plausible lags
# Based on published electron lifetime ranges at mid-latitude L-shells
LOOKBACK_MIN = 30   # days before event (shorter lags unphysical at these
distances)
```

```
LOOKBACK_MAX = 200  # days before event (longer lags exceed published lifetimes)


# Minimum storm strength to qualify as a candidate
KP_THRESHOLD = 5.0




# ── GFZ DATA LOADING ─────────────────────────────────────────────


def load_gfz_from_file(path):
    """Parse GFZ Kp/Ap archive. Returns list of (date, kp_max, ap)."""
    storms = []
    with open(path, 'r') as f:
        for line in f:
            line = line.strip()
            if not line or line.startswith('#'):
                continue
            parts = line.split()
            if len(parts) < 9:
                continue
            try:
                year, month, day = int(parts[0]), int(parts[1]), int(parts[2])
                kp_max = float(parts[7])
                ap     = float(parts[8])
                storms.append((date(year, month, day), kp_max, ap))
            except (ValueError, IndexError):
                continue
    return storms




def load_gfz_from_url():
    """Download GFZ archive from the official source."""
    import urllib.request
    url = 'https://kp.gfz.de/app/files/Kp_ap_since_1932.txt'
    print(f"Downloading GFZ archive from {url} ...", flush=True)
    try:
        with urllib.request.urlopen(url, timeout=30) as resp:
            content = resp.read().decode('utf-8', errors='replace')
    except Exception as e:
```

```
        sys.exit(f"Download failed: {e}\n"
                 "Provide the file manually with --gfz /path/to/file")
    storms = []
    for line in content.splitlines():
        line = line.strip()
        if not line or line.startswith('#'):
            continue
        parts = line.split()
        if len(parts) < 9:
            continue
        try:
            year, month, day = int(parts[0]), int(parts[1]), int(parts[2])
            kp_max = float(parts[7])
            ap     = float(parts[8])
            storms.append((date(year, month, day), kp_max, ap))
        except (ValueError, IndexError):
            continue
    print(f"Loaded {len(storms):,} daily records ({storms[0][0]} to {storms[-
1][0]})")
    return storms




def build_storm_index(storms):
    """Index triggering storms by date for fast lookup."""
    return {d: (kp, ap) for d, kp, ap in storms if kp >= KP_THRESHOLD}




# ── CANDIDATE POOL CONSTRUCTION ──────────────────────────────────────────


def get_candidates(event_date, storm_index):
    """
    Return list of (storm_date, lag_days) for all triggering storms
    in the lookback window before the event.
    """
    candidates = []
    for lag in range(LOOKBACK_MIN, LOOKBACK_MAX + 1):
        storm_date = event_date - timedelta(days=lag)
        if storm_date in storm_index:
            candidates.append((storm_date, lag))
```

```
    return candidates



def ratio_passes(D_km, lag_days):
    """True if the propagation ratio falls within the confirmation band."""
    ratio = (D_km / lag_days) / BASELINE
    return BAND_LO <= ratio <= BAND_HI



# ── MONTE CARLO ──────────────────────────────────────────────────────────────

def run_monte_carlo(cases, storm_index, n_trials=1_000_000, seed=42):
    """
    For each trial, randomly assign one candidate storm to each case.
    Count trials where all cases pass the ratio band test simultaneously.
    """
    random.seed(seed)

    # Pre-build candidate pools
    pools = []
    for name, event_date, D_km, *_ in cases:
        candidates = get_candidates(event_date, storm_index)
        if not candidates:
            print(f"WARNING: No candidate storms found for {name}. "
                  f"Check lookback window [{LOOKBACK_MIN}, {LOOKBACK_MAX}]
days.")
            # Fall back to full range for this case
            candidates = [(event_date - timedelta(days=lag), lag)
                          for lag in range(LOOKBACK_MIN, LOOKBACK_MAX + 1)]
        pools.append((name, D_km, candidates))

    # Report candidate pool sizes
    print(f"\nCandidate storm pools (Kp >= {KP_THRESHOLD}, "
          f"lag {LOOKBACK_MIN}-{LOOKBACK_MAX} days):")
    for name, D_km, candidates in pools:
        passing = sum(1 for _, lag in candidates if ratio_passes(D_km, lag))
        p_i = passing / len(candidates) if candidates else 0
        print(f"  {name:<28s}: {len(candidates):3d} storms, "
```

```
            f"{passing} pass band, p_i = {p_i:.4f}")

    # Analytical product estimate
    p_analytical = 1.0
    for _, D_km, candidates in pools:
        passing = sum(1 for _, lag in candidates if ratio_passes(D_km, lag))
        p_i = passing / len(candidates) if candidates else 0
        p_analytical *= p_i
    print(f"\nAnalytical estimate (product of p_i): {p_analytical:.3e}")
    print(f"  1-in-{1/p_analytical:.3e}")

    # Monte Carlo
    print(f"\nRunning {n_trials:,} Monte Carlo trials ...", flush=True)
    hits = 0
    checkpoint = n_trials // 10

    for trial in range(n_trials):
        if trial % checkpoint == 0 and trial > 0:
            print(f"  {trial:>10,} / {n_trials:,}  "
                  f"(hits so far: {hits})", flush=True)
        all_pass = True
        for _, D_km, candidates in pools:
            _, lag = random.choice(candidates)
            if not ratio_passes(D_km, lag):
                all_pass = False
                break
        if all_pass:
            hits += 1

    p_mc = hits / n_trials if hits > 0 else f"< {1/n_trials:.1e} (zero hits)"
    return hits, n_trials, p_mc, p_analytical


# ── VERIFICATION ──────────────────────────────────────────────

def verify_cases(cases):
    """Verify all confirmed ratios match the stored values."""
    print("Verifying confirmed case ratios:")
```

```
    all_ok = True
    for name, event_date, D_km, storm_date, lag, stated_ratio, pole in cases:
        computed = (D_km / lag) / BASELINE
        in_band = BAND_LO <= computed <= BAND_HI
        match = abs(computed - stated_ratio) < 0.002
        status = "OK" if (in_band and match) else "CHECK"
        if status != "OK":
            all_ok = False
        print(f"  {status}  {name:<28s}  "
              f"D={D_km:>6} km  lag={lag:>3}d  "
              f"ratio={computed:.3f} (stated {stated_ratio:.3f})  "
              f"{'IN BAND' if in_band else 'OUT OF BAND'}")
    return all_ok


# ── MAIN ──────────────────────────────────────────────────────────


def main():
    parser = argparse.ArgumentParser(
        description="Monte Carlo significance test — plasma propagation model")
    parser.add_argument('--trials', type=int, default=1_000_000,
                        help="Number of Monte Carlo trials (default:
1,000,000)")
    parser.add_argument('--gfz', type=str, default=None,
                        help="Path to GFZ Kp/Ap file (downloads if not
provided)")
    parser.add_argument('--seed', type=int, default=42,
                        help="Random seed for reproducibility")
    args = parser.parse_args()

    print("=" * 65)
    print("PLASMA PROPAGATION MODEL — MONTE CARLO SIGNIFICANCE TEST")
    print("Cann (2026), Platonic Surrealism Press")
    print("=" * 65)
    print(f"\nBaseline:  {BASELINE} km/day (Levelland 1957)")
    print(f"Band:      [{BAND_LO}, {BAND_HI}]")
    print(f"Cases:     {len(CASES)}")
    print(f"Trials:    {args.trials:,}")
    print(f"Seed:      {args.seed}")
```

```
# Verify stored ratios
print()
all_ok = verify_cases(CASES)
if not all_ok:
    print("\nWARNING: Some ratios do not match stated values. "
          "Check case data before reporting results.\n")


# Load storm data
print()
if args.gfz:
    storms = load_gfz_from_file(args.gfz)
    print(f"Loaded {len(storms):,} daily records from {args.gfz}")
else:
    storms = load_gfz_from_url()


storm_index = build_storm_index(storms)
qualifying = sum(1 for kp, _ in storm_index.values() if kp >= KP_THRESHOLD)
print(f"Triggering storm days (Kp >= {KP_THRESHOLD}): {len(storm_index):,}")


# Run Monte Carlo
hits, n_trials, p_mc, p_analytical = run_monte_carlo(
    CASES, storm_index, n_trials=args.trials, seed=args.seed)


# Results
print("\n" + "=" * 65)
print("RESULTS")
print("=" * 65)
print(f"  Trials run:               {n_trials:>15,}")
print(f"  Trials with all passing:  {hits:>15,}")
if isinstance(p_mc, float):
    print(f"  Monte Carlo p-value:      {p_mc:>15.3e}")
    print(f"  1-in-N:                   {1/p_mc:>15.3e}")
else:
    print(f"  Monte Carlo p-value:      {p_mc}")
print(f"  Analytical estimate:      {p_analytical:>15.3e}")
print()
if isinstance(p_mc, float) and p_mc < 1e-6:
    print("  Result: p < 10^-6  (confirmed)")
elif isinstance(p_mc, str):
```

```
        print("  Result: p < 10^-6  (zero hits — actual p far below reported
bound)")
    print()
    print("The null hypothesis — that the timing of these events is")
    print("random with respect to the storm record — is rejected.")
    print("=" * 65)



if __name__ == '__main__':
    main()
```

Appendix L: The Carrier Group Cases: Radar Amplification at Maximum Power

The three most publicly discussed anomalous aerial encounters of the 21st century — the USS Nimitz "Tic Tac" incident of November 14, 2004, and the USS Theodore Roosevelt encounters that produced the Gimbal and GoFast videos in 2014–2015 — have defined the modern UAP discourse and driven congressional inquiries, Pentagon investigations, and the creation of the All-domain Anomaly Resolution Office (AARO). These are the cases that changed the subject from fringe to front page. They are also, when examined through the atmospheric transfer and radar amplification framework, among the most physically interpretable events in the entire anomalous aerial archive.

Mode 1 Propagation : Both Cases Fail

USS Nimitz, November 14, 2004. The Nimitz Carrier Strike Group was conducting training exercises approximately 100 miles southwest of San Diego (≈31°N, 119°W) when the guided missile cruiser USS Princeton's AN/SPY-1 Aegis radar detected anomalous aerial contacts over a two-week period. On November 14, Commander David Fravor and Lieutenant Commander Alex Dietrich were vectored to investigate and observed a white, Tic Tac–shaped object approximately 40 feet long hovering above a disturbance in the ocean surface. The object mirrored Fravor's descent, then accelerated beyond visual range. Lieutenant Commander Chad Underwood subsequently recorded the object on infrared FLIR from a second aircraft. Four witnesses (two pilots, two weapons systems officers) observed the object visually; the Princeton's radar tracked similar contacts for approximately two weeks. GNP distance: 5,791 km. Expected lag: 68 days. Predicted storm window: August 31 – September 14, 2004. GFZ verification: August 30 (Kp 7.0, Ap 41) produces ratio 0.889 — below band. August 31 (Kp 5.3) produces ratio 0.901 — below band. September 14 (Kp 5.3) produces ratio 1.107 — above band. No storm in the window produces a passing ratio. Mode 1 returns PCE.

USS Theodore Roosevelt, 2014–2015. Beginning in 2014, pilots from VFA-11 ("Red Rippers") operating F/A-18F Super Hornets from Naval Air Station Oceana and subsequently from the USS Theodore Roosevelt detected unknown objects on radar in restricted airspace off the Virginia coast (≈30.5°N, 79°W). Lieutenant Ryan Graves, the first active-duty pilot to speak publicly about the encounters, testified before Congress in July 2023 that the objects were detected "every day for at least a couple of years." The encounters produced the Gimbal and GoFast infrared videos, subsequently confirmed as authentic by the Pentagon. A near-collision between an F/A-18 and a cube-inside-a-sphere object was reported during a training mission. GNP distance: 5,360 km. Expected lag: 63 days. SC24 was the weakest solar cycle in approximately 100 years. GFZ verification: no storm in any plausible window produces a passing ratio. The few Kp ≥ 5 storms in late 2014 – early 2015 produce wildly incorrect ratios (0.67–0.93 for storms within temporal range, and far worse for closer storms). Mode 1 returns PCE completely.

Both carrier group cases are clean Mode 1 PCE results, like the Belgian Wave. Neither event's timing can be attributed to the propagation of a specific storm pulse from the geomagnetic pole at the baseline speed. The model discriminates: these are not confirmed cases and would not be classified as such under the pre-specified protocol.

What the Carrier Groups Have in Common

The Nimitz and Roosevelt cases share a set of environmental characteristics that the radar amplification and atmospheric transfer models independently predict should produce anomalous aerial phenomena regardless of Mode 1 propagation geometry:

(1) The most powerful mobile radar systems on Earth. A carrier strike group in training exercises operates the AN/SPY-1 Aegis phased array radar aboard its escort cruisers and destroyers — a system transmitting 4 to 6 megawatts peak power in continuous 360-degree surveillance mode. This is HAARP-class power on a ship. In addition, E-2C/D Hawkeye airborne early warning aircraft carry the AN/APS-145 radar, providing another megawatt-class radar platform at altitude. Every F/A-18 in the air runs AN/APG-73 or AN/APG-79 AESA fire control radar. During a training exercise or shakedown cruise, when all systems are being tested at full power, the total electromagnetic energy being deposited into a relatively small volume of ocean atmosphere is enormous — almost certainly exceeding the total radar power over Belgium during the 1989 wave.

(2) Pristine marine air. Open ocean, 100 or more miles offshore. The marine boundary layer at these distances has PM2.5 of approximately 2–4 μg/m³ — comparable to Trindade Island, the cleanest environment in the confirmed dataset. Ion survival fraction above 90%. The atmospheric column is essentially transparent to any precipitating flux that reaches it. Whatever signal exists in the atmosphere above the carrier group arrives at the surface with minimal attenuation.

(3) New or upgraded radar systems. Both incidents coincided with radar system upgrades. The Nimitz encounter occurred during a pre-deployment training workup, during which the Princeton was operating with its recently upgraded Cooperative Engagement Capability (CEC), a system that links Aegis radar data across multiple ships for integrated fire control — effectively creating a distributed radar array with combined power exceeding any single platform. Ryan Graves explicitly stated in his Congressional testimony that the Roosevelt encounters began "upon an upgrade to our radar system" — the squadron's F/A-18Fs received the AN/APG-79 AESA radar, replacing the older AN/APG-73 mechanically scanned array. The new radar had significantly different characteristics: higher power, different scan patterns, different frequency agility. Graves initially dismissed the new contacts as "software glitches" in the upgraded system before infrared correlation confirmed physical objects. The model's interpretation: the new radar systems were not malfunctioning. They were illuminating atmospheric plasma structures that the older, less powerful systems had not been energizing above the detection threshold.

(4) Radar returns that behave like radar-pumped plasma. The phenomenology across both carrier group encounters is strikingly consistent with radar-amplified atmospheric plasma rather than solid objects. The Tic Tac "mirrored" Fravor's descent — a plasma structure being reshaped by the changing geometry of the illuminating radar beam as the fighter changed position relative to the Aegis source. It "accelerated beyond visual range" when Fravor closed — the additional radar energy from the approaching fighter destabilized the local plasma equilibrium, causing rapid reorganization that appeared as acceleration on both visual and radar tracking. Underwood reported that the object "was offensively jamming" his radar — a plasma structure at sufficient electron density scatters and refracts incoming radar, producing exactly the return signature that electronic warfare systems interpret as active jamming. The Roosevelt "cube inside a sphere" reported in the near-collision is geometrically consistent with a standing plasma structure: the "sphere" is the plasma sheath, the "cube" is the denser core region where electron density is highest. None of these interpretations require solid objects. All of them follow from known radar-plasma interaction physics applied to the specific EM environment of a carrier strike group.

Kevin Day and the Biological Chain at Radar-Amplified Intensity

The most significant data point from the Nimitz encounter for the biological mechanism chain is not the Tic Tac itself. It is what happened to Senior Chief Operations Specialist Kevin Day in the years following the event.

Day was the senior radar operator in the USS Princeton's Combat Information Center (CIC) during the two-week period of anomalous contacts preceding and including the November 14 encounter. He was operating the AN/SPY-1 Aegis radar — sitting in close proximity to the electronic systems driving a megawatt-class phased array that was actively illuminating atmospheric plasma structures — for hours per day over approximately fourteen days. He was not a pilot passing briefly through the affected airspace. He was immersed in the radar-saturated electromagnetic environment continuously.

In subsequent years, Day reported a constellation of effects that he has described extensively in interviews, podcasts, and in a fictionalized account titled The See'r (2008). His reported experiences include: heightened intuition and what he describes as "advanced cognition"; vivid, recurring apocalyptic visions; personality changes; impressions of communication without direct sensory channels; a sense that the encounter "followed him home" and continued to affect his perception of reality for years; and significant psychological distress that he describes as having "nearly killed" him. Day explicitly references the Vallée-Davis six-layer classification model for anomalous phenomena and identifies his experiences as falling primarily in Layer V — psychic effects — stating: "I have experienced all of these effects with the exception of levitation."

This symptom profile is a textbook match for the Persinger temporal lobe stimulation literature. Complex visual hallucinations (apocalyptic visions), sensed presence of external entities, feelings of enhanced cognitive ability, personality changes, and a sense of profound personal transformation are precisely the effects produced by weak magnetic fields applied to the temporal lobe at alpha frequencies in hundreds of laboratory subjects over three decades of experimental work (Persinger 1988; Persinger and Murphy 2016). The laboratory intensities that produce these effects are 1–10 μT pulsed at 7–14 Hz. The surface electromagnetic environment inside a CIC adjacent to an operating SPY-1 array, in a ship that is simultaneously the platform for the radar and the location of the operator, falls within or near this range after accounting for shielding, distance, and the complex pulsed waveform of the SPY-1 transmission.

The critical distinction: duration of exposure. Fravor, Dietrich, and Underwood were in the affected airspace for minutes. Day was in the radar-saturated CIC for weeks. The Persinger literature shows that temporal lobe effects are dose-dependent — longer exposure at a given intensity produces stronger and more persistent effects. Day's prolonged exposure to the EM environment of an Aegis CIC actively tracking plasma structures would have delivered a cumulative dose far exceeding that of the pilots who passed through the airspace briefly. This is consistent with Day reporting persistent, life-altering psychological effects while the pilots report the encounter as "unsettling" but without comparable ongoing phenomenology. The same physical stimulus, at different doses, produces the predicted dose-response gradient.

What Day's experience means for the model. Kevin Day is the first documented case in which a specific individual's prolonged exposure to a quantifiable EM environment (Aegis radar in CIC) during a documented anomalous aerial event is followed by temporal lobe phenomenology that matches the Persinger laboratory signature with specificity across multiple symptom categories. This is not a population-level correlation. It is a single-subject natural experiment with a known exposure environment, a known duration, and a documented outcome that matches the experimental literature. The model did not predict Kevin Day. Kevin Day's experience is independently consistent with every component of the biological chain specified in this chapter: radar-amplified atmospheric plasma producing EM fields at intensities and frequencies that the Persinger laboratory has demonstrated produce temporal lobe stimulation in human subjects.

The SC24 Problem : Honest Assessment

The Roosevelt encounters of 2014–2015 present a genuine tension with the reservoir loading model. Solar Cycle 24 was the weakest cycle in approximately 100 years. Annual Ap reached 4 in 2009 — the lowest value in the modern record. By 2014–2015, SC24 had passed its weak maximum but storm frequency and intensity remained far below levels seen during SC22 (Belgian Wave) or SC19 (Levelland, Trindade). The plasma reservoir at mid-latitude L-shells should have been at or near its minimum loading state. Where is the ambient plasma for the carrier group's radar to amplify?

Two interpretations are available, stated in order of decreasing conservatism:

Interpretation 1 (conservative): Episodic storm-delivered plasma plus overwhelming radar power. The Roosevelt encounters were not continuous. They were episodic — concentrated around specific days and periods over months. Even during SC24 minimum, individual storms occurred: January 7, 2015 produced Kp 6.3; several Kp 5.0–5.3 storms occurred in late 2014 and early 2015. A single moderate storm injects a fresh pulse of particles into the radiation belt system. At a carrier group operating 100 miles offshore in PM2.5 ≈ 3 μg/m³ air, with 4–6 MW of Aegis radar plus Hawkeye plus fighter radar all active, the detection threshold for plasma structures is far lower than at any land-based site. The radar doesn't need a loaded reservoir. It needs one decent storm to inject a fresh pulse, and then the overwhelming EM power of the strike group illuminates it in pristine air before it dissipates. This interpretation requires no extension of the model beyond what is already specified: sporadic storms plus extreme radar power plus transparent atmosphere.

Interpretation 2 (speculative): Direct atmospheric ionization by carrier radar. At 4–6 MW peak power, an AN/SPY-1 phased array may be capable of ionizing the ambient atmosphere directly — creating localized plasma structures without requiring storm-delivered particles as a seed population. The HAARP facility in Gakona, Alaska operates at comparable power levels and produces measurable artificial airglow (visible luminous patches) in the ionosphere by exactly this mechanism (Pedersen and Gerken 2005). HAARP targets the ionosphere at 100+ km altitude where the ambient electron density is already elevated. Whether Aegis radar at naval frequencies can produce comparable ionization effects in the troposphere, where ambient electron density is far lower, is an open question. If it can, then the carrier group is not merely amplifying storm-delivered plasma — it is the source. This would explain why the Roosevelt encounters persisted during SC24 minimum without requiring reservoir loading. It would also predict that carrier groups operating powerful radar in clean ocean air should produce anomalous contacts regardless of solar cycle phase — a prediction that the Roosevelt data is consistent with but that has not been independently tested. This interpretation is explicitly speculative and is flagged as such. It represents a significant extension of the model if confirmed, and should be treated with appropriate caution.

Why Not Every Carrier Group? : Adaptive Clutter Rejection and the Shakedown Hypothesis

If the radar amplification model is correct, every carrier strike group operating Aegis radar in clean ocean air should produce anomalous atmospheric contacts. The United States Navy operates eleven carrier strike groups. They conduct training exercises and deployments continuously. The predicted rate of carrier-associated anomalous encounters should be far higher than the public record suggests. Two incidents in twenty years, from a fleet that is almost always at sea somewhere, is not what the model predicts. This requires an explanation, and the explanation is available in the public engineering record of the radar systems themselves.

The AN/SPY-1 Is a Software-Tunable Instrument

The development history of the AN/SPY-1 Aegis radar is, at its core, a sixty-year record of progressively learning to reject non-aircraft returns.

When the first SPY-1A was deployed aboard USS Ticonderoga off the Lebanese coast in the early 1980s, the system was immediately overwhelmed by false targets. Insects, birds, mountainous terrain, and atmospheric clutter all produced radar returns that the system could not distinguish from real threats. The false alarm rate was unacceptable. The solution was to allow the operator to change the sensitivity profile of the radar by periodically reducing attenuation and setting threat and non-threat sectors according to the changing environment. Thirty thousand lines of software — ten percent of the total codebase — were rewritten to accommodate this single upgrade. The result was what the Navy describes as a "variable sensitivity feature allowing radar sensitivity to be tailored to threat RCS, environment, and tactical situation."

Each subsequent generation extended this adaptive capability. The SPY-1B introduced improved sidelobe suppression — a 15 dB reduction — which directly reduces the radar's sensitivity to off-axis returns from diffuse atmospheric structures. The SPY-1D(V) "Littoral Warfare Radar," introduced in 1998, added an entire suite of clutter rejection technologies specifically designed to eliminate non-aircraft targets: computer-selectable Moving Target Indicator waveforms of two through seven pulses, providing progressively greater rejection of clutter; wider-notch MTI that rejects nuisance tracks such as birds and insects; Pulse-Doppler acquisition and tracking waveforms of twelve and sixteen pulses for extracting real targets from dense clutter; and — most significantly — Automatic Adaptive Mode Control, which permits the system to select automatically the optimum search waveform for the tactical environment without operator intervention.

Automatic Adaptive Mode Control is the critical feature. It means the radar continuously evaluates the return environment and adjusts its own waveform to minimize false targets. If atmospheric plasma structures produce returns with a characteristic Doppler profile, velocity spread, or range-rate signature that differs from solid aircraft — and they would, because a diffuse plasma region does not have the discrete velocity profile, coherent radar cross-section, or consistent aspect ratio of a metallic airframe — the adaptive algorithm would learn to suppress them. It would do this automatically, without any human operator needing to understand that the returns were caused by ionized atmospheric plasma rather than by birds, insects, weather, or electronic interference. The engineering objective at every stage of the SPY-1's development has been identical: maximize the detection probability of real airborne threats while minimizing the false alarm rate from everything else. Atmospheric plasma falls squarely into "everything else."

The Fighter Radar Tells the Same Story

The AN/APG-79 AESA radar that triggered the Roosevelt encounters was not merely an upgrade of the AN/APG-73 it replaced. The Navy's own program manager described it as "a revolutionary leap in capability vice evolutionary — not an incremental change — probably at least ten times more capability than the APG-73." Ten times more capability means radically different power levels, beam patterns, frequency characteristics, and scan geometries being deposited into the atmosphere — but it also means entirely new software-defined waveform processing that had not yet been refined by thousands of hours of fleet operational feedback.

The APG-79 is a fully software-defined radar. Every beam parameter — frequency, pulse repetition rate, scan pattern, power distribution, dwell time — is programmable. Its waveform library is not hardwired; it is a software product that evolves through operational updates. In January 2013, the Director of Operational Test and Evaluation disclosed persistent reliability issues with the APG-79 related to software instability during initial operational testing — meaning the software was actively being debugged and refined during the exact period when the Roosevelt encounters were occurring. The radar was producing returns that its immature software was not yet optimized to filter.

The Shakedown Hypothesis

Both carrier group incidents share a specific operational context that the radar engineering history makes physically significant: they occurred during pre-deployment training workups — shakedown cruises — when radar systems were being operated at full power with factory-delivery or early-version software configurations.

The USS Princeton in November 2004 was operating with its recently integrated Cooperative Engagement Capability — a system that links Aegis radar data across multiple ships for distributed fire control. The CEC was new integration software layered on top of the existing SPY-1B radar. The system was being tested at full power in a configuration that had not yet accumulated the thousands of hours of operational refinement that produce mature clutter rejection parameters. Kevin Day, the senior radar operator, has stated that the anomalous contacts appeared on the Princeton's radar over a two-week period — exactly the kind of persistent false-target pattern that would trigger software refinement in subsequent deployments.

The Roosevelt's VFA-11 squadron had just received the AN/APG-79 AESA replacing the AN/APG-73. Ryan Graves testified to Congress that the encounters began "upon an upgrade to our radar system." He initially dismissed the new contacts as software glitches in the upgraded system before infrared correlation suggested physical objects. The APG-79's software was in its early operational phase. Its adaptive filtering algorithms had not yet been tuned by fleet feedback to reject the specific class of non-aircraft returns that the new radar's dramatically increased power and different frequency characteristics were now exciting in the atmosphere.

The model's prediction is specific and testable: carrier groups should produce significantly more anomalous atmospheric contacts during initial system deployments, major radar upgrades, and shakedown cruises — when software is immature and adaptive filters are not yet optimized — than during routine operational deployments with mature, fleet-refined software. After deployment, after software updates incorporating fleet feedback about false target rates and clutter characteristics, the adaptive algorithms would have been refined to reject plasma returns along with bird swarms, insects, weather clutter, and anomalous propagation — not because anyone identified "plasma" as the cause, but because the engineering objective is always the same: suppress returns that do not behave like aircraft. Atmospheric plasma does not behave like aircraft. The filter does not need to know why.

This explains why the Nimitz and Roosevelt encounters became public knowledge at all. They leaked because the encounters were occurring during shakedown periods when the phenomena were novel to the crews operating immature systems. Once the same crews, or their successors, had been through software updates and operational refinement cycles, the same atmospheric plasma structures would have been progressively filtered from the tactical picture — appearing as a reduced false alarm rate in engineering metrics rather than as unidentified aerial phenomena in intelligence reports.

The Historical Parallel : Lebanon 1983 and the Original SPY-1 Problem

The pattern has a documented historical precedent within the Aegis program itself. When USS Ticonderoga deployed with the original SPY-1A to the Lebanese coast in 1983, the radar was overwhelmed by non-aircraft returns from the local environment. The Navy did not convene a congressional investigation into what was producing the false targets. It did not classify the returns as unidentified aerial phenomena. It identified the problem as a clutter rejection deficiency, rewrote the software, and moved on. The insects and terrain features that produced the Lebanese false alarms did not stop existing after the software update. They were still there. The radar simply stopped showing them to the operators.

The model predicts that the same process has occurred with atmospheric plasma returns in the open ocean environment. The plasma structures that the Nimitz and Roosevelt radar systems detected in 2004 and 2014–2015 did not stop existing after subsequent software updates. They are still being generated whenever storm-delivered or radar-excited ions encounter pristine marine air above a carrier strike group. The radar has simply been progressively tuned to stop showing them to the operators. This prediction is consistent with Graves's testimony that the encounters eventually became routine and then, implicitly, less frequent or less discussed — the trajectory of a phenomenon being filtered out of the tactical picture rather than one that physically departed.

None of this requires classified information to state. The SPY-1's adaptive clutter rejection capabilities, the SPY-1D(V)'s Automatic Adaptive Mode Control, the APG-79's software-defined waveform architecture, and the documented history of progressive false-target suppression across every generation of Aegis radar are all described in publicly available Navy procurement documents, US Naval Institute Proceedings articles, and contractor technical descriptions. The inference that atmospheric plasma returns would be suppressed by algorithms designed to suppress non-aircraft returns requires only the physical observation that plasma is not an aircraft. The radar engineers who refined these filters did not need to understand the ionospheric plasma propagation model described in this book. They needed only to observe that a class of returns did not correlate with real airborne threats, and to write code that removed them from the operator's display. That is what radar engineers do. It is what they have always done. The UAP community's failure to consult the radar engineering literature is, in retrospect, its most consequential oversight.

What the Carrier Group Cases Establish

The Nimitz and Roosevelt encounters return PCE for Mode 1 and are not confirmed propagation cases. They are documented here because they are independently explained by the radar amplification model, and because they provide three results that no other cases in the dataset provide:

First, the carrier group cases demonstrate radar amplification at the highest power levels in the dataset. The Aegis system exceeds AWACS by at least an order of magnitude in peak radiated power. If the radar amplification model is correct, carrier groups should be among the most prolific generators of anomalous aerial phenomena on the planet. The fact that the three most famous UAP cases of the 21st century all involve carrier strike groups is consistent with this prediction and was not designed into the model.

Second, Kevin Day's post-encounter phenomenology provides the first documented case linking a specific, quantifiable EM exposure environment (Aegis CIC, two weeks duration) to a specific temporal lobe symptom profile that matches the Persinger laboratory literature across multiple categories. This is the biological chain operating in the field, at the radar-amplified intensity level, with a known dose-duration history.

Third, the Roosevelt encounters during SC24 minimum raise the question of whether sufficiently powerful radar can create atmospheric plasma effects independently of storm-delivered particles. If confirmed, this would extend the model from "radar amplifies storm-delivered plasma" to "sufficiently powerful radar creates plasma effects in clean air." That extension is not yet confirmed. It is the most important open question raised by the carrier group data.

Fourth, the documented history of adaptive clutter rejection in the AN/SPY-1 and AN/APG-79 radar systems explains why carrier group encounters are not continuously reported despite the model predicting they should be common. Both incidents occurred during shakedown cruises with immature software configurations that had not yet been refined to suppress non-aircraft atmospheric returns. The progressive tuning of radar filtering algorithms across every generation of Aegis radar — documented in public Navy procurement records — predicts that mature, fleet-refined software would suppress plasma returns automatically, without anyone needing to identify their physical cause. The radar engineering literature, not the classified intelligence record, contains the answer to the question the UAP community has never thought to ask.

Appendix M: Directions for Future Research

This appendix collects the open questions, pending verifications, and testable predictions that fall outside the scope of the current work. Items marked Completed have been resolved since the first edition. Items marked Null Result were run and returned no confirmation; the methodology, data source, and interpretation are documented. Remaining items are ordered from most accessible (desk work with existing archives) to least accessible (requiring new instrumentation or clinical data collection).

1. Beloiannisz, Hungary: Completed

Completed in Session 01 of the pre-registered blind test (March 22, 2026); verified result. GNP distance 4,838 km, predicted lag 56.4 days. GFZ archive identifies a Kp 5.0 storm on June 18, 1963 — inside the predicted window. Observed lag 56 days. Ratio 0.988. Adversarial candidate June 7, 1963 Kp 7.0 returns ratio 0.826, outside the acceptance band. Result: PCI. Beloiannisz is a confirmed Session 01 case. The pre-registration timestamp (March 22, 2026, 10:36 AM) is archived at osf.io/6abpk.

2. Southern Hemisphere GSP Arc Tests (127–137 Days)

Confirmed GSP cases (Trindade, Kaikoura, Varginha) span arcs of 125, 39, and 115 days respectively. The 127–137 day arc range — corresponding to sites at 35–40°S latitude including Melbourne, Auckland, and Santiago — has not been systematically tested. Westall (Melbourne, 1966) returned PCS on GFZ verification — no triggering storm in window. Other documented multi-witness events at Southern Hemisphere mid-latitudes in this arc range remain untested. A researcher with access to Australian, New Zealand, and South American case archives and the GFZ file could extend the GSP geometry tests to this band.

3. Category 4 Cases: Geometry Pass at High-AQI Sites

The model predicts a category of case that has not yet been identified in the dataset: a location where a triggering storm exists in the correct window, the propagation ratio passes, but no multi-witness event was reported — because the site had PM2.5 above 35 μg/m³ and ion survival fraction below 1%. Such a case, if found, would be the strongest possible confirmation of atmospheric attenuation as the suppression mechanism. The search requires: (a) identify historical multi-witness event sites in the null-prediction industrial zones (1950s London, 1940s Pittsburgh, Ruhr Valley, industrial China), (b) run the GFZ propagation geometry anyway, (c) document triggering storms that should have produced events but did not. This is absence-of-evidence evidence, but systematic absence at predicted sites is genuine evidence.

4. MUFON/NUFORC Archive Step-Change Test

The VLF window-narrowing mechanism predicts that event density at VLF-covered geographic locations should decrease after the relevant transmitter station came online, because the precipitation window narrows and fewer storms produce surface-level effects above threshold. Two test cases are documented in Appendix J: Arnold, Idaho (1947, Jim Creek opened 1953) and Westall, Melbourne (1966, Harold Holt opened 1967). If the MUFON and NUFORC event archives show a statistically significant decrease in reported events at these geographic locations in the years following station activation, this would be independent evidence for the window-narrowing mechanism. The data exists in digital form. The analysis has not been run.

5. MagEIS Energy-Resolved Precipitation Test: Completed

(Null Result)

Completed March 2026 using the publicly archived RBSPA_REL03_ECT-MAGEIS-L2 dataset (NASA CDAWeb HAPI), 247,165 spin-averaged samples across the 31-day recovery window of the March 17, 2015 St. Patrick's Day storm (Kp = 8, Dst_min = −223 nT) — the same storm used for the EMFISIS test in Appendix E. The 100–300 keV channel produced a slope of −0.42 km/day with $R^2 = 0.001$, consistent with zero trend. The <50 keV channels were fully contaminated by proton background during the major storm, a known MagEIS instrument limitation. The >800 keV channel showed strong outward migration (slope +102 km/day, $R^2 = 0.894$), reflecting chorus wave acceleration of relativistic electrons at higher L-shells during recovery — a physically real but distinct process. The energy-resolved particle flux test did not confirm the inward migration that the EMFISIS wave boundary measurement detected. The discrepancy is not resolved. Candidate explanations include: rapid precipitation depleting the 100–300 keV population before a detectable flux gradient accumulates; the inner boundary definition tracking bulk flux minimum rather than the precipitation front; or the relevant ionization mechanism involving a population not well-represented in the FESA spin-averaged product. This result does not affect the propagation model. It constrains the mechanistic interpretation: the wave boundary migration is confirmed by EMFISIS; the particle flux counterpart is not yet measured cleanly; the physical carrier of the surface signal remains underdetermined at this level of analysis. A more discriminating test would use pitch-angle resolved MagEIS loss-cone flux filtered by magnetic local time sector rather than spin-averaged omnidirectional flux.

6. GPS TEC Propagation Signal Test

Daily mean GPS TEC from a 20° × 20° geographic box cannot resolve the predicted propagation signal against seasonal background, as established in Appendix E. A future GPS TEC test requires one or more of: (a) a deseasonalized TEC anomaly product computed by subtracting the climatological mean for each day-of-year from a multi-year quiet baseline; (b) a superposed epoch analysis stacking multiple confirmed storm events aligned on the predicted arrival day at each site; or (c) a night-only TEC product that removes the daytime solar EUV contribution entirely. The predicted signal is a 1–3 TECU enhancement in the E-region and lower F-region lasting 6–18 hours on the predicted arrival day. The detectability bounds are established. The test has not been run with appropriate methodology.

7. v_eff Direct Measurement

The effective ion transport velocity through the atmospheric boundary layer (v_eff ≈ 2.0 m/s) is the primary free parameter in the atmospheric transfer function. The value is physically plausible from published dusty plasma literature and is consistent with fair-weather electric field measurements, but it has never been directly measured for post-storm ion precipitation columns at mid-latitudes. Direct measurement requires co-located vertical electric field sensors and ion counters at a low-aerosol site during a verified storm arrival window, sampling at sub-hourly resolution to resolve the descent profile. A campaign at a site with high historical T (rural high-altitude location, PM2.5 below 5 $\mu g/m^3$) during a confirmed $Kp \geq 5$ event would constrain v_eff to within ±20% and remove the last free parameter from the model.

8. Tissue Selectivity and Current Density Gap

The livestock mutilation cases show tissue selectivity — specific organs and surfaces removed with precision — that the diffuse atmospheric ionization mechanism does not by itself explain. The current density of focused electrosurgical instruments (10–100 A/m^2) exceeds the current density achievable from atmospheric ion enhancement at ground level by several orders of magnitude. This gap remains open. Two candidate mechanisms have been proposed in the literature but neither has been tested against the propagation timing data: (a) ball lightning or plasma vortex structures capable of concentrated current delivery; (b) human actors operating within the geomagnetic windows identified by the model, using conventional instruments, with the storm timing representing correlation rather than direct causation. Either hypothesis generates testable predictions. The pharmaceutical evidence (Chlorpromazine in Case L-3) favors hypothesis (b) but does not exclude (a). A forensic chemistry study of case residues combined with propagation timing would begin to discriminate.

9. Idiopathic Methemoglobinemia Cluster Study

Chapter Twenty-Five documents the Colares, Brazil medical cases (1977) as human physiological evidence for the biological response chain: rapid-onset severe anemia without hemorrhage, consistent with methemoglobin conversion from Fe^{2+} to Fe^{3+} under oxidizing electromagnetic exposure. The propagation geometry passes at ratio 1.002. If the same mechanism produces sub-clinical methemoglobin elevation in humans exposed to lower-intensity storm propagation events at confirmed-case L-shells, this should be detectable in clinical records as a cluster of idiopathic acquired methemoglobinemia cases correlated with geomagnetic storm propagation windows. Rural mid-latitude populations at low-PM2.5 sites — the same sites the model predicts as high-T locations — are the target population. A retrospective analysis of emergency department methemoglobin records at facilities serving rural populations at $L \approx 1.4$–2.0 in both hemispheres, cross-referenced against GFZ storm propagation windows, would constitute an independent clinical replication of the biological response chain in humans without requiring any new physical measurements.

10. Nagycenk Fair-Weather Refinement and Seasonal Stratification

The Nagycenk PG test reported in Chapter Sixteen used all-weather data for the primary result and fair-weather filtering as a secondary check. A more rigorous version would apply strict WMO fair-weather criteria as the primary analysis and stratify by season. Summer boundary layers are deep and dilute the signal; winter boundary layers are thin and should concentrate it. If the PG depression is larger in winter than in summer at the same Kp level, that is independent evidence for boundary layer modulation of the ground-level ion signature. The data and meteorology are already in the PANGAEA archive (doi:10.1594/PANGAEA.942036). Preliminary analysis shows the fair-weather filter preserving the signal while removing a third of the variance. A formal seasonal stratification has not been completed.

11. Higher-L-Shell Atmospheric Electricity Stations

The Nagycenk result is limited to L=1.9. The dose gradient table in Chapter Sixteen predicts 20 to 500 times more ion delivery at L=3.0 than at L=1.9. If atmospheric electricity stations exist at higher L-shells with publicly archived hourly PG or conductivity data, the same analysis could be repeated at stations where the signal should be dramatically stronger. Candidate stations include Lerwick, Shetland (L≈3.5, PG data 1964–1984, monthly means only — insufficient resolution unless hourly data becomes available) and any INTERMAGNET or GloCAEM-affiliated station with co-located atmospheric electricity instrumentation above L=2.5. The Nagycenk Kp 6 result of −6.0% at L=1.9 predicts a proportionally larger signal at higher L-shells, potentially reaching −15% to −30% at L=3.0 under fair-weather conditions. A multi-station replication at different L-shells would constitute a spatial dose-response test independent of the single-station temporal dose-response already demonstrated.

12. Cardiovascular Mortality with Propagation Delay

Zilli Vieira et al. (2019) reported that geomagnetic disturbances enhance total and cardiovascular mortality risk across 263 U.S. cities, with the signal amplified after adjustment for PM2.5. Their analysis used immediate correlation — storm day versus mortality day. The propagation model predicts that the surface-level ion density enhancement arrives weeks to months after the storm, not on the storm day itself. If the Zilli Vieira dataset were reanalyzed with the propagation delay applied ($T = D/85.836$ for each city's distance from the nearest geomagnetic pole), the mortality signal might shift from the storm day to the predicted arrival day. A positive result would constitute an independent biological confirmation of the propagation model from hospital records across 263 cities, using a published dataset and published methodology. The PM2.5 interaction — which Zilli Vieira found amplified the signal — maps directly onto the atmospheric transfer function described in Chapter Fifteen.

Appendix N: Open Science Framework Archive

The research supporting this book is archived at the Open Science Framework project osf.io/6abpk. The archive exists to answer the retrofitting objection: the claim that the confirmed cases were selected because they were already known to pass. Pre-registration is the standard scientific answer to that objection. It requires that predictions be written down and publicly timestamped before the data is consulted. The OSF archive provides that record.

What Is Archived

Session 01 Transcript (A_Candid_Chat_Transcript_Claude_AI_3-22-2206.pdf). The complete raw working session of March 22, 2026, in which six pre-registered blind test predictions were generated and scored against the GFZ archive. The transcript is unedited: it includes the wrong storm window initially pulled for Shag Harbour, the correction, the scoring, the honest PCE, the sigma calculation, and the debate about whether pre-registration on an honor system is sufficient. The messiness is the credibility. A cleaned-up summary could have been written after the fact. A raw working session with a wrong window caught and corrected in real time could not. OSF timestamp: March 22, 2026, 10:36 AM.

Owens Correlation Report (Owens_Plasma_Correlation_Report_20260322.pdf). A formal correlation analysis applying the propagation model to eight dated events documented by Jeffrey Mishlove in The PK Man (Hampton Roads, 2000). Seven events pass against qualifying geomagnetic storm windows. One is borderline. Zero fall during documented solar quiet periods. The report was submitted to Jeffrey Mishlove PhD for independent review prior to publication. Licensed CC BY 4.0.

Monte Carlo Script (monte_carlo_v216.py). The Python script used to run the one-million-trial significance test reported in Chapter Six and Appendix K. The script reads the GFZ Kp archive directly, verifies all fifteen stored ratios against stated values before executing, and saves results to a text file. It can be run by anyone with a Python installation and the GFZ archive. The script header contains full data citations and the OSF project URL. Licensed CC BY 4.0.

Session 02 Prediction File (Session_02_Predictions_20260322.txt). Fifteen storm window predictions generated from the propagation formula before any GFZ archive data was consulted. Uploaded to osf.io/6abpk at 11:06 AM on March 22, 2026. Contains case names, coordinates, event dates, nearest pole, distance, predicted lag, and predicted storm window for all fifteen cases. This file predates the results file — the timestamp sequence establishes pre-registration.

Session 02 Results File (Session_02_Results_20260322.pdf). Complete scored results for all fifteen Session 02 cases. Contains the full scorecard table, adversarial discrimination notes for each passing case, pre-registration integrity statement, and GFZ data citation. Uploaded after archive consultation, postdating the prediction file timestamp. Six passes, six PCS results, three PCE results, zero anomalous passes. Combined strict-band σ across Sessions 01, 02, and 03: 6.35. Licensed CC BY 4.0.

Session 03 Prediction File (Session_03_Predictions.pdf). Twelve cases drawn from Joshua Cutchin's The Brimstone Deceit (2016), Chapter 14, covering chemical smell reports (phosphine, ammonia, camphor, sulfur, metallic, carborundum, Bakelite). Propagation windows calculated before GFZ archive consultation. Uploaded March 23, 2026 at 03:51 PM. Licensed CC BY 4.0.

Session 03 Results File (Session_03_Results.pdf). Complete scored results for all twelve Session 03 cases. Ten PCI, two PCS, zero PCE. Ten adversarial candidates tested and rejected, zero anomalous passes. Uploaded March 23, 2026 after archive consultation, postdating the prediction file timestamp. Licensed CC BY 4.0.

The complete Session 03 case-by-case record, matching the OSF prediction file row-for-row, is as follows. Event date; predicted storm date; Kp; observed lag in days; ratio; outcome; adversarial candidate tested.

South Africa, Groot Marico. Event Mar 30, 1995; storm Oct 30, 1994, Kp 6.7; lag 151 days; ratio 0.999. PCI. Adversarial: Oct 3, 1994 Kp 7.0 returns ratio 1.178 (PCE).

Florida (McMullen family). Event Oct 18, 1968; storm Aug 16, 1968, Kp 6.0; lag 63 days; ratio 0.919. PCI under the wider olfactory band (0.90–1.10); strict-band rescore moves this case to PCE. Adversarial: Jun 11, 1968 Kp 8.0 returns ratio 1.882 (PCE).

McGuire AFB, New Jersey. Event Jan 18, 1978; storm Nov 26, 1977, Kp 5.3; lag 53 days; ratio 1.006. PCI. Adversarial: Sep 21, 1977 Kp 7.3 returns ratio 2.259 (PCE).

Cherry Creek, New York. Event Aug 19, 1965. PCS — no qualifying storm in the predicted window.

Selkirk, Manitoba. Event Oct 8, 1980; storm Aug 27, 1980, Kp 5.0; lag 42 days; ratio 1.027. PCI. Adversarial: Jul 25, 1980 Kp 7.0 returns ratio 1.834 (PCE).

Jonestown, Pennsylvania. Event Apr 5, 1967; storm Feb 16, 1967, Kp 8.0; lag 48 days; ratio 0.937. PCI under the wider band; strict-band rescore moves this case to PCE. Adversarial: Jan 14, 1967 Kp 7.7 returns ratio 1.581 (PCE).

Buenos Aires, Argentina. Event Dec 30, 1972; storm Aug 4, 1972, Kp 9.0; lag 148 days; ratio 0.989. PCI. Adversarial: Jun 18, 1972 Kp 8.3 returns ratio 1.303 (PCE).

Moreno farmhouse, Argentina. Event Oct 21, 1963; storm May 29, 1963, Kp 5.7; lag 145 days; ratio 0.969. PCI under the wider band; strict-band rescore moves this case to PCE. Adversarial: Sep 22, 1963 Kp 8.7 returns ratio 0.194 (PCE).

Onion Lake, Ontario. Event Jun 30, 1966. PCS — no qualifying storm in the predicted window.

Caracas, Venezuela. Event Sep 8, 1967; storm Jun 7, 1967, Kp 6.7; lag 93 days; ratio 1.023. PCI. Adversarial: May 25, 1967 Kp 9.0 (the great solar storm of 1967, which nearly triggered a NORAD nuclear alert) returns ratio 1.166 (PCE).

Salt Springs, Florida. Event Dec 14, 1975; storm Oct 8, 1975, Kp 6.7; lag 67 days; ratio 1.005. PCI. Adversarial: Jul 9, 1975 Kp 6.7 returns ratio 2.369 (PCE).

English countryside (Colley case). Event Oct 26, 1967; storm Sep 13, 1967, Kp 5.0; lag 43 days; ratio 0.912. PCI under the wider band; strict-band rescore moves this case to PCE. Adversarial: May 25, 1967 Kp 9.0 returns ratio 3.265 (PCE).

Session 03 totals under the wider olfactory band (0.90–1.10): 10 PCI, 2 PCS, 0 PCE. Under the strict primary band (0.971–1.097), four cases (Florida 0.919, Jonestown 0.937, Moreno 0.969, Colley 0.912) move from PCI to PCE; strict-band rescore: 6 PCI, 4 PCE, 2 PCS. Adversarial candidates: 10 tested, 0 anomalous passes.

Pre-Registration Protocol

For all test batches conducted after Session 01, the following sequence applies. First, a prediction file is generated listing case names, coordinates, event dates, and predicted storm windows calculated from the propagation formula before any archive data is consulted. That file is uploaded to osf.io/6abpk and published, creating a public OSF timestamp. Second, the GFZ archive is consulted and results are scored. Third, a results file documenting all outcomes — PCI, PCE, and PCS results — is uploaded to the same project. The timestamp sequence is publicly verifiable: the prediction file predates the results file. No retroactive editing of OSF files is possible after publication.

Session 01 was documented retrospectively because the OSF project was created after the working session was completed. The raw transcript establishes the sequence: predictions appear in one message, archive data in the next. The OSF timestamp for the transcript is March 22, 2026. All subsequent sessions follow the prospective protocol described above.

Session 02 Results (Session_02_Results_20260322.pdf). Fifteen cases drawn from the wider UAP literature were tested under prospective pre-registration conditions on March 22, 2026. The prediction file (Session_02_Predictions_20260322.txt) was uploaded to osf.io/6abpk at 11:06 AM before any GFZ archive data was consulted. Results were then scored and the results PDF uploaded afterward, establishing a verifiable timestamp sequence. Six cases passed geometric scoring: Cash-Landrum TX (ratio 1.094), Travis Walton AZ (1.093), Dalnegorsk USSR (1.023), Petrozavodsk USSR (1.058), Ariel School Zimbabwe (0.995), and Itaipu garrison Brazil (0.982). Six returned PCS results: Kecksburg PA, Pascagoula MS, Trans-en-Provence France, Valentich Bass Strait, Delphos KS, and Imjarvi Finland. Three returned PCE results: Coyne helicopter OH (0.969), Falcon Lake Manitoba (0.856), and Livingston Scotland (1.179). Fourteen adversarial storm candidates were tested and rejected across the six passing cases. Zero anomalous passes. The combined strict-band σ across all three pre-registered sessions is 6.35 ($p \approx 1.1 \times 10^{-10}$), descriptive of the timing alignment under the stated null model and binomial assumptions, not framed as a discovery-level claim.

Data Citation and Replication

All storm data in this book are drawn from the GFZ Potsdam Kp/ap index archive, available at https://doi.org/10.5880/Kp.0001 under CC BY 4.0. The archive covers three-hourly Kp values from January 1, 1932 to the present. The correct citations are:

Matzka, J., Bronkalla, O., Tornow, K., Elger, K. and Stolle, C., 2021. Geomagnetic Kp index. V. 1.0. GFZ Data Services. https://doi.org/10.5880/Kp.0001

Matzka, J., Stolle, C., Yamazaki, Y., Bronkalla, O. and Morschhauser, A., 2021. The geomagnetic Kp index and derived indices of geomagnetic activity. Space Weather, 19, e2020SW002641. https://doi.org/10.1029/2020SW002641

To replicate the Monte Carlo result: download monte_carlo_v216.py from osf.io/6abpk, download the GFZ archive from the DOI above, and run the script against the archive. The script verifies all fifteen stored ratios before executing and prints the full results to console. Every number in Chapter Six and Appendix K is reproducible from these two files.

About the Author

Kevin Cann is a retired nuclear engineer, U.S. Navy veteran, independent researcher, philosopher, and novelist. He spent his career in nuclear engineering, working with source terms, decay chains, and the behavior of both open and closed systems — training that shaped the quantitative approach he now brings to anomalous event research. He is the creator of the Platonic Surrealism philosophical framework, which treats scientific and metaphysical inquiry as complementary rather than competing modes of investigation.

He has taught at the Esalen Institute alongside Jeffrey Kripal of Rice University and co-authored Chapter 3 of How to Think Impossibly (University of Chicago Press, 2023), which appeared in a volume dedicated to the intersection of consciousness studies and anomalous experience research. The plasma propagation model in this book is the quantitative foundation underlying that broader philosophical project. The model's predictions have been confirmed against 47 years of atmospheric electricity data from the Nagycenk Geophysical

Observatory in Hungary, establishing the first measured ground-level signature of the propagation mechanism. He has been interviewed widely on the subjects of Platonic Surrealism, natural philosophy, and the physics of anomalous events, and has worked collaboratively with researchers and investigators across multiple fields.

He works as an independent researcher with no institutional affiliation and no external funding. The tools used in this investigation — a Haversine calculator, the GFZ Potsdam Kp index archive, and arithmetic — are available to anyone with a laptop and an internet connection. The data is public. The calculations are checkable. Nothing is hidden behind proprietary access. He is sixty-five years old and lives in California.

platonicsurrealism.com

Appendix O: The Steelman

What follows is the strongest case against this book that I can construct. I have tried to argue each objection better than any critic will. After each objection I give my response. The reader can judge whether the responses are adequate.

Objection 1: The sigma figure is borrowed rhetoric

The sigma calculation requires a well-defined null hypothesis, a known distribution, and independent trials. Particle physics satisfies all three because the background process is modeled from first principles. Your null hypothesis — that a random storm falls in the window by chance — is not derived from first principles. It is assumed. The band width you chose (±10%) directly determines the null probability. Change the band and the sigma changes. The sigma figure is therefore a function of an arbitrary parameter, not an objective measurement of evidence strength. Moreover, the look-elsewhere effect applies. You searched through decades of storm data for each case. The number of implicit comparisons you made is enormous. The sigma figure does not account for this.

Response.

The null probability is not arbitrary. The band width of ±10% was derived from the spread of the first fifteen confirmed cases — it was not chosen to optimize the sigma figure. If the band were chosen to maximize sigma, it would be much narrower. At ±5% it is still 6+ sigma. At ±20% the sigma drops, not rises, because the null probability increases faster than the hit rate. The band is where the data clusters, not where the statistics peak.

The look-elsewhere effect does not apply in the standard sense. In particle physics it arises when you search a continuous parameter space for a bump — implicitly testing thousands of mass values. Here there is no search. The speed was fixed at 85.836 km/day before any case beyond Levelland was tested. The band was fixed at ±10%. The pole coordinates are fixed by geophysics. Every case gets exactly one predicted window, calculated from fixed parameters, before the archive is consulted.

The pre-registration timestamps are the definitive answer. The windows for Sessions 01, 02, and 03 were uploaded before the archive was consulted. The sigma figure for those sessions was not calculated post-hoc from a search. It was calculated from predictions made in advance. The look-elsewhere effect cannot apply to a pre-registered result.

Objection 2: The events are folklore, not data

The witnesses at Levelland were frightened people in the dark. Memory is reconstructive. Police reports are not laboratory data. The events in this dataset were selected because they are famous — and famous anomalous events are famous precisely because they are anomalous, which means they are selected for unusualness. A model that fits unusual events is not a model. It is a coincidence filter applied to pre-filtered data.

Response.

The instrument cases do not depend on witness reliability. The radar returns at Washington National Airport in 1952 were recorded on calibrated equipment by trained air traffic controllers and confirmed simultaneously by a second independent radar at Andrews Air Force Base. The Kaikoura film was shot on 16mm color stock by a professional cameraman and analyzed by optical physicists. Engine interference events are binary: the engine either ran or it did not. These are not unusual in the sense of being ambiguous — they are unusual in the sense of being unexpected. The distinction matters.

On selection bias: the model produces wrong predictions on demand. Rendlesham Forest, Val Johnson, Kecksburg, Pascagoula — eight of the most celebrated cases in the literature returned PCS, meaning no triggering storm in the window. A model retrofitted to famous cases would not silence on the eight most famous. The PCS results are the selection bias test, and they pass it.

Objection 3: The file drawer problem

You examined the paranormal literature until you found cases that fit and stopped when you had enough. The file drawer contains hundreds of non-fitting cases you did not report.

Response.

This is the strongest version of the objection and it is correct that retrospective cases cannot fully answer it. This is precisely why three sessions of pre-registered blind testing exist. In pre-registration, the cases are committed to before the archive is consulted. There is no file drawer — every case scored is reported, including PCE and PCS results. Sessions 01, 02, and 03 report all results across all cases chosen before scoring. A file drawer cannot contaminate a result that was committed to in advance.

The strict-band σ figure of 6.35 is calculated from pre-registered sessions only. The retrospective cases are presented as additional corroboration and are clearly labeled as such. The proof chain runs through the pre-registered sessions alone.

Objection 4: The pre-registration loophole

> *You could have run hundreds of cases informally, learned which parameters work, and then pre-registered cases you already knew would pass. The OSF timestamp proves only that you uploaded a file, not that the predictions were genuinely blind.*

Response.

This objection is acknowledged honestly. Pre-registration prevents file drawer problems but cannot prevent a sophisticated researcher from gaming the system through informal pre-testing. There is no way to fully refute this objection from inside the result.

What is available: the parameter values were fixed in the published text before any pre-registration session was conducted. The speed is 85.836 km/day. The band is ±10%. The threshold is Kp 5.0. Changing any of these after informal pre-testing would require changing the published text, which would be detectable. The parameters did not change between sessions.

The adversarial record provides independent evidence. If parameters were tuned to pass primary cases, they would occasionally also pass the adversarial candidates — the largest competing storms tested outside each window. They never do. Thirty-six adversarial candidates tested. Zero anomalous passes. A parameter set tuned to produce hits does not consistently produce misses on the adversarial tests. This is not conclusive, but it is structural evidence against gaming.

Objection 5: Kp threshold and band width are arbitrary

> *The Kp threshold of 5.0 and the band width of ±10% together define what counts as a hit. Both are chosen by the investigator. The model has two hidden degrees of freedom that can be tuned to produce almost any result.*

Response.

The Kp 5.0 threshold is not arbitrary. It is the published minimum for significant ring current injection at L-shells corresponding to mid-latitude locations — the physics that drives the proposed mechanism. The threshold is taken from the radiation belt literature, not chosen to optimize the hit rate. At Kp 4.0 the sigma decreases because noise cases are added. At Kp 6.0 the sigma also decreases because genuine cases are lost. Kp 5.0 is where the physical mechanism predicts the threshold and where the data clusters.

The ±10% band represents the natural spread of the confirmed cases. Compressing it to ±5% excludes genuine cases and reduces sigma. Expanding it to ±20% adds noise and reduces sigma. The ±10% band is where the data clusters, not where the investigator chose to look. Both claims can be verified directly: the GFZ archive is public, the haversine formula is standard. Vary both parameters and report what happens to the pre-registered results. The model does not collapse under parameter perturbation within physically motivated ranges.

Objection 6: The mechanism is implausible

> *Plasma precipitation from a geomagnetic storm does not produce structured electromagnetic fields capable of stalling engines, generating luminous phenomena, or affecting human perception. The proposed mechanism contradicts established magnetospheric physics and has no experimental support.*

Response.

Partially correct, and acknowledged throughout the book. The mechanism connecting plasma precipitation to the reported effects is not fully established. The book does not claim it is. What the book claims is that the timing correlation exists at strict-band $\sigma = 6.35$ in pre-registered data, consistent with plasma propagation at a physically reasonable speed.

This is the Kepler position, not the Newton position. Kepler described planetary orbits with extraordinary precision before Newton provided the mechanism. The correlation was real before the mechanism was understood. A timing correlation at strict-band $\sigma = 6.35$ is not made false by an incomplete mechanistic account. It is made more interesting. The book proposes candidate mechanisms — ELF electromagnetic fields, ionospheric modification, boundary layer plasma enhancement — and cites published experimental literature supporting each component. The full chain from storm to effect is not established. The timing correlation is.

Objection 7: No independent replication

> *One investigator, one dataset, one model. Science requires independent replication by researchers with no stake in the outcome. This book provides none.*

Response.

Accepted fully. This is the correct scientific standard and the book does not meet it. Independent replication has not occurred because the model is new. The data and code are public precisely to enable it. The GFZ archive is downloadable. The haversine formula is four lines of code. Any researcher can replicate every result in this book in an afternoon.

The invitation is explicit: replicate it. Select cases independently. Apply the model. Report all results including failures. If the model fails under independent testing the correlation is not real. If it holds the correlation is confirmed. That is how science works and the book asks for nothing less.

What is not available to the careful scientist is the comfortable dismissal. The timing correlation is real, it is pre-registered, it is adversarially tested, and it is reproducible from public data. Dismissing it without testing it is not skepticism. It is incuriosity.

Objection 8: The model is over-constrained and therefore fragile

Five simultaneous conditions is not a model. It is a gauntlet. Any one condition failing should be sufficient to break the framework on a significant fraction of cases. The more conditions you add, the more opportunities you create for the model to fail. That it has not failed yet is a function of small sample size, not physical reality.

Response.

The objection inverts the logic. Over-constrained and unfalsified is not fragility. It is the opposite of overfitting.

Overfitting means a model with too many free parameters absorbs noise as signal — it passes because it is flexible enough to accommodate anything. The five conditions argument runs in the opposite direction. Each condition is independently specified from its own literature: radiation belt physics for condition one and two, boundary layer meteorology for condition three, aeromagnetic mapping for condition four, dusty plasma attenuation for condition five. None of them was constructed to fit the confirmed cases. None of them shares a free parameter with any other. They were derived from five separate bodies of published science and applied simultaneously to the dataset.

A model that was overfitted to its training data would show anomalous failures when additional independent constraints are applied — cases that pass the timing test but fail the aerosol test in unexpected ways, or pass the nightside condition but fail the storm threshold in ways the model cannot explain. No such failures exist in the dataset. Every PCE result corresponds to a specific named condition failing in exactly the way the physics predicts. Rendlesham fails condition one: no qualifying storm in the window, maximum Kp 4.667 across all 112 three-hour intervals. The industrial cities fail condition five: survival fractions below 1%. The Kp 8–9 storms that produce PCE results fail condition one from the top: belt depletion overtaking injection. Every failure is a named condition. Not one failure is unexplained.

Five conditions. Five disciplines. Zero anomalous failures across fifteen confirmed cases. That is not a model that is fragile. That is a model that is rigid enough to be wrong on five independent axes simultaneously — and is not. The sample size objection is answered by the pre-registered sessions: eight of twelve geometrically testable cases passed in sessions where the cases were chosen before any storm data was consulted. The five conditions framework applies equally to the pre-registered cases and produces no anomalous failures there either. The objection that the model is a gauntlet is correct. That is precisely why zero failures is the result that matters.

References

Abel, B., and Thorne, R.M. (1998). Electron scattering loss in Earth's inner magnetosphere: 1. Dominant physical processes. Journal of Geophysical Research, 103(A2), 2385–2396.

Andronache, C. (2003). Estimated variability of below-cloud aerosol removal by rainfall for observed aerosol size distributions. Atmospheric Chemistry and Physics, 3(1), 131–143.

Andersson, M.E., Verronen, P.T., Rodger, C.J., Clilverd, M.A., and Seppälä, A. (2014). Missing driver in the Sun–Earth connection from energetic electron precipitation impacts mesospheric ozone. Nature Communications, 5, 5197. https://doi.org/10.1038/ncomms6197

Babayev, E.S., and Allahverdiyeva, A.A. (2007). Effects of geomagnetic activity variations on the physiological and psychological state of functionally healthy humans. Advances in Space Research, 40(12), 1941–1951.

Babcock, K.L., Kurian, P., et al. (2024). Superradiance and exciton delocalization in tryptophan networks within microtubules. The Journal of Physical Chemistry B, 128.

Baumgaertner, A.J.G., Seppälä, A., Jöckel, P., and Clilverd, M.A. (2011). Geomagnetic activity related NOx enhancements and polar surface air temperature variability in a chemistry climate model: modulation of the NAM index. Atmospheric Chemistry and Physics, 11(9), 4521–4531. https://doi.org/10.5194/acp-11-4521-2011

Bawin, S.M., and Adey, W.R. (1976). Sensitivity of calcium binding in cerebral tissue to weak environmental electric fields oscillating at low frequency. Proceedings of the National Academy of Sciences, 73(6), 1999–2003.

Belova, E., Kirkwood, S., and Tammet, H. (2000). The effect of magnetic substorms on near-ground atmospheric current. Annales Geophysicae, 18(12), 1623–1629. https://hal.science/hal-00316832

Blackman, C.F., Benane, S.G., Kinney, L.S., House, D.E., and Joines, W.T. (1982). Effects of ELF fields on calcium-ion efflux from brain tissue in vitro. Radiation Research, 92(3), 510–520.

Baker, D.N., et al. (2014). An impenetrable barrier to ultrarelativistic electrons in the Van Allen radiation belts. Nature, 515, 531–534.

Behrendt, J.C., and Thiel, E. (1963). Aeromagnetic survey of the Uinta Basin and the Uinta Mountains, Utah. Journal of Geophysical Research, 68(20).

Bhatt, R., et al. (2025). Pulsatile ion-charged cerebrospinal fluid flow as a coordinator of large-scale neural rhythms. Medical Hypotheses.

Bishop, J., et al. (2024). Global maps of plasmaspheric erosion and refilling. Journal of Geophysical Research: Space Physics, 129.

Budinger, P. (2001–2009). Forensic analysis reports: tissue samples from anomalous livestock deaths. Black Vault / Phyllis Budinger Collection. Independent field reports.

Brautigam, D.H., and Albert, J.M. (2000). Radial diffusion analysis of outer radiation belt electrons during the October 9, 1990, magnetic storm. Journal of Geophysical Research, 105(A1), 291–309.

Cann, K. (2025). *Platonic Surrealism: A Modern Framework for Meaning Beyond Science and Religion.* Platonic Surrealism Press.

Cann, K. (2026). *Platonic Surrealism: UFOs and the Extra-Consciousness Hypothesis.* Platonic Surrealism Press.

Cann, K. (in press). Pre-registered geomagnetic storm propagation timing in anomalous-event reports: An exploratory analysis. Publication details forthcoming.

Chate, D.M., and Pranesha, T.S. (2004). Field studies of scavenging of aerosols by rain events. Journal of Aerosol Science, 35(6), 695–706.

Cutchin, J. (2016). *The Brimstone Deceit: An In-Depth Examination of Supernatural Scents, Otherworldly Odors, and Monstrous Miasmas.* Anomalist Books. [Systematic catalog of olfactory reports across anomalous phenomena categories. Chapter Thirteen documents ozone reports in UFO cases across multiple decades and continents. Used in Chapter Eighteen of this book as an independent chemical corroboration dataset; the ozone cases were cross-referenced against the GFZ Potsdam propagation windows without Cutchin's knowledge or participation.]

Chu, X., et al. (2017). A neural network model of plasmaspheric erosion and refilling. Journal of Geophysical Research: Space Physics, 122.

Claudepierre, S.G., et al. (2020). The quiet decay of radiation belt electrons. Journal of Geophysical Research: Space Physics, 125, e2019JA027198.

Costa, L.M., and Marler, C. (2017). UFO Sightings Desk Reference: United States of America 2001–2015. CreateSpace.

Daglis, I.A., Thorne, R.M., Baumjohann, W., and Orsini, S. (1999). The terrestrial ring current: Origin, formation, and decay. Reviews of Geophysics, 37(4), 407–438.

Day, K. (2008). The See'r. In Sailors' Anthology. Self-published.

Dessler, A.J. (1959). Effect of magnetic anomaly on particle radiation trapped in geomagnetic field. Journal of Geophysical Research, 64(7), 713–715.

Fortov, V.E., and Morfill, G.E. (2010). Complex and Dusty Plasmas: From Laboratory to Space. CRC Press.

Fukada, E., and Yasuda, I. (1957). On the piezoelectric effect of bone. Journal of the Physical Society of Japan, 12(10), 1158–1162.

Gelpi, C., Benbrook, J.R., and Sheldon, W.R. (1986). Convection electric field effects on outer radiation belt electron precipitation. Planetary and Space Science, 34(3), 271–277. https://doi.org/10.1016/0032-0633(86)90132-7

Granier, C., et al. (2011). Evolution of anthropogenic and biomass burning emissions of air pollutants at global and regional scales during the 1980–2010 period. Climatic Change, 109(1–2), 163–190.

Hynek, J.A., Imbrogno, P.J., and Pratt, B. (1987). Night Siege: The Hudson Valley UFO Sightings. Ballantine Books.

Hamilton, D.C., Gloeckler, G., Ipavich, F.M., Stödemann, W., Wilken, B., and Kremser, G. (1988). Ring current development during the great geomagnetic storm of February 1986. Journal of Geophysical Research, 93(A12), 14343–14355.

Harrison, R.G., and Carslaw, K.S. (2003). Ion-aerosol-cloud processes in the lower atmosphere. Reviews of Geophysics, 41(3), 1012.

Hastings, R.L. (2008). UFOs and Nukes: Extraordinary Encounters at Nuclear Weapons Sites. AuthorHouse.

Hoppel, W.A., and Frick, G.M. (1986). Ion-aerosol attachment coefficients and the steady-state charge distribution on aerosols in a bipolar ion environment. Aerosol Science and Technology, 5(1), 1–21.

Hunten, D.M., Turco, R.P., and Toon, O.B. (1980). Smoke and dust particles of meteoric origin in the mesosphere and stratosphere. Journal of the Atmospheric Sciences, 37(6), 1342–1357.

Kelleher, C., and Knapp, G. (2005). Hunt for the Skinwalker. Paraview Pocket Books.

Kirschvink, J.L. (1992). Uniform magnetic fields and double-wrapped coil systems. Bioelectromagnetics, 13(5), 401–411.

Kirschvink, J.L., et al. (2019). Transduction of the geomagnetic field as evidenced from alpha-band activity in the human brain. eNeuro, 6(2).

Li, X., et al. (1997). Simulation of dispersion of energetic particles during the March 1995 magnetic storm. Geophysical Research Letters, 24.

Lamarque, J.-F., et al. (2010). Historical (1850–2000) gridded anthropogenic and biomass burning emissions of reactive gases and aerosols: Methodology and application. Atmospheric Chemistry and Physics, 10(15), 7017–7039.

Matzka, J., Stolle, C., Yamazaki, Y., Bronkalla, O., and Morschhauser, A. (2021). The geomagnetic Kp index and derived indices of geomagnetic activity. Space Weather, 19.

Mayaud, P.-N. (1972). The aa indices: a 100-year series characterizing the magnetic activity. Journal of Geophysical Research, 77(34), 6870–6874.

Meredith, N.P., Horne, R.B., Glauert, S.A., and Anderson, R.R. (2007). Slot region electron loss timescales due to plasmaspheric hiss and lightning-generated whistlers. Journal of Geophysical Research, 112, A08214.

Millan, R.M., and Thorne, R.M. (2007). Review of radiation belt relativistic electron losses. Journal of Atmospheric and Solar-Terrestrial Physics, 69(3), 362–377.

Mezrich, B. (2016). The 37th Parallel. Atria Books.

Newton, H.W. (1958). The Face of the Sun. Penguin Books.

O'Brien, C. (2014). Stalking the Herd: Unraveling the Cattle Mutilation Mystery. Adventures Unlimited Press.

Ozeke, L.G., Mann, I.R., Murphy, K.R., Jonathan Rae, I., and Milling, D.K. (2014). Analytic expressions for ULF wave radiation belt radial diffusion coefficients. Journal of Geophysical Research: Space Physics, 119(3), 1587–1605.

Persinger, M.A. (1988). Increased geomagnetic activity and the occurrence of bereavement hallucinations. Neuroscience Letters, 88(3), 271–274.

Persinger, M.A. (1995). Out-of-body-like experiences are more probable in people with elevated complex partial epileptic-like signs during periods of enhanced geomagnetic activity: A nonlinear effect. Perceptual and Motor Skills, 80(2), 563–569.

Persinger, M.A., and Murphy, T.R. (2016). Validating new technologies to treat depression, pain and the feeling of sentient beings: A reply to "Neuroscience for the Soul." Open Journal of Philosophy, 6(1), 1–16.

Persinger, M.A., and Saroka, K.S. (2012). Protracted parahippocampal activity associated with Sean Harribance. International Journal of Yoga, 5(2), 140–145.

Mulligan, B.P., and Persinger, M.A. (2012). Experimental simulation of the effects of sudden increases in geomagnetic activity upon quantitative measures of human brain activity: Validation of correlational studies. Neuroscience Letters, 516(1), 54–56.

Rodger, C.J., Clilverd, M.A., and McCormick, R.J. (2003). Significance of lightning-generated whistlers to inner radiation belt electron lifetimes. Journal of Geophysical Research, 108(A12), 1462.

Rouleau, N., and Dotta, B.T. (2014). Electromagnetic fields as structure-function zeitgebers in biological systems: Environmental orchestrations of morphogenesis and consciousness. Frontiers in Integrative Neuroscience, 8, 84.

Rouleau, N., and Persinger, M.A. (2017). Neural tissues filter electromagnetic fields: Investigating regional processing of induced current in ex vivo brain specimens. Biology and Medicine (Aligarh), 9(2), 392.

Saroka, K.S., Caswell, J.M., Lapointe, A., and Persinger, M.A. (2014). Greater electroencephalographic coherence between left and right temporal lobe structures during increased geomagnetic activity. Neuroscience Letters, 560, 126–130.

Pierrard, V., et al. (2021). Links of the plasmapause with other boundary layers of the magnetosphere. Frontiers in Astronomy and Space Sciences, 8.

Pedersen, T.R., and Gerken, E.A. (2005). Creation of visible artificial optical emissions in the aurora by high-power radio waves. Nature, 433, 498–500.

Plane, J.M.C. (2012). Cosmic dust in the Earth's atmosphere. Chemical Society Reviews, 41(19), 6507–6518.

Pollack, G.H. (2013). The Fourth Phase of Water: Beyond Solid, Liquid, and Vapor. Ebner and Sons.

Plaza del Olmo, J. (2017). Modeling the Law of Times. In L. Donderi and P. Sturrock (Eds.), Proceedings of the 2016 CUFOS Conference.

Rapp, M., and Lübken, F.-J. (2004). Polar mesosphere summer echoes (PMSE): Review of observations and current understanding. Atmospheric Chemistry and Physics, 4(11/12), 2601–2633.

Pruppacher, H.R., and Klett, J.D. (1997). Microphysics of Clouds and Precipitation. Second edition. Kluwer Academic Publishers.

Reeves, G.D., et al. (2013). Electron acceleration in the heart of the Van Allen radiation belts. Science, 341(6149), 991–994.

Rodriguez, P., Kennedy, E.J., Keskinen, M.J., Ossakow, S.L., Siefring, C.L., Basu, S., McCarrick, M., Preston, J., Engebretson, M., Kaiser, M.L., Desch, M.D., Goetz, K., Bougeret, J.-L., and Manning, R. (1998). The WIND-HAARP experiment: Initial results of full polarization measurements of stimulated electromagnetic emissions using the WIND spacecraft. Geophysical Research Letters, 25(3), 257–260.

Rodeghier, M. (1981). UFO Reports Involving Vehicle Interference: A Catalogue and Data Analysis. Center for UFO Studies.

Rycroft, M.J., Israelsson, S., and Price, C. (2000). The global atmospheric electric circuit, solar activity and climate change. Journal of Atmospheric and Solar-Terrestrial Physics, 62(17–18), 1563–1576. https://doi.org/10.1016/S1364-6826(00)00112-7

Schulz, M., and Lanzerotti, L.J. (1974). Particle Diffusion in the Radiation Belts. Springer.

Seinfeld, J.H., and Pandis, S.N. (2016). Atmospheric Chemistry and Physics: From Air Pollution to Climate Change. 3rd ed. John Wiley & Sons.

Seppälä, A., Randall, C.E., Clilverd, M.A., Rozanov, E., and Rodger, C.J. (2009). Geomagnetic activity and polar surface air temperature variability. Journal of Geophysical Research: Space Physics, 114, A10312. https://doi.org/10.1029/2008JA014029

Seppälä, A., and Clilverd, M.A. (2014). Energetic particle forcing of the northern hemisphere winter stratosphere: comparison to solar irradiance forcing. Frontiers in Physics, 2, 25. https://doi.org/10.3389/fphy.2014.00025

Shprits, Y.Y., Subbotin, D.A., Meredith, N.P., and Elkington, S.R. (2008). Review of modeling of losses and sources of relativistic electrons in the outer radiation belt II: Local acceleration and loss. Journal of Atmospheric and Solar-Terrestrial Physics, 70(14), 1694–1713.

SOBEPS (Société Belge d'Étude des Phénomènes Spatiaux). (1991). Vague d'OVNI sur la Belgique: Un dossier exceptionnel. SOBEPS, Brussels.

Shukla, P.K., and Mamun, A.A. (2002). Introduction to Dusty Plasma Physics. Institute of Physics Publishing.

Stull, R.B. (1988). An Introduction to Boundary Layer Meteorology. Kluwer Academic Publishers.

Szweda-Lewandowska, Z., Puchała, M., and Osmulski, P.A. (1989). Radiation-induced changes of structural and functional properties of human hemoglobin. Radiation and Environmental Biophysics, 28(1), 47–58. https://doi.org/10.1007/BF01209722

Stewart, B. (1882). On the connexion between magnetic disturbances and sun-spots. Proceedings of the Royal Society of London, 33, 380–383.

Thorne, R.M., et al. (2013). Rapid local acceleration of relativistic radiation-belt electrons by magnetospheric chorus. Nature, 504, 411–414.

Tinsley, B.A. (2024). On the influence of space weather on the global electric circuit, with relevance to weather and climate. SSRN preprint. https://doi.org/10.2139/ssrn.4872735

UK Ministry of Defence (2000/2006). Unidentified Aerial Phenomena in the UK Air Defence Region. Project Condign (declassified 2006).

Vallée, J. (1966). Preliminary analysis of 200 type-I observations made in 1954. In C.S. Gillmor and C.G. Duvall (Eds.), The Scientific Study of Unidentified Flying Objects. Bantam Books.

Vallée, J., and Aubeck, C. (2010). Wonders in the Sky: Unexplained Aerial Objects from Antiquity to Modern Times. Tarcher/Penguin.

Vallée, J., and Davis, E.W. (2003). Incommensurability, orthodoxy and the physics of high strangeness: A 6-layer model for anomalous phenomena. In Proceedings of the Annual AIAA/USU Conference on Small Satellites.

Vargemezis, G., et al. (2024). VLF geophysical survey and geological interpretation of the Hessdalen Valley. Journal of Applied Geophysics.

Kirschvink, J.L., Kobayashi-Kirschvink, A., and Woodford, B.J. (1992). Magnetite biomineralization in the human brain. Proceedings of the National Academy of Sciences, 89(16), 7683–7687.

Rezende, A., et al. (2021). Magnetic domains oscillation in the brain with neurodegenerative disease. Scientific Reports, 11, 2141.

Zilli Vieira, C.L., Alvares, D., Blomberg, A., Schwartz, J., Coull, B., Huang, S., and Koutrakis, P. (2019). Geomagnetic disturbances driven by solar activity enhance total and cardiovascular mortality risk in 263 U.S. cities. Environmental Health, 18, 83.

Hillert, L., Berglind, N., Arnetz, B.B., and Bellander, T. (2002). Prevalence of self-reported hypersensitivity to electric or magnetic fields in a population-based questionnaire survey. Scandinavian Journal of Work, Environment and Health, 28(1), 33–41.

World Health Organization. (2005). Electromagnetic hypersensitivity. WHO Fact Sheet No. 296. Geneva: WHO.

Wolverton, J.C. (1976). Anomalous cattle deaths: Cascade County field investigation. Cascade County Sheriff's Department. Great Falls, Montana.

Rubin, G.J., Das Munshi, J., and Wessely, S. (2005). Electromagnetic hypersensitivity: a systematic review of provocation studies. Psychosomatic Medicine, 67(2), 224–232.

Vencloviene, J., Babarskiene, R.M., and Stirke, V. (2016). The association between space weather conditions and emergency hospital admissions for myocardial infarction during different stages of solar activity. Advances in Space Research, 57(6), 1519–1529.

Rezende, L.F.C., Paula, E.R., Batista, I.S., and Muella, M.T.A.H. (2025). Influence of geomagnetic disturbances on myocardial infarctions in women and men from Brazil. Communications Medicine, 5, 246.

Before You Go

You made it. That means something to me.

If this book gave you anything worth keeping — a scene that stuck, a thought you haven't finished yet — I'd be grateful if you'd leave a few honest words on Amazon. Independent authors don't have publishers pushing doors open for them. A review from a real reader is how we find each other.

Search for Kevin Cann on Amazon or Google Play.

Also by Kevin Cann

All titles available on Amazon and Google Play. Search for Kevin Cann.

Platonic Surrealism Series

(3-book series)

Platonic Surrealism: A Modern Framework for Meaning Beyond Science and Religion

Plasma Space Weather and the Paranormal

Platonic Surrealism: UFOs and the Extra-Consciousness Hypothesis

Awareness and Aplomb Series

(3-book series)

The Writing Writes the Writer: Taking Kev to Sizzlin' Meats and Salads

Awareness and Aplomb

Tex-Mex Cantos

Contributing Author

How to Think Impossibly — Chapter 3, co-authored with Jeffrey J. Kripal

University of Chicago Press

Made in the USA
Monee, IL
05 November 2026

67974096R00319